Advances in Environmental Science

D.C. Adriano and W. Salomons, Editors

Acidic Precipitation

Volume 4

Soils, Aquatic Processes, and Lake Acidification

Edited by S.A. Norton, S.E. Lindberg, and A.L. Page

Springer-Verlag
New York Berlin Heidelberg
London Paris Tokyo Hong Kong

Volume Editors:

S.A. Norton
Department of Geological Sciences
University of Maine
Orono, ME 04469
USA

S.E. Lindberg
Environmental Sciences Division
Oak Ridge National Laboratory
Oak Ridge, TN 37831
USA

A.L. Page
Department of Soil and Environmental Sciences
University of California, Riverside
Riverside, CA 92521
USA

Library of Congress Cataloging-in-Publication Data

Soils, aquatic processes, and lake acidification/volume editors, S.A. Norton, S.E. Lindberg, and A.L. Page.
p. cm.—(Advances in environmental science; vol. 4)
ISBN 978-1-4612-8782-7
1. Acid pollution of rivers, lakes, etc. 2. Soil acidification.
3. Acid deposition—Environmental aspects. I. Norton, Stephen A. (Stephen Allen), 1940- . II. Lindberg, S.E. III. Page, A.L. (Albert Lee), 1927- . IV. Series: Advances in environmental science; v. 4.
TD427.A27S65 1989
628.1′683-dc20 89-11466

Printed on acid-free paper

Typeset by McFarland Graphics and Design, Dillsburg, Pennsylvania.

9 8 7 6 5 4 3 2 1

ISBN 978-1-4612-8782-7 ISBN 978-1-4612-4456-1 (eBook)
DOI 10.1007/978-1-4612-4456-1

Preface to the Series

In 1986, my colleague Prof. Dr. W. Salomons of the Institute for Soil Fertility of the Netherlands and I launched the new *Advances in Environmental Science* with Springer-Verlag New York, Inc. Immediately, we were faced with a task of what topics to cover. Our strategy was to adopt a thematic approach to address hotly debated contemporary environmental issues. After consulting with numerous colleagues from Western Europe and North America, we decided to address *Acidic Precipitation,* which we view as one of the most controversial issues today.

This is the subject of the first five volumes of the new series, which cover relationships among emissions, deposition, and biological and ecological effects of acidic constituents. International experts from Canada, the United States, Western Europe, as well as from several industrialized countries in other regions, have generously contributed to this subseries, which is grouped into the following five volumes:

Volume 1 *Case Studies*
(D.C. Adriano and M. Havas, editors)

Volume 2 *Biological and Ecological Effects*
(D.C. Adriano and A.H. Johnson, editors)

Volume 3 *Sources, Deposition, and Canopy Interactions*
(S.E. Lindberg, A.L. Page, and S.A. Norton, editors)

Volume 4 *Soils, Aquatic Processes, and Lake Acidification*
(S.A. Norton, S.E. Lindberg, and A.L. Page, editors)

Volume 5 *International Overview and Assessment*
(T. Bresser and W. Salomons, editors)

From the vast amount of consequential information discussed in this series, it will become apparent that acidic deposition should be seriously addressed by many countries of the world, in as much as severe damages have already been inflicted on numerous ecosystems. Furthermore, acidic constituents have also been shown to affect the integrity of structures of great historical values in various places of the world. Thus, it is hoped that this up-to-date subseries would increase the "awareness" of the world's citizens and encourage governments to devote more attention and resources to address this issue.

The series editors thank the international panel of contributors for bringing this timely series into completion. We also wish to acknowledge the very insightful input of the following colleagues: Prof. A.L. Page of the University of California, Prof. T.C. Hutchinson of the University of Toronto, and Dr. Steve Lindberg of the Oak Ridge National Laboratory.

We also wish to thank the superb effort and cooperation of the volume editors in handling their respective volumes. The constructive criticisms of chapter reviewers also deserve much appreciation. Finally, we wish to convey our appreciation to my secretary, Ms. Brenda Rosier, and my technician, Ms. Claire Carlson, for their very able assistance in various aspects of this series.

Aiken, South Carolina

Domy C. Adriano
Coordinating Editor

Preface to *Acidic Precipitation*, Volume 4 (*Advances in Environmental Science*)

Acidic precipitation and its effects have been the focus of intense research for over two decades. Initially, research centered on the acidity status and acidification of surface waters and consequent impact on the status of sports fisheries; evidence suggested impacts on fisheries in Sweden and Norway, and in North America, in eastern Ontario, Quebec, and in the Adirondack Mountains of New York. More recently, research has become more process oriented with a major objective being greater understanding of dose–response relationships between atmospheric loading of acidifying material and lake acidity.

This volume of the subseries *Acidic Precipitation* emphasizes acid-neutralizing processes and the capacity of terrestrial and aquatic systems to assimilate acidifying substances and, conversely, the ability of systems to recover after acid loading diminishes.

Reuss and Walthall review the various interactive processes between atmospheric inputs and soils. They focus on the ability of the soil to neutralize incoming acidity as well as the ability of the soil to buffer solution chemistry after atmospheric loading of acidifying substances is reduced. Prasittikhet and Gambrell describe the origin of acid sulfate soils. These soils, acidic end members of a continuum, yield insight into the processes one expects in acute acidification of podsolic soils or as a result of drainage and oxidation of soils (e.g., bogs) that have stored reduced sulfur. Bertsch describes the historical development of analytical and computational methods for the determination of Al speciation in natural waters. He then reviews toxicity relationships between Al species and aquatic organisms (primarily fish) and terrestrial plants. The chapter closes with a discussion of Al speciation in various compartments of the environment. Jeffries reviews regional snowpack chemistries for both strong acid content and trace metal content and relates this chemistry to the metamorphism of snow, snowmelting, and impacts of snowmelt on surface water quality. Norton and colleagues discuss the ability of stream sediments to interact with acidic waters, based on experimental and empirical data. They also review the evidence for the role of lake sediment in neutralizing lake water via cation release and sulfur reduction. Porcella and associates review the state of the art for mitigation of acidity, largely using liming, in streams and lakes. They include a discussion of effects on trace metals and biological impacts related to the neutralization. Havas describes three case studies of the rapid chemical and biological recovery of lakes from a chronically acidic status: an artificially acidified lake (Lake 223), two lakes formerly under the strong influence of the now-closed Coniston smelter in Ontario, and moderately acidified lakes in the Sudbury, Ontario, region. Charles and coworkers review the paleolimnological data (diatoms and chrysophytes) related to the reconstruction of pH, alkalinity, dissolved labile Al, and dissolved organic carbon in lakes historically subjected to atmospheric deposition of acidic or acidifying material.

Orono, Maine — *Stephen A. Norton*

Oak Ridge, Tennessee — *Steven E. Lindberg*

Riverside, California — *Albert L. Page*

Contents

Contributors

Richard W. Battarbee, Palaeoecology Research Unit, Department of Geography, University College London, 26 Bedford Way, London WC1H 0AP, UK

Paul M. Bertsch, Division of Biogeochemistry, University of Georgia, Savannah River Ecology Laboratory, P.O. Drawer E, Aiken, SC 29801, USA

P.A. Bukaveckas, Department of Biology, Indiana University, Bloomington, IN 47405, USA

Donald F. Charles, U.S. Environmental Protection Agency, Environmental Research Laboratory, 200 S.W. 35th Street, Corvallis, OR 97333, USA

J.V. Depinto, Department of Civil and Environmental Engineering, Clarkson University, Potsdam, NY 13676, USA

C.T. Driscoll, Department of Civil Engineering, Syracuse University, Syracuse, NY 13244, USA

Robert P. Gambrell, Laboratory for Wetland Soils and Sediments, Louisiana State University, Baton Rouge, LA 70803, USA

S.P. Gloss, Wyoming Water Resources Center, University of Wyoming, Laramie, WY 82071, USA

Magda Havas, Environmental and Resource Studies, Trent University, Peterborough, Ontario K9J 7B8, Canada

Arne Henriksen, Norwegian Institute for Water Research, Box 333, 0314 Oslo, Norway

Dean S. Jeffries, Rivers Research Branch, National Water Research Institute, Department of Environment, P.O. Box 5050, Burlington, Ontario L7R 4A6, Canada

Jeffrey S. Kahl, Department of Environmental Protection, Augusta, ME 04333, USA

Stephen A. Norton, Department of Geological Sciences, University of Maine, Orono, ME 04469, USA

D.B. Porcella, Ecological Studies Program, Electric Power Research Institute, Palo Alto, CA 94303, USA

Jirapong Prasittikhet, Division of Soils, Department of Agriculture, Bangkok, Thailand

Ingemar Renberg, Department of Ecological Botany, Umeå University, S-90187 Umeå, Sweden

John O. Reuss, Department of Agronomy, Colorado State University, Fort Collins, CO 80525, USA

C.L. Schofield, Department of Natural Resources, Cornell University, Ithaca, NY 14853, USA

John P. Smol, Department of Biology, Queen's University, Kingston, Ontario K7L 3N6, Canada

Herman van Dam, Research Institute for Nature Management, P.O. Box 46, 3956 ZR Leersum, The Netherlands

Paul Mark Walthall, Department of Agronomy, Louisiana State University Agricultural Center, Baton Rouge, LA 70803, USA

Richard F. Wright, Norwegian Institute for Water Research, Box 333, 0314 Oslo, Norway

T.C. Young, Department of Civil and Environmental Engineering, Clarkson University, Potsdam, NY 13676, USA

Soil Reaction and Acidic Deposition

John O. Reuss* and Paul Mark Walthall*†

Abstract

This chapter discusses the major chemical processes by which acidic deposition interacts with soils. The focus is on forest soils, as the effects of acidic deposition on soils used for production of food and fiber are generally small compared to effects of agricultural practices such as nitrogen fertilizer applications and liming.

Buffering mechanisms considered include aluminum buffering, silicate mineral buffering, cation exchange, organic buffering, and the effect of anion immobilization processes such as nitrate uptake and sulfate adsorption. The effects of acidic inputs on capacity factors such as exchange acidity, exchangeable base content, and sulfate adsorption capacity are considered, as are related natural processes such as acidification due to accumulation of bases in biomass. Particular attention is paid to intensity effects, such as the effect of increased concentration of anions associated with strong acids on the chemical composition of the soil solution, as they are likely to be highly nonlinear with respect to the capacity factors. These include pH, aluminum mobilization, and loss of alkalinity in the soil solution, which in turn may result in acidification of drainage waters.

Due to the variety and complexity of processes, results of field and laboratory experiments are likely to be highly variable. A few examples are discussed in relation to the mechanisms that appear to be controlling the response in each case.

I. Introduction

Acidic deposition affects the chemical composition of the soil system, including solid and adsorbed phases, soil solution, and ultimately the drainage water. A discussion of the effect of acidic deposition on soil reaction must deal with a complex set of processes. These processes may interact in a manner such that a relatively small change in soil reaction as measured by the pH of the soil solution

*Department of Agronomy, Colorado State University, Fort Collins, CO 80525, USA.

†Current address: Louisiana State University Agricultural Center, Baton Rouge, LA 70803, USA.

may be accompanied by very significant changes in solution chemistry. Among the more important effects are mobilization of monomeric forms of aluminum and loss of acid-neutralizing capacity (ANC or "alkalinity") in the soil solution, leading to acidification of drainage waters.

Responses of individual systems will be highly variable, depending on the intensity and duration of deposition and the properties of the soil. Simply applying acid to soils and measuring the response in terms of "soil acidity," or acidity of runoff or leachates can be expected to give results that are highly variable and difficult to interpret. Another approach sometimes used is to evaluate the H^+ ion input attributable to acidic deposition in terms of the total input resulting from natural or ecosystem processes (Andersson et al., 1980; Richter, 1986; Binkley and Richter, 1987). Although the results are often interesting, this approach considers only the capacity factors of soil acidity and the effect on the chemical composition of the solution, that is, the intensity effects, may be very different.

We believe the most useful approach is that of a *chemical system*: a model or framework describing the various chemical processes that occur when acidic deposition impinges on a soil, and interactions that occur among these processes. Although quantitative models describing these processes are available (Reuss, 1978, 1980, 1983; Cosby et al., 1985; Reuss and Johnson, 1985, 1986), in this chapter we will confine our discussion to the conceptual model. Furthermore, acidic deposition effects are not likely to have a significant effect on the acidity of soils that contain free carbonates or of agricultural soils where amendments are commonly used to mitigate the much greater acidification potential of routine agricultural practices, particularly the use of nitrogenous fertilizers. Therefore, we have focused the discussion on effects in acidic forest soils.

II. Buffering Mechanisms in Soil

A variety of pH-buffering mechanisms occur in soils. For example, Ulrich (1983) lists the following buffer mechanisms with approximate pH ranges:

Calcium carbonate buffer (pH > 8 to 6.2)
Silicate buffers (pH 6.2 to 5.0)
Cation exchange buffer (pH 5.0 to 4.2)
Aluminum buffer (pH 4.2 to 2.8)
Iron buffer (pH 3.8 to 2.4)

These buffer ranges may be considered as a general guide, but in practice the ranges often overlap and are dependent on the method chosen for pH measurement. As a result of the overlap in buffer ranges, titration curves for soils often resemble those for weak polybasic acids (Russell, 1961), and individual buffer ranges may not be clearly expressed. Although we shall not discuss the carbonate buffers in detail, we will consider the buffering effects resulting from anion immobilization processes and buffering due to organic acids.

A. The Aluminum Buffer

1. Buffering by Aluminum Trihydroxides

The Al buffer is most commonly described in terms of the reaction of an aluminum trihydroxide with H^+

$$3H^+ + Al(OH)_3 \rightleftharpoons Al^{3+} + 3H_2O \quad (1)$$

for which the equilibrium expression may be written as

$$\frac{(Al^{3+})}{(H^+)^3} = K^o \quad (2)$$

Commonly accepted values of the equilibrium constant, log K°, are approximately 8.04 for gibbsite and 9.66 for amorphous $Al(OH)_3$ (Lindsay, 1979). From a plot of Al^{3+} activity versus pH for these two forms (Figure 1–1), the pH of a system containing gibbsite will be buffered near pH 4.0, whereas amorphous $Al(OH)_3$ will buffer the system at about pH 4.5. When these pH levels are reached, further inputs of H^+ will be largely consumed in the dissolution of the aluminum trihydroxide mineral. As a result the Al^{3+} concentration will increase rapidly, and the pH change will be small. Microcrystalline gibbsite (not shown) will cause the system to buffer at an intermediate value.

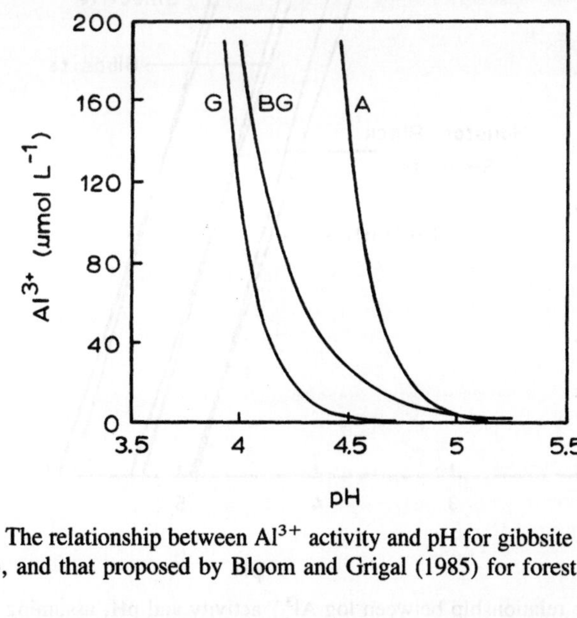

Figure 1–1. The relationship between Al^{3+} activity and pH for gibbsite (G), amorphous $Al(OH)_3$ (A), and that proposed by Bloom and Grigal (1985) for forest surface soils in Minnesota.

2. Aluminum Buffering by Aluminosilicates

Although the solubility of gibbsite or amorphous aluminum hydroxide is often used to estimate the activity of Al^{3+} in solution, soil solutions that are undersaturated with respect to the solubility of either of these two phases have been reported (Bloom and Grigal, 1985; Driscoll et al., 1984; Hooper and Shoemaker, 1985; Nordstrom and Ball, 1986). Figure 1–2 shows the pH–Al relationships for several aluminosilicates as well as gibbsite and amorphous $Al(OH)_3$. The imposed conditions were $\log(H_4SiO_4^o) = -4.0$ (quartz); $\log(Mg^{2+}) = -4.0$; $\log(K^+) = -4.5$; and $\log(Fe^{3+}) = 3.54–3pH$, amorphous $Fe(OH)_3$. The mineral illite (Lindsay, 1979) yields a pH buffer very near that of amorphous $Al(OH)_3$, but

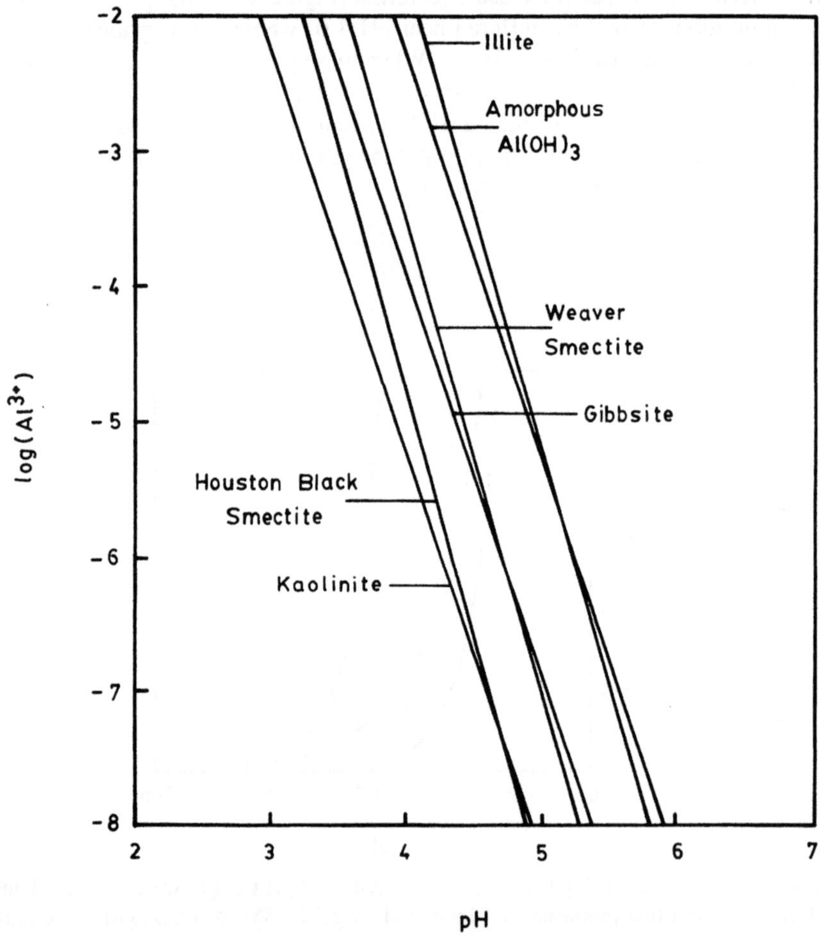

Figure 1–2. The relationship between log Al^{3+} activity and pH, assuming control by a number of common soil minerals.

would not be expected to play a major role under acidic conditions as the K interlayer would be rapidly removed and the mineral would alter to a more stable secondary phase due to the low activity of K^+ generally found in acidic forest soils. The secondary minerals smectite (Carson et al., 1976) and kaolinite (Lindsay, 1979) provide a lower pH buffer (Figure 1–2) than the $Al(OH)_3$ phases; where these dominate, soil pH values may drop below the buffer range of the $Al(OH)_3$ phases. Van Breeman and Wielemaker (1974) reported that the pH of acidic sulfate soils studied in Thailand appeared to be buffered by the conversion of smectite to kaolinite, partially explaining why pH values of these soils, when drained for agricultural purposes, did not drop to values below 3.3 to 4.0 upon oxidation of FeS_2 and formation of H^+ and $SO_4^=$ ions. This buffer range is illustrated in Figure 1–2. Hydroxy-interlayered vermiculites and smectites have been observed to follow the solubility of the $Al(OH)_3$ interlayer rather than the 2:1 layer structure, and to provide a pH buffer near that of amorphous $Al(OH)_3$ (Kittrick, 1983). Through the course of soil weathering under well-drained conditions, one would expect loss of silica, destruction of the 2:1 and 1:1 aluminosilicates, and precipitation of one of the $Al(OH)_3$ phases, effects that have been observed in many of the soils in the southeastern United States.

Many aluminosilicate minerals contain base cations, and the weathering of these minerals may play a role in buffering against pH decrease. The occurrence of such minerals in acidic soils depends on the parent material and the amount of weathering that has taken place. In relatively young acidic soils, substantial amounts of these aluminosilicates may be present, and base cations will be released as they weather. We will use the Ca aluminosilicate mineral known as *anorthite* as an example:

$$\underset{\text{(anorthite)}}{CaAl_2Si_2O_8} + 8H^+ = Ca^{2+} + 2Al^{3+} + 2H_4SiO_4^o \qquad (3)$$

The weathering of this mineral releases 1 mole of Ca^{2+} for every 2 mol of Al^{3+}, or an equivalent (charge) Al^{3+}/Ca^{2+} ratio of 3. If the replacement of the base cations removed by leaching due to acidic deposition were to depend on the weathering of anorthite, only 1 mol of charge can be furnished as Ca^{2+} for every 3 mol of charge released as Al^{3+}. If the reaction is not rate limited and no secondary reactions occur, the effect of acidic deposition must be either reduction of the reservoir of exchangeable bases or mobilization of Al^{3+} in solution. As the amount of Ca^{2+} released would be less than the release of the acidic Al^{3+} ion, the system would eventually become acidic.

In these systems, the release of Fe^{3+} on weathering will store acidity in the same manner as the release of Al^{3+}. Table 1–1 shows the $(Al^{3+} + Fe^{3+})$/base cation ratios of some of the common materials that might be weathered in response to acidic deposition inputs. In virtually all cases, substantially more Al^{3+} (or Fe^{3+}) will be released than base cations, so if the breakdown of these minerals is controlling the release of base cations, we would expect an increase in Al^{3+} and a decrease in bases, even if the process were not rate limited. If the process is rate limited, the sum of base cations plus Al^{3+} released will be less than the amount of

Table 1-1. $(Al^{3+} + Fe^{3+})$/base cation release ratios of several silicate minerals.

Mineral	Formula	$(Al^{3+} + Fe^{3+})$/base cation charge ratios
Albite	$NaAlSi_3O_8$	3
Orthoclase	$KAlSi_3O_8$	3
Muscovite	$KAl_3Si_3O_{10}(OH)_2$	9
Anorthite	$CaAl_2Si_2O_8$	3
Actinolite	$Ca_2Fe_5Si_8O_{22}$	3.75
Cordierite	$Mg_2Al_4Si_5O_{18}$	3

acidic input, so that eventually the acidity in solution would remain in the form of H^+ rather than Al^{3+}.

In most cases, secondary reactions will occur, so it is unlikely that effects can be predicted simply on the basis of the release ratios. If we combine the reverse reaction from Equation 1 for the precipitation of gibbsite with Equation 3, we obtain Equation 4.

$$\underset{\text{(anorthite)}}{CaAl_2Si_2O_8} + 2H^+ \longrightarrow Ca^{2+} + \underset{\text{(gibbsite)}}{2Al(OH)_3} + 2H_4SiO_4^o \tag{4}$$

Similarly, by combining the reverse of Equation 5 for the dissolution of kaolinite with Equation 3 we obtain Equation 6.

$$\underset{\text{(kaolinite)}}{Al_2Si_2O_5(OH)_4} + 6H^+ \longrightarrow 2Al^{3+} + 2H_4SiO_4^o + H_2O \tag{5}$$

$$\underset{\text{(anorthite)}}{CaAl_2Si_2O_8} + 2H^+ \longrightarrow Ca^{2+} + \underset{\text{(kaolinite)}}{Al_2Si_2O_5(OH)_4} \tag{6}$$

Which of these reactions might occur depends on a number of factors, particularly the $H_4SiO_4^o$ concentration that is maintained in solution. If neither the dissolution nor reprecipitation reaction is rate limited, the replacement of bases removed will be complete, and neither the soil nor the drainage water is likely to be acidified. However, if the reprecipitation reactions are rate limited, acidification will occur, because in the original dissolution reaction Al^{3+} release is in excess of Ca^{2+} release. Whether or not the dissolution reactions are rate limited would seem to depend on the supply of the primary mineral and the physical accessibility to weathering. Even if the dissolution reactions are not rate limited, it is entirely possible that the net $(Al^{3+} + Fe^{3+})$/base cation release ratios may be substantially affected by the rate of the reprecipitation reactions. A possible signature of such reactions would be a narrowing of the $(Al^{3+} + Fe^{3+})$/base cation release ratios, and this narrowing would increase as reaction time increases. Conversely, the ratio should tend to widen with a short reaction time or with the rapid removal of silicon or monosilic acid, thus decreasing the probability of forming kaolinite or 2:1 clay minerals.

Another type of reaction that may tend to decrease the (Al^{3+} + Fe^{3+})/base cation release ratio is that of incongruent dissolution (Huang and Keller, 1971; Galloway et al., 1983); that is, base cations at or near the surface of the lattices may be released by a process of H^+ diffusing in and replacing the base cations. This process would result in the conversion of the structure of the outer surfaces from that of a mineral such as anorthite to that of kaolinite, without the occurrence of complete dissolution and reprecipitation. This type of reaction should be characterized by rapid release of bases on exposure to acid, with a switch from a low to higher (Al^{3+} + Fe^{3+})/base cation release ratio on repeated exposure.

3. The H^+-Al^{3+} Relationship

If Al buffering is controlled by an Al trihydroxide, the activity of Al^{3+} will be proportional to the third power of the H^+ activity (Equation 2), and the slope of the lines in a logarithmic plot such as Figure 1–2 will be 3.0. If the Al buffering is controlled by other minerals, the exponent associated with the H^+ activity may be somewhat different. For example, Lindsay (1979) gives the reaction of illite with H^+ as

$$K_{0.6}Mg_{0.25}Al_{2.3}Si_{3.5}O_{10}(OH)_2 + 8H^+ + 2\,H_2O \rightarrow 0.6K^+ + 0.25\,Mg^{2+} + 2.3Al^{3+} + 3.5H_4SiO_4^o \quad (7)$$

If H^+ were to increase while the concentrations of K^+, Mg^{2+}, and $H_4SiO_4^o$ remained constant, the Al^{3+} activity would vary with the (8/2.3) or approximately the 3.5 power of H^+. In practice, lack of equilibrium and kinetic considerations may obscure the precise H^+-Al^{3+} relationship. Although the Al trihydroxide model may well be appropriate for B and some C horizons in forest soils, some investigators have found quite different relationships, particularly for surface soils. For example, Bloom and Grigal (1985) have suggested that for the top 30 cm of forest soils in Minnesota, the overall relationship may be described by

$$\log(Al^{3+}) = 2.60 - 1.66\,\text{pH} \quad (8)$$

This indicates an exponent of 1.66 in Equation 2 above, substantially less than the exponent of 3.0 for Al trihydroxides. This relationship suggests a somewhat less abrupt buffering process than would be required by an exponent of 3.0, with the buffer curve generally lying between that of gibbsite and amorphous $Al(OH)_3$ in the range of interest (Figure 1–1). Even though the buffering is less abrupt than would be the case if the exponent were near 3.0, this relationship still indicates a powerful buffer system in which pH values below 4.0 would only occur in the presence of very high levels of Al^{3+}. Cronan and others (1986) have proposed a system for organic horizons whereby the slope and intercept of the log Al^{3+} versus pH relationship can be estimated using pH, Cu-extractable Al, and the titratable carboxyl content of the humus.

The Al buffering mechanism provides a very large buffering capacity in most mineral soils. In systems where this is the dominant pH buffering mechanism, the effects of acidic deposition on the pH of the soil solution are likely to be small.

Paradoxically, however, the most deleterious effects are likely to occur in systems where the other buffering mechanisms are exhausted and significant Al buffering occurs.

When stressed with mineral acid inputs, the concentration of SO_4^{2-}, and to some extent NO_3^-, in solution will increase. Charge balance requires that the total charge associated with cations in solution must increase by an equal amount. Buffering by the Al mechanism means that in the solution the charge previously associated with H^+ in the mineral acids is now associated with Al^{3+} or other monomeric Al species. The inorganic forms of Al, particularly Al^{3+}, are toxic to many plants. Furthermore, Al mobilization results in loss of ANC in the solution and acidic Al-rich drainage waters (Reuss and Johnson 1985, 1986). Although the two are obviously interdependent, aluminum mobilization is likely to be a much more important factor in determining possible deleterious effects of acidic deposition in terrestrial and associated aquatic ecosystems than is change in soil pH.

B. The Cation Exchange Buffer

Many readers accustomed to pH ranges commonly maintained in agricultural soils may be surprised at the relatively low pH range (4.2 to 5.0) suggested by Ulrich (1983) for ion exchange buffering. These buffer ranges refer to pH values measured in salt solutions, which are generally lower than those measured in water (see section IV, C). A graph showing pH versus Ca + Mg saturation plotted from data reported by Clark and Hill (1964) for 48 forest soil samples from eastern Canada (Figure 1–3) illustrates the existence of the exchange buffer and is consistent with the suggested pH range. As pointed out by Clark and Hill, this

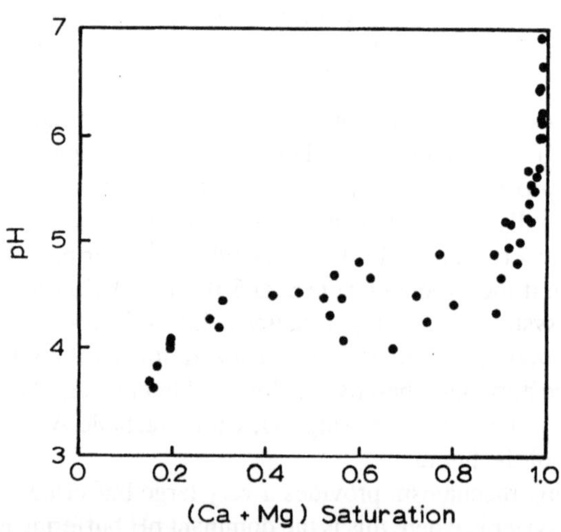

Figure 1–3. The relationship between Ca + Mg saturation and pH (0.01 M $CaCl_2$) for 48 forest soil samples from eastern Canada. (From the data of Clark and Hill, 1964.)

relationship is clearly expressed only if Ca + Mg saturation is calculated using cation exchange capacity (CEC) determined at the pH of the soil (unbuffered CEC). Inclusion of variable charge, due to use of standard methods involving CEC determined at pH 7.0 or above obscures the relationship.

The cation exchange buffer is complex, as many different cations may be involved. The general principles relevant to the effects of acidic deposition, however, can be understood by considering ion exchange relationships as they apply to simplified systems involving only the H^+, Al^{3+}, and Ca^{2+} ions.

1. Ion Exchange Equations

An extensive literature is available concerning ion exchange relationships in soils, most of which is beyond the scope of this review. Therefore, we shall discuss only a couple of the more common relationships as they relate to ion exchange buffering.

Consider the exchange of cations A^{a+} and B^{b+} with an exchanger X

$$bA^{a+} + BX = aB^{b+} + AX \tag{9}$$

One of the most common equations used to describe the exchange equilibrium for ions of unequal valence is

$$K_s \frac{(A^{a+})^b}{(B^{b+})^a} = \frac{(AX)^b}{(BX)^a} \tag{10}$$

where the parentheses denote activity. AX and BX are the fractions of the adsorbed ions A and B, respectively, and K_s is the exchange or "selectivity" coefficient. If these fractions are expressed as fractions of the total adsorbed moles (mole fractions), the equation is that proposed by Vanselow (1932), whereas if expressed as a fraction of the moles of charge on the exchanger (equivalent fractions), the equation is that of Gaines and Thomas (1953). One other formulation commonly used to describe exchange processes in soil is that of Gapon (1933).

$$K_{gp} \frac{(A^{a+})^{1/b}}{(B^{b+})^{1/a}} = \frac{(AX)}{(BX)} \tag{11}$$

where K_{gp} is the Gapon selectivity coefficient. The implications of these relationships when applied to the effects of acidic deposition on acidic soil systems are discussed in the following sections.

2. Ca^{2+}-H^+ Exchange

Schofield and Taylor (1955a, 1955b) called attention to the fact that the pH of acidic soils measured in dilute $CaCl_2$ solutions increased as the $CaCl_2$ solutions were made more dilute, but that the quantity (pH − ½ pCa) remained relatively constant, where pCa is defined as the negative log of the Ca^{2+} (or Ca^{2+} + Mg^{2+}) activity. They referred to this constant as the *lime potential*. The principle of the constancy of this parameter has been utilized, although in somewhat different ways, in at least two models designed to describe quantitatively the effects of acidic deposition on soils (Reuss, 1978, 1980; Bloom and Grigal, 1985). Although the constancy has been explained in a number of ways, perhaps the simplest is the application of Equation 6 to Ca^{2+}-H^+ exchange

$$\frac{(Ca^{2+})}{(H^+)^2} = \frac{(CaX)}{K(HX)^2} \tag{12}$$

where K is a constant specific to the exchanger. If the total number of each of the ions on the exchange is much greater than that in solution, CaX and HX remain relatively constant as solution strength changes. Thus, for a specific pair of values of CaX and HX, the right side of Equation 12 can be considered as constant. By taking the square root and changing to logarithmic form we find

$$pH - ½\,pCa = ½\log\left\{\frac{(CaX)}{K(HX)^2}\right\} \tag{13}$$

This tells us that the lime potential is a function of the relative amount of H^+ and Ca^{2+} on the exchange, and the exchange coefficient. The significance of this relationship to ion exchange buffering is that at any particular Ca + Mg saturation, a doubling of H^+ in solution requires an increase of a factor of four in the Ca^{2+} activity. This is illustrated in Figure 1–4 for the relevant range of lime potentials. At lime potentials above 2.50, pH values could drop below 4.0 only if solution strengths are very high, that is, Ca^{2+} activities > 1,000 μmol L^{-1}.

3. Ca^{2+}-Al^{3+} Exchange

Whereas the Ca^{2+}-H^+ exchange model is useful in understanding the ion exchange buffer relationships, most exchange acidity in soils is thought to exist in

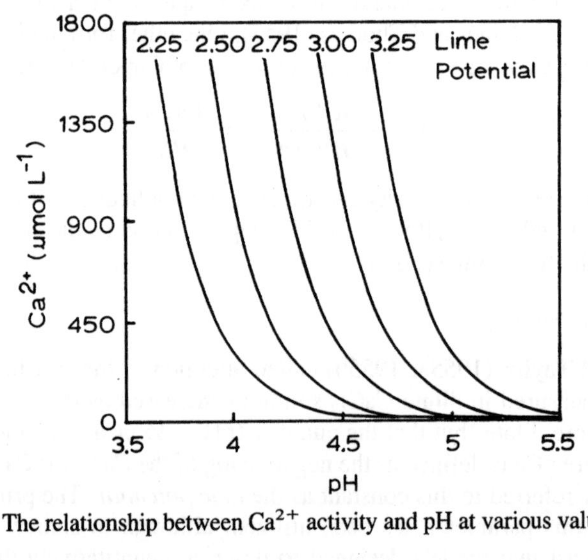

Figure 1–4. The relationship between Ca^{2+} activity and pH at various values of the lime potential.

the form of Al species, particularly Al^{3+} (Coleman and Thomas, 1967). Furthermore, the ion exchange buffering systems and the Al buffering processes occur simultaneously, and Al may be brought into solution by exchange processes as well as by the dissolution-precipitation processes described above. For the Ca^{2+}-Al^{3+} exchange system, Equation 10 may be written

$$K_p \frac{(Al^{3+})^2}{(Ca^{2+})^3} = \frac{(AlX)^2}{(CaX)^3} \tag{14}$$

For a simple system in which the cations are Ca^{2+} and Al^{3+} and the anions are $SO_4^{2-} + NO_3^-$, the charge balance principle requires that

$$2[Ca^{2+}] + 3[Al^{3+}] = 2[SO_4^{2-}] + [NO_3^-] \tag{15}$$

where the brackets indicate molar concentration. Furthermore, in this system $CaX = (1 - AlX)$. For any total anion concentration and value of the exchange constant, Equations 14 and 15 can be solved simultaneously for Ca^{2+} and Al^{3+} if appropriate adjustments are made for the difference between concentration and activities (Reuss, 1983). Isotherms showing the relationship between the Ca (or Ca + Mg) saturation and the fraction of total solution charges, consisting of Al species at three total solution concentrations, are shown in Figure 1–5.

Both the shape of these isotherms and the effect of solution concentration are important to an understanding of the effects of acidic deposition. The shapes indicate that, within the range of exchange coefficients found in most soils, the divalent ions (Ca^{2+} and Mg^{2+}) dominate the solution even though Al^{3+} may dominate the exchange. As the basic cations are depleted, usually in the range of

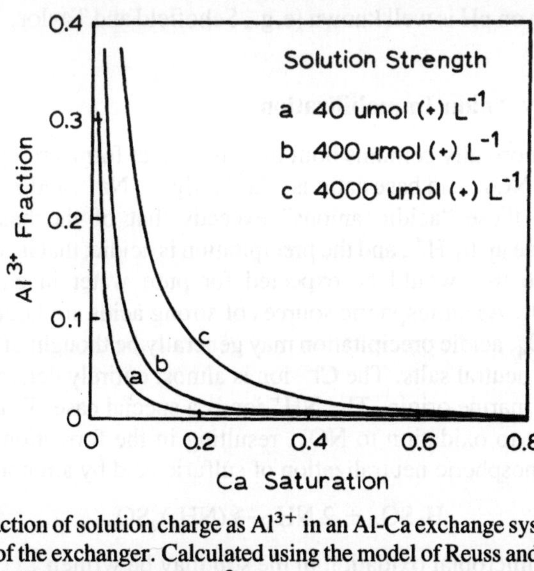

Figure 1–5. Fraction of solution charge as Al^{3+} in an Al-Ca exchange system as a function of Ca saturation of the exchanger. Calculated using the model of Reuss and Johnson (1985). The selectivity coefficient is taken as 10^2.

10% to 25% (Ca + Mg) saturation, the switch to Al domination in solution is quite abrupt. As total solution strength increases, the Al^{3+} in solution increases, as does the value of the Ca + Mg saturation at which Al^{3+} in solution becomes significant. The importance of these effects to acidic deposition effects will be discussed in later sections.

The Ca^{2+}-Al^{3+} exchange model is not incompatible with the Ca^{2+}-H^+ exchange concepts. Rearranging Equation 2 gives

$$Al^{3+} = K^o(H^+)^3 \tag{16}$$

Substitution for Al^{3+} in Equation 14, converting to logarithmic form, and rearranging (Reuss, 1980) gives

$$pH - \tfrac{1}{2}\,pCa = \tfrac{1}{3}\log K^o + \tfrac{1}{6}\log K_s + \tfrac{1}{6}\log\frac{(CaX)^3}{(AlX)^2} \tag{17}$$

which is similar to the relationship proposed by Turner and Clark (1964). Thus, the constancy of pH − ½ pCa for any given Ca + Mg and Al saturation is preserved, and the lime potential is shown to be a function of the solubility of Al in the system K^o, the Ca^{2+}-Al^{3-} exchange coefficient (K_s), and the relative degrees of Ca + Mg and Al saturation. An expression relating pH − ½ pCa to these parameters may also be derived by combining Equation 2 and the Gapon relationship for Ca^{2+}-Al^{3+} exchange.

Some investigators (e.g., McBride and Bloom, 1977) have questioned the applicability of equations such as Equation 14 to the Ca^{2+}-Al^{3+} exchange system. Although a detailed discussion of this problem is beyond the scope of this chapter, the general shape of the isotherms shown in Figure 1–5 is confirmed by the data of McBride and Bloom (1977) and of Pleysier and others (1979), and the effect of solution strength on pH is well known (e.g., Schofield and Taylor, 1955a, 1955b).

C. Buffering by Anion Immobilization

Wet deposition normally contains both the anions that form strong mineral acids, Cl^-, SO_4^{2-}, and NO_3^-, and base cations, Ca^{2+}, Mg^{2+}, Na^+, and K^+. If the charge associated with these "acidic anions" exceeds that of the base cations, the difference is made up by H^+, and the precipitation is acidic; that is, the pH is below the value of 5.6 that would be expected for pure water in equilibrium with atmospheric CO_2. As atmospheric sources of strong acids are largely confined to HNO_3 and H_2SO_4, acidic precipitation may generally be thought of as a mixture of these acids with neutral salts. The Cl^- ion is almost entirely derived from neutral salts, usually of marine origin. The NH_4^+ ion is a special case. When it enters the soil, it is subject to oxidation to NO_3^-, resulting in the formation of nitric acid. Consider the atmospheric neutralization of sulfuric acid by ammonia.

$$H_2SO_4 + 2\,NH_3 \rightarrow (NH_4)_2SO_4 \tag{18}$$

The subsequent microbial oxidation in the soil may be written

$$(NH_4)_2SO_4 + 4\,O_2 \rightarrow 2\,HNO_3 + H_2SO_4 + 2\,H_2O \tag{19}$$

Thus, the addition of 1 mol of $(NH_4)_2SO_4$ to the soil has the same potential for acidifying the soil as the original 1 mol of atmospheric H_2SO_4 plus 2 mol of HNO_3. Similarly, the addition of 1 mol of NH_4NO_3 has the same acidifying potential as 2 mol of HNO_3. Neutralization of atmospheric acids by NH_3 actually increases the potential to acidify the system. A more complete discussion of these effects may be found in Reuss and Johnson (1986).

1. Nitrogen Immobilization

Immobilization of strong acidic anions in the soil has the effect of removing them from the system and effectively neutralizing some or all of the accompanying acidity. This immobilization may occur as a result of uptake by plants or soil organisms or by chemical processes of anion adsorption. For example, consider the fate of HNO_3 in precipitation that enters the soil or that is formed by the oxidation of NH_4^+ followed by uptake of nitrate

$$H^+ + NO_3^- \text{ (soil)} \longrightarrow HOH + NO_3^- \text{ (plant)} \tag{20}$$

If the uptake of NO_3^- by the plant is not accompanied by the uptake of cations of equivalent charge, the release of OH^- ions (or HCO_3^-) by the plant maintains charge balance within the plant and neutralizes the soil solution relative to the HNO_3 input. The acidity created within the plant is subsequently neutralized when the NO_3^- is reduced and the nitrogen assimilated into proteins. Thus, nitrogen from atmospheric deposition that enters the soil and is subsequently taken up by plants has no effect on the acidity of the soil or the drainage water. Nitrogen inputs, as either HNO_3 or NH_4^+ salts, do not affect the acidity of the soil or drainage water unless the input exceeds the capacity of the ecosystem to incorporate the nitrogen within the biomass, or unless accompanied by a mobile conservative anion such as SO_4^{2-}. Moderate amounts of nitrogen enhance the forest growth, at least on nitrogen-deficient sites, without having a direct acidifying effect on either soil or drainage water. In the long term, the annual input that can be incorporated into the biomass without causing "leakage" of NO_3^- is much less than the annual uptake, as most of the uptake is recycled in throughfall and litterfall. Quantifying acceptable levels of nitrogen deposition is difficult and undoubtedly site specific. Nilsson (1986) has tentatively suggested that the "critical load" for Scandinavia may be in the neighborhood of 0.5 g N m^{-2}. This value is probably low for rapidly growing forest sites. However, N deposition levels above 2 g m^{-2} are reported in southern Sweden and 3 to 4 g m^{-2} in central Europe (Grennfelt and Hultberg, 1986). There is little doubt that such levels will result in acidification of soils and drainage waters (van Breemen et al., 1982; van Breemen and Jordens, 1983).

2. Sulfur Immobilization

Similarly, to the degree that $SO_4^=$ from H_2SO_4 is incorporated into biomass, inputs of this acid will not affect the acidity of soil or drainage water. Because plants use much less sulfur than nitrogen, this mechanism of immobilization is less important in the case of sulfate. Microbial biomass contains large amounts of sulfur, and microbial immobilization may yet prove to be a significant sink for atmospherically deposited sulfur (Fitzgerald et al., 1982). To date, however, most attention has been given to immobilization by chemical processes; as suggested by Johnson and others (1985), we believe that these are generally more important than microbial immobilization.

Under reducing conditions, sulfate may also be lost from the system by microbial reduction to H_2S or immobilized by the formation of metal sulfides. For example, in a forested catchment with extensive bog areas in Nova Scotia, Ogden (1982) found 33% and 40% decreases in sulfate output relative to input during two summers, and this difference was plausibly attributed to sulfate reduction. These high apparent levels of reduction were not observed in the winter, nor were they observed in two other catchments where bog areas were less prevalent. The actual significance of sulfate reduction as a pathway for either immobilization or loss of SO_4^{2-} is difficult to estimate. In the northeastern United States and eastern Canada, Kramer and others (1986) observed a close relationship between wet deposition inputs and lake sulfate, suggesting that sulfate generally tends to be conservative in these systems.

Sulfur may also be immobilized in some soils by adsorption on the chemically active soil surfaces and possibly by the precipitation of aluminosulfate minerals. The occurrence of chemical adsorption of sulfate by soils is well known, particularly in soils containing substantial amounts of hydrous oxides of Fe and Al. Furthermore, it is well recognized that adsorption of SO_4^{2-} from solution is often accompanied by an increase in pH, which leads to the simplest concept of buffering by this process, that is, the simple exchange of SO_4^{2-} for OH^- groups on the clay surfaces (Chao et al., 1965; Rajan, 1979).

$$H_2SO_4 + 2\,OH\text{-}X \rightleftharpoons SO_4^{=}X + 2\,H_2O \qquad (21)$$

To the extent this simple anion exchange mechanism is accurate, the buffering effect is obvious; 2 mol of OH^- are released for each mole of SO_4^{2-} adsorbed. However, at high levels of SO_4^{2-} adsorption, Rajan (1979) noted an increase in negative charge associated with the clays. This suggests that at some sites less than 2 mol of OH^- were displaced per mole of SO_4^{2-} adsorbed, although the residual charge on the SO_4^{2-} was free to contribute to the cation exchange capacity. This results in a somewhat more complex effect on buffering and a possible feedback mechanism between sulfate adsorption and cation exchange capacity, thus affecting the ion exchange buffer. The degree to which this occurs and the possible significance to sulfate adsorption effects on acidic deposition cannot be reliably evaluated at present.

Several points must be clearly understood concerning sulfate adsorption. First, the process is concentration dependent; that is, the adsorption capacity varies with the SO_4^{2-} concentration so that there is no fixed sulfate adsorption capacity associated with a particular soil. If SO_4^{2-} inputs increase due to acidic deposition, the adsorption capacity increases by an amount dependent on the slope of the isotherm. A new equilibrium between SO_4^{2-} inputs and soil solution concentration will not be attained until the corresponding increase in adsorption capacity is satisfied. This "buffering" is essentially a delay mechanism.

Second, the adsorption capacity is highly soil dependent. There is some evidence that Spodosols, which are common in the northeastern United States, Canada, and Scandinavia, adsorb much less sulfate than do the Ultisols and related soils commonly found in the southeastern United States (Johnson and Todd, 1983).

Finally, the reversibility of SO_4^{2-} adsorption may be a major factor in determining the recovery of soils and waters from acidic deposition effects. If the process is completely reversible, elevated levels of SO_4^{2-} in solution may persist long after deposition ceases. Although the process is at least partially reversible in the laboratory, whether it will prove to be more or less so over longer time periods is not known. Section III contains further discussion of the implications of these processes.

Several authors (e.g., Nordstrom, 1982; Nilsson and Bergkvist, 1983; Prenzel, 1983) have suggested that sulfur contained in acidic deposition might be immobilized in the form of minerals containing both aluminum and sulfate. Minerals in this category include alunite [$KAl_3(OH)_6(SO_4)_2$], basaluminite [$Al_4(OH)_{10}SO_4$], and jurbanite [$Al(OH)SO_4$]. Below about pH 4.5, alunite might theoretically be more stable than gibbsite, if SO_4^{2-} exceeds about 100 μmol/L. Sulfate concentrations in excess of this value may be observed in soil solutions and groundwater in areas of high deposition (e.g., Lee, 1985; Nilsson and Bergkvist, 1983; Johnson et al., 1985). We are not aware of any evidence that either precipitation of alunite or basaluminite is controlling the concentrations of SO_4^{2-} or Al_{3+} in catchments influenced by acidic deposition. There have been reports, however, that suggest the precipitation of jurbanite or a similar mineral (Nilsson and Bergkvist, 1983; Prenzel, 1983; Khanna et al., 1987, Weaver et al., 1985). Unfortunately, most of the evidence reported is in the form of stability diagrams plotted using pH (or pOH) on both axes, a method that tends to distort the diagram in a fashion that makes reliable interpretation very difficult. Probably the most convincing data concerning control of SO_4^{2-} by jurbanite are those of Weaver and others (1985). The point is particularly important in that precipitation of jurbanite would also exert a control on Al^{3+} activity. Nordstrom (1982) suggests that adsorbed SO_4^{2-} may be converted to one or more of these mineral forms on aging. Resolution of this question is difficult, as much of the observed behavior may logically be interpreted by either a precipitation or an adsorption mechanism.

D. Organic Buffering

Organic matter interacts with soil acidity in a variety of ways, some of which may not be buffering reactions per se. Therefore, our discussion of organic matter reactions in acidic soils is not strictly limited to buffering reactions. Organic matter may be present in a wide variety of chemical forms and size fractions. Of most interest here are short-chain organic acids in the solution phase and colloidal material in which carboxyl and phenol groups may dissociate, providing cation exchange capacity.

The colloidal material has a high charge density and may contribute the major fraction of the observed CEC, particularly in organic horizons. Its contribution to buffering may be considered as a form of ion exchange buffering. It may occur as a direct protonation of the dissociated carboxyl or phenolic groups, while basic cations are displaced in the process. However, the CEC is reduced so that charges that are protonated in this manner contribute to the pH-dependent charge of the system.

The dissolved organic acids play a major role in forest soils. The pH of forest soils may be below 3.5 just below the organic horizons, that is, H^+ exceeds 300 μmol L^{-1}. This requires a similar concentration of negative charge, which in natural forest soil is furnished almost entirely by organic anions. If these acids persist in the system, their direct protonation or dissociation provides a buffer that may be of considerable importance, particularly in terms of the chemistry of the drainage water. Of even more interest, however, is the fate of these anions.

As suggested above, Al trihydroxide minerals or layer silicates with hydroxy interlayers are unlikely to persist in soils below about pH 4.0. Conversely, due to Al buffering, pH values are unlikely to be below 4.0 in the presence of these minerals. Many organic anions have the ability to complex Al, a property that is important in reducing the toxic effects observed from Al in monomeric forms (Hue et al., 1986). As these very acidic waters dominated by organic acids percolate downward, they encounter soil layers where more soluble forms of Al occur, the pH rises due to Al buffering, and the Al brought into solution is complexed by the organic anions. When the solubility of these organoaluminum complexes is exceeded, they precipitate, a process that appears to be of considerable importance in the formation of the humic B_h horizon characteristic of Spodosols (podzolic soils). These processes are apparent in the data of Cronan and Aiken (1985), who found an average of 112 μmol(−) L^{-1} organic charge in leachates from the O/A horizons of soils from the Adirondack region of New York, whereas in leachates from the B horizon charge associated with organic anions averaged only 29 μmol L^{-1}. In addition to precipitation as humic complexes, organic acids may be consumed by microbial processes and lost from the system. Whether by precipitation or by microbial breakdown, the loss of the organic anions results in the loss of the organic acidity.

III. Capacity Effects of Acidic Deposition

The concepts of capacity and intensity are very useful when considering the effects of acidic deposition on soils and drainage waters. In this context we consider capacity effects to be those that deal with total amounts or quantities, such as the quantity of H^+ input per year or the quantity of base cations that are present or that may be lost as a result of acidic deposition. The "intensity" effects such as pH or the concentrations of various constituents present in soil solutions or drainage waters are dealt with in section IV.

A. Exchange Acidity

The concept of exchange acidity provides us with one of our simplest capacity measures of soil acidity. Exchange acidity may be determined by measuring the acidity, largely in the form of H^+ and Al species, using a buffered (usually pH 8.2) extractant such as $BaCl_2$-triethanolamine or by a neutral salt extractant such as KCl. The former includes potential exchange acidity to the pH of the buffer, and the latter includes only that acidity that may be released at the pH of the soil. In acidic soils with little or no pH-dependent charge, the two may be very similar, and the buffered extractant may give substantially higher values (a factor of two or more) in soils where pH-dependent or "variable" charge predominates.

As an example, let us consider a soil system 0.3 m in depth, having a cation exchange capacity (CEC) of 15 cmol(+) kg^{-1} and a bulk density of 1.11 g cm^{-3}, with 20% of the CEC consisting of exchangeable bases and 80% of exchange acidity. For convenience we will assume all fixed charge, so that buffered or unbuffered measurements would give similar results. Such a system would contain about 40 mol(+) m^{-2} of exchange acidity and 10 mol (+) m^{-2} of exchangeable bases (Figure 1–6).

We will further consider 1 m annually of acidic rainfall with an H^+ concentration of 63 μmol L^{-1} (pH 4.2) and dry deposition in an amount equal to wet deposition, for a total annual deposition of about 0.125 mol H^+ per square meter.* The annual input is very small compared to the exchange acidity (about 0.4%), and we would not expect significant changes in exchange acidity to occur unless very heavy deposition levels were encountered over long periods of time. Similar calculations have led to the conclusion that acidic deposition is unlikely to have a significant effect on soils and surface waters (Richter, 1983). For reasons discussed later, we consider such conclusions as inappropriate; however, it is clear that exchange acidity is unlikely to be sensitive to acidic deposition effects.

B. Exchangeable Bases

Depletion of exchangeable bases due to acidic deposition has been a major concern of most investigators. The shape of the ion exchange buffer curve (Figure 1–3) and the Ca^{2+}-Al^{3+} exchange isotherm (Figure 1–5) suggest that in most soils pH changes and Al mobilization are both likely to be small as long as base saturation exceeds about 15% or 20% of the charge present at soil pH. Ulrich (1987) also gives 15% base saturation as the threshold value at which Al^{3+} in solution increases significantly, and the range is consistent with that suggested by Reuss

*For comparison, Hultberg (1985) reported 0.106 mol m^{-2} at Lake Gjårdsjön in southern Sweden; Johnson and others (1985) reported 0.16 mol m^{-2} at Walker Branch near Oak Ridge, Tennessee; Johnson and others (1981) reported 0.096 mol m^{-2} (wet only) at Hubbard Brook, New Hampshire, and Khanna and others (1987) reported input of as much as 80 kg S ha^{-1} (0.5 mol H^+ m^{-2} if expressed as H_2SO_4) for spruce stands at Solling, FRG.

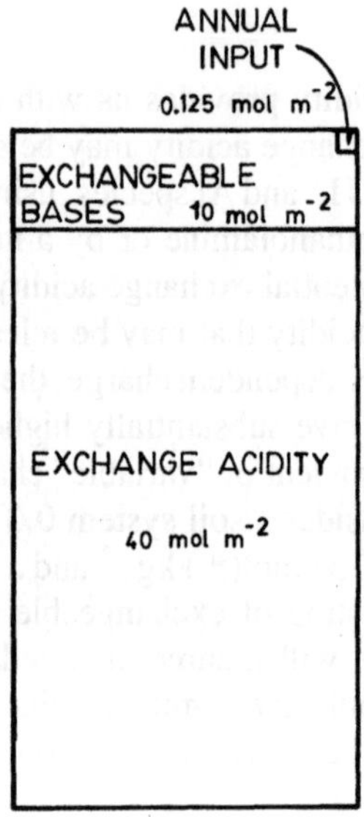

Figure 1–6. Relative amounts of H^+ input, exchangeable acidity, and exchangeable bases. See text for assumptions concerning deposition levels and soil parameters.

and Johnson (1986), as determined using a chemical equilibrium model. In the range in which the ion exchange buffer dominates, buffering of strong acidic inputs occurs as a result of removal of H^+ from solution by exchange and/or mineral dissolution, whereas charge balance is maintained as base cations on the exchange are brought into solution. The base cations are then subject to leaching as the soil solution percolates downward. This process accelerates the loss of base cations. In our example above, the exchangeable bases total 10 mol(+) m^{-2}, and the annual input of H^+ (i.e., the excess of strong acidic anions minus base cations) is 0.125 mol m^{-2}. A simplistic calculation that assumes 100% replacement and no sources of replenishment suggests that complete depletion would require some 80 years. Reduction from 20% base saturation to 15% would require 20 years and to 10% would take 40 years. These calculations suggest that in acidic (low base saturation) soils, the time frame for deleterious effects due to base depletion may be much shorter than that suggested by the total exchange acidity calculation in the previous section. The time required for serious depletion may be as short as a couple of decades on shallow soils with low CEC and low base saturation and could extend to centuries in deep soils that are more base rich. Although such projections require many assumptions and are subject to substantial errors, they do give a useful sense of the time scales that are likely to be involved.

Acidic deposition is not the only process that results in base cation depletion. Other processes include leaching in association with HCO_3^- formed by internal generation of H_2CO_3 and immobilization due to biomass accumulation and/or loss due to biomass removal. Several authors have pointed out that base depletion due to accumulation in forest biomass may be of the same magnitude as the accelerated base loss due to acidic deposition. For example, Johnson and others (1985) found base accumulation in the annual biomass increment to be 0.09 and 0.11 mol(+) M^{-2} for yellow poplar and chestnut oak sites, respectively, in eastern Tennessee. For comparison, input from acidic deposition was of 0.14 mol H^+ m^{-2} deposition, and 0.06 mol m^{-2} were leached in association with HCO_3^-.

Such comparisons are useful in assessing the capacity effects associated with various processes, but caution should be exercised. The effects of base depletion due to acidic deposition on soil solution composition may be quite different from that of HCO_3^- leaching, due to effects associated with the anions present in the acidic deposition.

C. CEC and Base Saturation

The true capacity factors defining base status are the total exchangeable bases and the exchange acidity as described above. However, because of the importance of base saturation in determining acidic deposition effects, and because CEC is often considered as a major factor in determining possible sensitivity of soils to acidic deposition effects (McFee, 1980), we shall briefly discuss the role of these parameters.

CEC is the total number of negative charges available for exchange per unit of soil, and is usually expressed on a weight basis. The number of charges available for exchange varies with pH in many soils. Most traditional measurements of CEC are carried out in buffered systems at either pH 7.0 or 8.2 and include both pH-dependent (variable) charge and non-pH-dependent (permanent) charge. Base saturation is a ratio, the amount of exchangeable bases, expressed as a fraction of this total charge. If we are concerned with the total capacity of the system to retain exchangeable bases, the appropriate measure of CEC is one that includes variable charge. However, if our goal is to determine whether or not acidic deposition is likely to result in mobilization of Al, the base saturation should be calculated using the CEC value determined at the in situ pH of the soil solution. A problem exists if we wish to make calculations to project the amount of base that must be removed to reach this threshold for Al mobilization, as we are often unable to predict the amount of pH-dependent charge that will be deactivated as base depletion proceeds.

High CEC alone does not necessarily indicate a high capacity to tolerate acidic deposition without deleterious effects. This sensitivity is determined by the amount of base that must be removed before the threshold base saturation is reached.

D. Sulfate Adsorption Capacity

As the mechanisms by which sulfate adsorption acts as a buffer have been discussed in section II, C, 2, only a brief discussion of the capacity aspects is given here. Several investigators (e.g., Hultberg, 1985) indicate that sulfate concentrations in lakes and streams are much less variable than is water flow. This suggests that a source-sink or buffer is operating in the system. This buffer is generally presumed to be a SO_4^{2-} adsorption mechanism. Using phosphate extractions of the B horizons of Spodosols from the northeastern United States, Fuller and others (1985) suggested a range of from 0.5×10^{-3} to 2.0×10^{-3} mol S kg^{-1}. This is similar to the values determined by Johnson and Todd (1983) for an Ultisol, a soil order generally considered to adsorb more SO_4^{2-} than Spodosols, although Fuller and others (1985) suggest the latter values may be low due to analytical problems. For 30 cm of B horizon, the values reported by Fuller and others would be equivalent to the S input in 0.16 to 0.64 mol H_2SO_4 m^{-2} (0.32 to 1.32 mol H^+ m^{-2}), or about 2.5 to 10 years input for our earlier illustration.

Other lines of evidence support the higher adsorption capacity of the more highly weathered soils. Rochelle and others (1987) examined the SO_4^{2-} retention of 40 intensively studied sites of <500 ha in the United States and Canada. Most sites south of the extent of Wisconsin glaciation show a net retention of SO_4^{2-}, whereas those north of this limit tend to have no net retention. Retention at Coweeta in North Carolina exceeded 80% of the input. Soil differences were also apparent. High net retention was characteristic of Ultisols, and low net retention was found in Spodosols. Smith and Alexander (1983) noted a tendency for positive trend slopes, that is, increasing SO_4^{2-}, for catchments in the southeastern United States, whereas the trend slope in the northeast actually tended to be negative. These results suggest that southeast catchments generally have not reached equilibrium with current levels of deposition, and at current levels in the northeastern United States and eastern Canada, there is likely to be little or no remaining buffering capacity due to the SO_4^{2-} adsorption mechanism. The possibility of further increases in SO_4^{2-} levels in streams in the southeast could have considerable implications for surface water acidification in this area.

IV. Intensity Effects

The capacity effects of acidic deposition mentioned above tend to accrue over relatively long time scales, typically decades or even centuries. Soil solution composition may change much more rapidly, and we shall refer to these changes as *intensity* effects. Although the nature and extent of these intensity responses depend on the current status of the capacity factors, this dependence is often highly nonlinear; that is, a small change in capacity factors may result in a relatively large effect on solution composition. In this section we shall consider effects related to neutral salt content, Al ion concentrations, and pH.

A. Neutral Salt Concentration

The solution concentration of the SO_4^{2-} and NO_3^- ions responds rapidly to inputs to H_2SO_4 or HNO_3. The magnitude of the change depends on the status of the factors that determine anion immobilization, as discussed above. In a typical aggrading forest in the northeastern United States or Canada, most NO_3^- is likely to be taken up and accumulated in the biomass, and thus changes in solution NO_3^- concentration are likely to be small. The SO_4^{2-} concentration will respond at a rate that depends largely on the sulfate adsorption properties of the soil and approach a new equilibrium concentration that depends on the level of input, the concentration that occurs due to evapotranspiration, and the sulfur-cycling characteristics of the system. Typical soil solution concentrations in areas affected by acidic deposition would appear to be in the range of 100 to 200 μmol L^{-1} (Table 1–2), although much higher values may be found in very high deposition areas. For example, Khanna and others (1987) report values ranging from about 10^{-4} to $10^{-3.2}$ molar (100 to 630 μmol L^{-1}).

If the exchange complex is well supplied with base cations, that is, above perhaps 15% to 20% base saturation (Figure 1–3), the system will most likely be within the ion exchange buffer range. In this case the negative charges on the SO_4^{2-} are balanced by base cations, primarily Ca^{2+} and Mg^{2+}, and the major response is an increase in the concentration of sulfate salts in the leachate. As discussed above, this increased solution concentration has obvious implications in relation to base depletion. The increased base cation concentration in solution may also have

Table 1-2. Sulfate concentrations (μmol L^{-1}) in soil solution in areas affected by acidic deposition.

	Walker Branch, TN[a]	
Horizon	Chestnut oak (Fullerton series)	Yellow poplar (Tarklin series)
02	172	173
A1	128	100
A2	165	174
B2	90	194
	Adirondack Mountains, NY[b]	
Horizon	Hardwood	Conifer
0	41	66
A	96	151
B	104	121
	Lake Gjårdsjön, Sweden[c]	
Depth (cm)	Concentration	
5	233	
15	137	
35	150	
55	182	

[a] From Johnson et al. (1985).

[b] From Mollitor and Raynal (1982).

[c] From Nilsson (1985).

significant implications in relation to the rate of nutrient uptake and biogeochemical cycling. Although the potential importance of this effect to ecosystem processes has been recognized (e.g., Johnson et al., 1985), there is yet very little evidence as to possible long-term ecosystem effects.

B. Al Mobilization Effects

If the increased concentration of the SO_4^- and possibly NO_3^- anions associated with acidic deposition inputs cannot be balanced by base cations through the ion exchange buffer mechanism, in well-drained mineral soils they will most likely be balanced by either H^+ or Al species. For example, in lysimeter experiments under heather *(Calluna vulgaris)* with a soil pH of 4.0 (1:2.5 H_2O), Stuanes and Abrahamsen (1980) found that an increase of 1 μmol(−) L^{-1} SO_4^{2-} resulted in an increase of 0.4 to 0.6 μmol(+) L^{-1} of (Al^{3+} + H^+). Under moor grass *Molina coerulea)* where the soil pH was 4.4 to 4.6, SO_4^{2-} in solution was almost entirely balanced by base cations. Apparently ion exchange buffering predominated in the moor grass system, whereas under heather a combination of decreased pH and aluminum buffering was observed.

As discussed earlier in relation to the Al buffer mechanism, Al species [Al^{3+}, $Al(OH)^{2+}$, and $Al(OH)_2^+$] are brought into solution by mineral dissolution and/or ion exchange processes. These processes provide an effective buffer against major changes in H^+ concentration, but this buffering is accompanied by an increased concentration of these Al species in solution. Such increases may have serious consequences in terms of the toxicity of monomeric Al species to some plants and of loss of alkalinity and toxic levels of Al species in the drainage waters.

Although mobilization of Al by acidic deposition may occur after an initial phase of cation depletion, cation depletion is not always a necessary precondition. Referring to Figure 1–5, we find that as solution strength increases, the Al in solution increases as well. This is not just a simple matter of the total Al in solution increasing in proportion to the total solution strength, as the fraction of total charge in solution comprised of Al species increases as well. The reason for this effect is not intuitively obvious. However, by taking the square root of both sides of Equation 14 and rearranging, we obtain

$$(Al^{3+}) = (Ca^{2+})^{3/2} \left(\frac{(AlX)}{K_s^{1/2}\,(CaX)^{3/2}} \right) \tag{22}$$

Equation 22 indicates that for any specific values of CaX and AlX, the activity of Al^{3+} will be proportional to the 3/2 power of the Ca^{2+} activity. As solution strength increases, the fraction of Al^{3+} in solution must also increase. The implication for acidic deposition effects is that for soils low in exchangeable bases, significant Al^{3+} mobilization may occur simply as a result of the increased solution strength associated with acidic deposition, without the necessity for futher depletion of the exchangeable base cations (Reuss and Johnson, 1985, 1986).

C. Effects of pH

1. Soil pH

Hydrogen ion activity or pH is one of the most commonly measured soil parameters; to the nonspecialist it is often the parameter of most interest. It may be measured in a variety of ways, each of which has certain advantages and limitations. Unfortunately, each of these methods gives a different pH value, and interpretation is often difficult.

The pH value most relevant to soil chemical and plant processes is the actual pH of the soil solution. Unfortunately, this parameter is both difficult to measure and variable over time. We have previously pointed out that solution strength affects chemical composition in a complex manner, and this is also true for the H^+ ion; changes in pH are not simply due to the effect of concentration or dilution. Taking the square root of Equation 12 and rearranging, we have

$$H^+ = (Ca^{2+})^{1/2} \frac{K(HX)}{(CaX)^{1/2}} \quad (23)$$

At a given level of Ca (or Ca + Mg) saturation, the H^+ activity is proportional to the square root of the Ca^{2+} activity. If one measures pH in a 1:2 soil-water suspension, the value obtained will usually be noticeably lower than that obtained in a 1:5 suspension. The difference, however, will be less than that expected due to simple dilution of the H^+ in solution because H^+ varies relatively less than does Ca^{2+}. Similarly, the pH of the soil solution at field capacity will be lower than that measured in a 1:2 soil-water suspension.

One of the problems with tracking field pH is that pH measured in water suspensions may fluctuate substantially over time. For example, consider a season of low to moderate rainfall when evapotranspiration exceeds precipitation, followed by a wet season during which most of the rainfall percolates through the soil. During the season of low rainfall, incoming anions (SO_4^{2-}, NO_3^-, and Cl^-) are not removed by deep percolation and thus tend to concentrate. This concentration will have the effect of lowering pH as measured in water. Similarly, as salts are removed during the wet season, the pH will increase. This dilution effect will tend to increase both soil solution pH and pH as measured in a soil-water suspension.

Because of these limitations of water pH measurements, many soil scientists prefer to measure pH in a salt solution, commonly 0.01 M $CaCl_2$. The pH values measured in this manner are commonly of the order of 0.8 to 1.0 unit lower than those measured in water. The relatively high $CaCl_2$ concentrations tend to mask changes in salt concentration in the soil, and thus measurements are much less variable than in the case of water suspensions. Furthermore, because Ca^{2+} activity is nearly constant in this $CaCl_2$ solution, pH measured in this manner reflects more directly the Ca + Mg saturation of the exchanger; that is, if Ca^{2+} is fixed, the right side of Equation 23 depends only on the Ca + Mg saturation. To a large extent the $CaCl_2$ pH is a measure of base status of the exchanger rather than a direct indication of the H^+ ion concentration of the soil solution. However, this is not strictly true for soils of low CEC and low base saturation, as in this case the Ca + Mg saturation may be increased substantially by exchange with the salt solution. The essential point for our present purpose is that changes in $CaCl_2$ pH brought about by acidic deposition effects will only occur as a result of changes in base status. They are unlikely to reflect changes that occur in the soil solution due to the effect of acidic deposition on the soil solution strength.

In addition to the measurement problem, there are other inherent difficulties in evaluating the effect of acidic deposition on soil reaction. Consider a hypothetical buffer curve such as Figure 1–7. In this case the Al buffer range and the ion exchange buffer range are clearly expressed. Soils with moderate to high base saturation are well within the ion exchange buffer range. In this situation even long-term exposure to acidic deposition would result mainly in somewhat elevated levels of neutral salts in the soils and drainage waters, whereas effects on soil reaction would be small. However, if the system is near the transition from the ion exchange to the Al buffer range, changes in pH of soil solution and drainage water would occur much more quickly. As base cations are further depleted, the system enters into the Al buffer range where further changes in pH are small, as H^+ inputs serve mainly to mobilize Al.

From our previous discussion, the Al buffer range is fixed by the solubility of Al in the system as represented by K^o. For example, amorphous $Al(OH)_3$ would tend to buffer the system near pH 4.5, whereas the buffering would probably occur near pH 4.0 if gibbsite were the controlling phase. Very low pH values in mineral soils

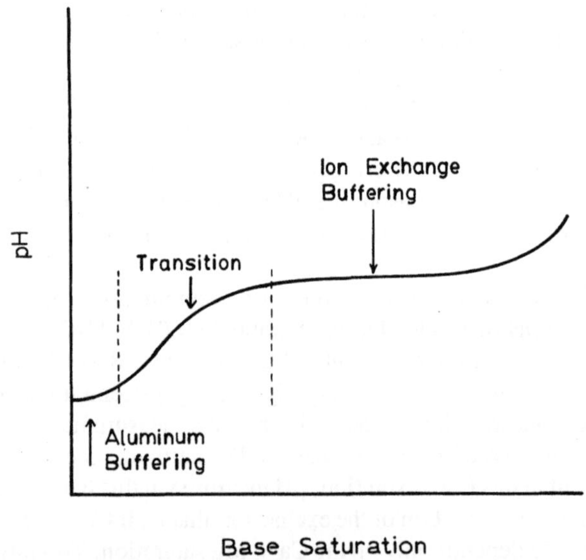

Figure 1–7. Idealized soil buffer curve illustrating the aluminum and ion exchange buffer ranges.

indicate that Al solubility is controlled by relatively insoluble aluminosilicates such as smectite, or that pH may be controlled by the solubility of Fe-containing minerals (Ulrich, 1983, 1987). Thus the lower end of the buffer curve may vary from near 4.5 to less than 3.5 in various soils. In the former case, there would likely be a marked overlap between the ion exchange buffer range and the Al buffer range, and the transition range where the pH of the system would respond quite rapidly to acidic deposition inputs would not be present. In this case the transition from neutral salt formation effects to Al mobilization would occur without a marked change in pH. Conversely, in the absence of relatively soluble Al materials, the transition range would be clearly expressed.

The ion exchange buffer range also varies somewhat, tending to be lower for high charge density exchangers such as montmorillonitic clays than for lower charge density materials such as kaolinite (Clark and Hill, 1964). In our formulations the pH of the ion exchange buffer range depends on the Ca^{2+}-Al^{3+} selectivity coefficient. Variations in this parameter also affect the expression of the transition range, although generally the magnitude of this effect is less than that for the solubility of Al.

2. Leachate pH

Another technique is the measurement of pH of leachates collected using various types of lysimeters (Mollitor and Raynal, 1982; Turner et al., 1985; Johnson et al., 1985). Although this method may also be useful, the pH of leachates may be very different from the pH of the soil solution. Carbon dioxide partial pressures in the soil are often well above atmospheric levels and result in the formation of basic bicarbonates, even in acidic soil solutions.

$$2\ CO_2 + 2\ H_2O \rightleftharpoons 2\ H^+ + 2\ HCO_3^- \qquad (24)$$

$$2\ H^+ + CaX \rightleftharpoons HX + Ca^{2+} \qquad (25)$$

$$2CO_2 + 2H_2O + CaX \rightleftharpoons HX + Ca^{2+} + 2HCO_3^- \qquad (26)$$

In a system subject to Al buffering reactions, the ANC or alkalinity of the solution is given by

$$ANC = [HCO_3^-] - [H^+] - \Sigma[Al^+] \qquad (27)$$

where the brackets indicate concentration and $\Sigma[Al^+]$ indicates the total positive charge associated with the Al species (the CO_3^{2-} can be neglected in acidic systems). The relationship between pH and ANC for leachates, given three different partial pressures of CO_2, is shown in Figure 1–8. At atmospheric CO_2 (about 0.03%), pH is very sensitive to ANC in the range of about −25 to +25 μmol L^{-1}. However, outside this range pH is quite insensitive to ANC. Furthermore, for positive or even moderately negative ANC values, if soil CO_2 levels are above those in the atmosphere, the pH will rise significantly on equilibration with the atmosphere (Reuss and Johnson, 1985). The formation of basic bicarbonates (Equations 19 through 21) depends on the CO_2 partial pressure

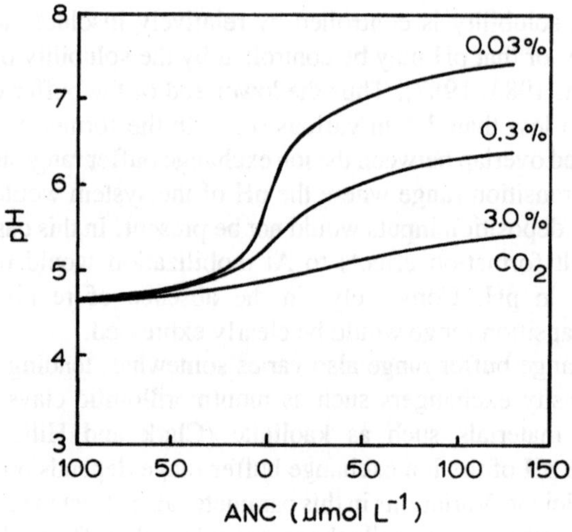

Figure 1–8. The relationship between pH and ANC in solutions draining from a soil for which the Al solubility is described by $3pH + \log(Al^{3+}) = 9.1$. Calculated using the model of Reuss and Johnson (1986).

in the soil. As soil CO_2 may be highly variable over time, the ANC of the leachates may also be highly variable. Furthermore, mobilization of Al species and/or increases in H^+ due to acidic deposition tend to decrease ANC. If the variations in ANC due to either changes in CO_2 partial pressure or changes in acidic inputs are such that ANC fluctuates in the range of about -25 to $+25$ μmol L^{-1}, major changes in leachate pH may be observed. Fluctuations outside this range, however, are likely to have relatively small effects on leachate pH.

Considering the complexity of the system, the difficulties in assessing acidic deposition effects by measurements of leachate pH are obvious. Under conditions where the mobilization of Al and decreases in solution pH result in a change from positive to negative ANC, the effect may be large. Under other circumstances, effects may be very small and/or highly inconsistent.

V. Discussion

In this section we first summarize the effects that are likely to occur as a result of acidic deposition inputs to acidic soils, as they relate to the processes discussed in the previous sections. We then discuss results or concepts that have been reported by other investigators within the framework of these processes.

A. Summary of Expected Effects

1. To the extent that anion immobilization occurs, there will be little or no effect on the system. The most common processes resulting in anion mobilization are

the biological uptake of nitrogen and adsorption of sulfate. Sulfate adsorption can be thought of as a delay mechanism, as the system will tend to reach a new steady state at which the SO_4^{2-} concentration of the soil solution and drainage water is in equilibrium with the input.

2. If the soils are well supplied with bases (base saturation above about 15% to 20% of the negative charge active at soil pH), the ion exchange mechanism will dominate. Increases in the concentration of anions that form strong acids brought about by acidic deposition will result mainly in an increased concentration of neutral salts in the soil solution and drainage water. Long-term exposure will result in base depletion due to export of cations. Soil solution pH will likely decrease slightly due to the higher concentration; this is a "salt" effect.
3. At very low base saturation, at least part of the increased charge on the strong acidic anions in solution will be balanced by either H^+ or Al species. These systems can be expected to respond by a decrease in pH of the solution phase, mobilization of inorganic Al species in solution, or both. The relative amounts of H^+ and Al species will depend to a large degree on the mineral phase(s) that control aluminum solubility. Where Al is controlled by relatively soluble materials (gibbsite, microcrystalline gibbsite, or amorphous aluminum trihydroxides), the major response will be an increase in Al species, with only a small change in solution pH.
4. Leachate pH may be highly dependent on CO_2 partial pressure in the soil. However, leachate pH is likely to be most affected in the range where the transition from ion exchange buffering to Al buffering occurs.

B. Discussion

1. ΔM/ΔH

Some authors (e.g., Wiklander and Andersson, 1972; Krug and Frink, 1983) have pointed out that leaching of base cations (ΔM) in response to acidic inputs may be substantially less than the change in H^+ from the input to the leachate (ΔH), that is, ΔM/ΔH is less than 1.0; in fact, Krug and Frink (1983) state that for soils below pH 5.0, ΔM/ΔH will be much less than 1.0. The mobile anion concept (Seip, 1980) indicates that cation removal requires a "mobile" anion moving in solution. Charge balance must be maintained in the solution. This principle suggests that if ΔM/ΔH is to be less than 1.0 either the anion in the acidic input is retained in the system, or the charge associated with base cations in the leachate is less than that associated with the acidic anion, in which case the charge balance must be maintained by either H^+ or Al species. In the latter case, if the system is in the Al buffer range, we would expect the major response to be Al mobilization, whereas in the transition range (Figure 1–7) the solution pH would be depressed. The essential point is that the reduction in cation depletion implied by a ΔM/ΔH value less than 1.0 is likely to be attained at the price of an increase in the solution concentration of the acidic cations, H^+ and Al species. Apparent values of ΔH/ΔM less than 1.0 that cannot be accounted for by Al mobilization may also be observed for organic horizons. These are discussed in the next section.

2. Canopy and Forest Floor Processes

The role of canopy and forest floor processes deserves special attention here, particularly in relation to the role of organic anions. The measured pH of throughfall may be either above or below that of the bulk precipitation. Decreases in pH and increased SO_2^{4-} are likely to be found in areas of conifers and high dry deposition (Khanna et al., 1987) and may be attributed to the effectiveness of the conifers as collectors of dry deposition. Under hardwoods, the pH of throughfall is often higher than that of bulk deposition under hardwoods (Mollitor and Raynal, 1982; Richter et al., 1983; Lindberg et al., 1979).

By a careful analysis of the processes involved, it becomes apparent that this increase in pH as a result of canopy processes does not decrease the capacity of the deposition to acidify the soil or drainage water. First, the apparent neutralization comes about due to exchange processes; H^+ is lost from solution while base cations increase, resulting in the formation of neutral salts. There is no increase in the supply of the base cations in the canopy, so this neutralization must come about at the expense of the return of base cations from the canopy to the forest floor by the normal pathways. The major pathways in the absence of acidic deposition are leaching in association with short-chain organic anions and return in the litterfall. If the anion in the acidic deposition is taken up by the plants, as is likely to be the case for moderate amounts of NO_3^-, there is no net effect on the acid-base status of the system. If the anion passes through the system into the subsoil or drainage water, as is likely to be the case with SO_4^{2-}, the capacity either to reduce base saturation by accelerated cation leaching or to mobilize the acid cations is unaffected by the apparent neutralization that occurred in the canopy. This is true because the total amount of bases returned to the forest floor is the same, regardless of whether they were leached from the canopy by acidic deposition or returned by the normal pathways.

Some investigators (e.g., Krug and Isaacson, 1984; James and Riha, 1986) have noted that buffering occurs in organic horizons in excess of what can be accounted for by exchange for either base cations or Al species. James and Riha (1986) suggested that in their experiments this may have been due to the presence of dissolved Ca bound to organic ligands in the control treatments, whereas in the acid treatments this became free Ca^{2+}, and the organic ligands were protonated.

Krug and Frink (1983) had proposed a somewhat similar mechanism whereby organic acids are protonated by acidic deposition. This process would essentially cause a switch from organic acidity to mineral acidity. Some mechanism such as this may indeed be operating in organic horizons. It must be understood, however, that any amelioration of acidity brought about by buffering by organic acids persists in the system only to the extent that the organic acids are preserved. In most mineral soils, these organic acids are removed at some point in the profile (Section II, D above). In forest soils not subject to acidic deposition, most acidity is organic and is lost with the removal of the organic anions. However, under acidic deposition, as the anions move downward, the strong acid-forming anions persist after the organic anions have been lost. As these mineral anions move through the system, they either accelerate base loss or maintain the acidic cations (H^+ or Al species) in solution by the mechanisms discussed in the previous system. The crucial point here is that acidity that appears to have been buffered by organic acids may reappear at a lower point in the profile when the organic anions are lost. Furthermore, even if the organic anions had persisted into the drainage water in the predeposition era, the net effect is a switch from organic to mineral acidity. Although the implications of this process are not completely understood, there seems to be little doubt that the mineral acids are much more likely to support toxic levels of monomeric Al than are organic acids (Hue et al., 1986).

3. Artificial Acidification Experiments

If one examines the reports in the literature of experiments in which acids have been applied to soils, followed by measurement of various chemical parameters of the leachates and the soils, the results appear highly inconsistent. In isolation, results from the various experiments appear to support very different conclusions concerning the likely effects of such additions. Although we can discuss only a few of these experiments, we find that in most instances what appear to be contradictory results are actually consistent with the above concepts. For example, using lysimeters watered with pH 3.0 H_2SO_4, Skeffington and Brown (1986) in the United Kingdom found that base cations in the leachate increased, but most of the cation flux was due to Al species (Figure 1-9). An increase in adsorbed SO_4^{2-} over the control was found, which was consistent with a decrease in SO_4^{2-} flux. Both control and treated lysimeters received a relatively large acidic input due to nitrification, which is a common problem in studies using even "undisturbed" lysimeters or cores, so that in this case substantial Al mobilization occurred in the control treatments.

Conversely, using time-averaged inputs of pH 3.16 and 3.40, Bergkvist (1986) found in Sweden that base cation flux increased while total Al flux actually decreased, apparently due to a decrease in mobile organic complexes. The obvious interpretation would be that in the Swedish case, the dominant process was ion exchange buffering, whereas Al buffering was dominant in the UK experiment. Unfortunately, Bergkvist (1986) did not report anions, so further interpretation is difficult.

The data of Lee and Weber (1982) illustrate the delay in response that occurs with soils having a high SO_4^{2-} adsorption capacity. In lysimeters treated with pH 3.0, 3.5, and 4.0 rainfall, sulfate enrichment was not observed in solutions from a depth of 20 cm until approximately 8, 10, and 12 months, respectively, after initiation of treatments. Nearly 3 years were required for SO_4^{2-} concentrations to increase to the value of the rainfall inputs. No sulfate enrichment was observed at a depth of 1 m in any plots when the experiment was discontinued after 38 months of treatment. After SO_4^{2-} enrichment was observed at 20 cm, the major effect on

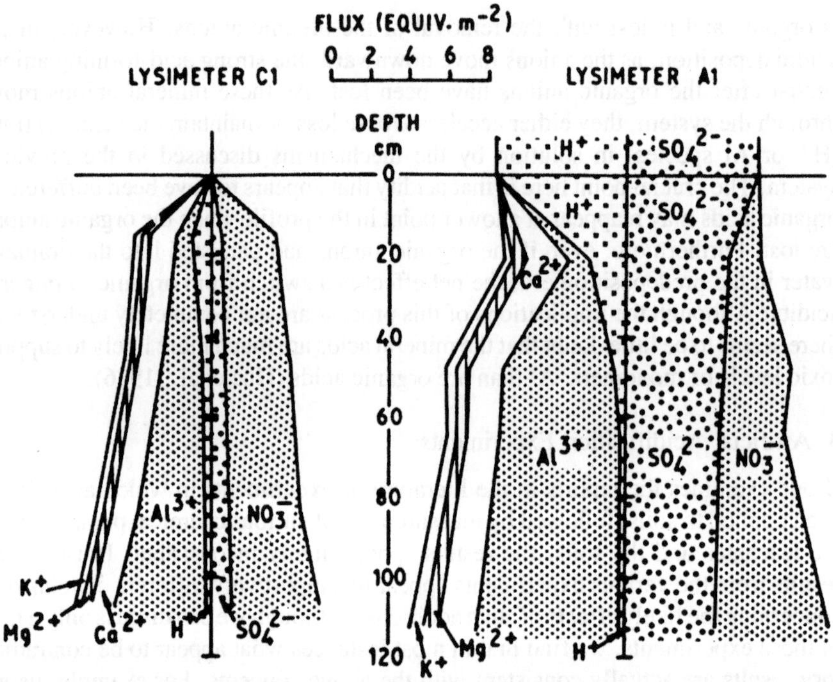

Figure 1–9. Fluxes of major cations in control (C1) and pH 3.0 acid treated lysimeters (A1). The width of any sector at a given depth represents the total flux of an ion that passed that point during the 4.75 years of the experiment. (From Skeffington and Brown, 1986. Copyright 1986 by D. Reidel Publishing Company. Reprinted with permission.)

solution composition was an increase in base cations, indicating ion exchange buffering. Some pH depression in the leachates was noted at high SO_4^{2-} concentrations. Unfortunately, Al was not measured.

C. Conclusion

From the above sections, it is apparent that the processes by which soils interact with acidic deposition are complex and that very different effects may be observed in different systems. As a result, experiments designed to measure these effects may yield very different results. If interpreted in isolation, results of different experiments appear to be contradictory. We have attempted here to provide a conceptual model by which these processes may be understood better and which can serve as a framework for interpretation of field and laboratory investigations. This model cannot be considered complete at the present time and very likely is not entirely correct in all aspects. Nonetheless, it is consistent with nearly all research results we have seen, and we are confident that in the main it is correct and will be confirmed by further investigations. Furthermore, it provides a framework for the formulation of testable hypotheses. Results from testing these hypotheses will allow future refinements and revisions that will improve our understanding of acidic deposition effects.

Acknowledgments

Contributions of the Colorado Agricultural Experiment Station, Project No. 6232. Although the research described in this article has been funded wholly or in part by the United States Environmental Protection Agency, Agreement No. CR812319, to Colorado State University, it has not been subjected to the agency's review and therefore does not necessarily reflect the views of the agency, and no official endorsement should be inferred.

References

Andersson, F., T. Fagerstrom, and I. Nilsson. 1980. *In* T. C. Hutchinson and M. Havas, eds. *Effects of acid deposition on terrestial ecosystems*. Plenum, New York. 319–334.
Bergkvist, B. 1986. Water Air Soil Pollut 34:901–916.
Binkley, D., and D. R. Richter. 1987. *In* A. MacFayden and E. D. Ford, eds. *Advances in ecological research,* vol. 16, 1–51. Academic Press, London.
Bloom, P. R., and D. F. Grigal. 1985. J Environ Qual 14:489–495.
van Breemen, N., P. A. Burrough, E. J. Velthorst, H. F. van Dobben, T. de Witt, T. B. Ridder, and H. F. R. Reignders. 1982. Nature 229:548–550.
van Breemen, N., and E. R. Jordens. 1983. *In* B. Ulrich and J. Pankrath, eds. *Effects of accumulation of air pollutants in forest ecosystems*, 171–182. Reidel, Boston.
van Breemen, N., and W. H. Wielemaker. 1974. Soil Sci Soc Amer Proc 38:61–66.
Carson, C. D., J. A. Kittrick, J. B. Dixon, and T. B. McKee. 1976. Clays and Clay Mineralogy 24:151–155.
Chao, T. T., M. E. Harward, and S. C. Fang. 1965. Soil Sci 99:104–107.
Clark, J. S., and H. G. Hill. 1964. Soil Sci Soc Amer Proc 28:490–492.
Coleman, N. T., and G. W. Thomas. 1967. *In* R. W. Pearson and F. Adams, eds. *Soil acidity and liming,* 1–34. American Society of Agronomy, Madison, WI.
Cosby, B. J., G. N. Hornberger, J. N. Galloway, and R. F. Wright. 1985. Water Resour Res 21:51–63.
Cronan, C. S., and G. R. Aiken. 1985. Geochim Cosmochim Acta 49:1697–1705.
Cronan, C. S., W. J. Walker, and P. R. Bloom. 1986. Nature 324:140–143.
Driscoll, C. T., J. P. Baker, J. J. Bisogni, and C. L. Schofield. 1984. *In* O. R. Bricker, ed. *Acid precipitation: Geological aspects,* 55–75. Butterworth, Boston.
Fitzgerald, J. W., T. C. Strickland, and W. T. Swank. 1982. Soil Biol Biochem 14:529–536.
Fuller, R. D., M. B. David, and C. T. Driscoll. 1985. Soil Sci Soc Am J 49:1034–1040.
Gaines, G. L., and H. C. Thomas. 1953. J Chem Phys 21:714–718.
Galloway, J. N., S. N. Norton, and M. R. Church. 1983. Environ Sci Technol 17:514A–545A.
Gapon, E. N. 1933. J General Chem USSR 3:144–163. (In Russian; original not seen.)
Grennfelt, P., and H. Hultberg. 1986. Water Air Soil Pollut 30:945–963.

Hooper, R. P., and C. A. Shoemaker. 1985. Science 229:463–465.
Huang, W. H., and W. D. Keller. 1971. Amer Mineralogist 56:1082–1095.
Hue, N. V., G. R. Craddock, and F. Adams. 1986. Soil Sci Soc Am J 50:28–34.
Hultberg, H. 1985. *In* F. Anderson and B. Olsson, eds. *Lake Gjårdsjön: An acid forest lake and its catchment* (Ecol Bull 37:133–157). Swedish Research Council, Stockholm.
James, B. R., and S. J. Riha. 1986. J Environ Qual 15:229–234.
Johnson, D. W., D. D. Richter, G. M. Lovett, and S. E. Lindberg. 1985. Can J For Res 15:773–782.
Johnson, D. W., and D. E. Todd. 1983. Soil Sci Soc Am J 47:792–800.
Johnson, N. M., C. T. Driscoll, J. S. Eaton, G. E. Likens, and W. H. McDowell. 1981. Geochim Cosmochim Acta 45:1421–1437.
Khanna, P. K., J. Prenzel, K. J. Meiwes, B. Ulrich, and E. Matzner. 1987. Soil Sci Soc Am J 51:446–452.
Kittrick, J. A. 1983. Clays and Clay Mineralogy 31:317–318.
Kramer, J. R., A. W. Anders, R. A. Smith, A. H. Johnson, R. B. Alexander, and G. Oehlert. 1986. *In Acid Deposition: Long-term trends,* 231–299. National Academy Press, Washington, DC.
Krug, E. C., and C. R. Frink. 1983. Science 221:520–525.
Krug, E. C., and P. J. Isaacson. 1984. Soil Sci 137:370–378.
Lee, J. R., and D. E. Weber. 1982. J Environ Qual 11:57–64.
Lee, Y. H. 1985. *In* F. Anderson and B. Olsson, eds. *Lake Gjårdsjön: An acid forest lake and its catchment* (Ecol Bull 37:109–119). Swedish Research Council, Stockholm.
Lindberg, S. E., R. C. Harriss, R. R. Turner, D. S. Shriner, and D. D. Huff. 1979. ORNL/TM-6674. Oak Ridge National Laboratory, Oak Ridge TN. 514 p.
Lindsay, W. L. 1979. *Chemical equilibria in soils.* John Wiley and Sons, New York. 449 p.
McBride, M. B., and P. R. Bloom. 1977 Soil Sci Soc Am J 41:1073–1077.
McFee, W. W. 1980. *In* D. S. Shriner, et al., eds. *Atmospheric deposition: Environmental and health effects,* 495–506. Ann Arbor Science, Ann Arbor, MI.
Mollitor, A. V., and D. J. Raynal. 1982. Soil Sci Soc Am J 46:137–141.
Nilsson, S. I. 1985. *In* F. Anderson and B. Olsson, eds. *Lake Gjårdsjön: An acid forest lake and its catchment* (Ecol Bull 37:120–132). Swedish Research Council, Stockholm.
Nilsson, S. I. 1986. *In* J. Nilsson, ed. *Critical loads for nitrogen and sulphur,* 211–221. Report from Nordic Working Group, Stockholm.
Nilsson, S. I., and B. Bergkvist. 1983. Water Air Soil Pollut 20:311–329.
Nordstrom, D. K. 1982. Geochim Cosmochim Acta 46:681–692.
Nordstrom, D. K., and J. W. Ball. 1986. Science 232:54–56.
Ogden, J. G., III. 1982. Water Air Soil Pollut 17:119–130.
Pleysier, J. L., A. S. R. Juo, and J. Herbillon. 1979. Soil Sci Soc Amer J 43:875–880.
Prenzel, J. 1983. *In* B. Ulrich and J. Pankrath, eds. *Effects of accumulation of air pollutants in forest ecosystems,* 157–170. Reidel, Boston.
Rajan, S. S. S. 1979. Soil Sci Amer J 43:65–69.
Reuss, J. O. 1978. *Simulation of nutrient loss from soils due to rainfall acidity.* US-EPA-600/3-78-053. Washington, DC. 45 p.
Reuss, J. O. 1980. Ecological Modelling 11:15–38.
Reuss, J. O. 1983. J Environ Qual 12:591–595.
Reuss, J. O., and D. W. Johnson. 1985. J Environ Qual 14:26–31.
Reuss, J. O., and D. W. Johnson. 1986. *Acid deposition and the acidification of soils and waters*. Ecological Studies Series 59. Springer-Verlag, New York. 119 p.
Richter, D. D. 1983. Environ Sci Technol 17:568–570.

Richter, D. D. 1986. Soil Sci Soc Am J 15:1584–1587.
Richter, D. D., D. W. Johnson, and D. E. Todd. 1983. J Environ Qual 12:263–270.
Rochelle, B. P., M. T. Church, and M. B. David. 1987. Water, Air Soil Pollut 33:73–83.
Russell, E. W. 1961. *Soil conditions and plant growth*. 9th ed. Longmans, London. 688 p.
Schofield, R. K., and A. W. Taylor. 1955a. J Soil Science 6:137–146.
Schofield, R. K., and A. W. Taylor. 1955b. Soil Sci Soc Amer Proc 19:164–167.
Seip, H. M. 1980. *In* D. Drablos and A. Tollan, eds. *Ecological impact of acid precipitation,* 358–366. SFNS project. Norwegian Institute for Water Research, Oslo.
Skeffington, R. A., and K. A. Brown. 1986. Water Air Soil Pollut 34:891–900.
Smith, R. A., and R. B. Alexander. 1983. U.S. Geological Survey Circular 910. Alexandria, VA.
Stuanes, A. O., and G. Abrahamsen. 1980. *Proc. international conference ecological impact acid precipitation,* 152–153. Norway SNSF Project, Oslo.
Turner, R. C., and J. S. Clark. 1964. Soil Sci 99:194–199.
Turner, R. S., A. H. Johnson, and D. C. Wang. 1985. J Environ Qual 14:314–323.
Ulrich, B. 1983. *In* B. Ulrich and J. Pankrath, eds. *Effects of accumulation of air pollutants in forest ecosystems,* 127–146. Reidel, Boston.
Ulrich, B. 1987. *In* E. D. Schulze and H. Z. Zwolfer, eds. *Ecological studies,* vol. 61, 11–49. Springer-Verlag, Heidelberg, FRG.
Vanselow, A. P. 1932. Soil Sci 33:95–113.
Weaver, G. T., P. K. Khanna, and F. Beese. 1985. Soil Sci Soc Amer J 49:746–750.
Wiklander, L., and A. Andersson, 1972. Geoderma 7:159–165.

Acidic Sulfate Soils

Jirapong Prasittikhet* and Robert P. Gambrell†

Abstract

Acidic sulfate soils are formed from potential acidic sulfate soils that are characterized by the accumulation of pyrite (FeS_2). Upon drainage of these soils, the pyrite oxidizes to produce sulfuric acid and hence acidic sulfate soils. The acid formed commonly decreases the soil's pH to less than 4, and sometimes the pH becomes as acid as 2. The acid produced has a major effect on chemical and microbial processes in soils. Substances detrimental to plant growth, such as Al^{3+}, Fe^{2+}, H_2S, and CO_2 are often generated in amounts toxic to plant growth.

In acidic sulfate soils, the potential for very acidic soil conditions to develop results from natural processes. Often, however, cultural disturbances of these potential acidic sulfate soils, such as drainage for agricultural purposes or other reasons, cause the chemical and microbial processes to occur that result in development of severe acidic conditions. Acidic sulfate soils occur worldwide in most climatic zones but are found primarily in tropical regions. Most acidic sulfate soils are found in coastal areas at low elevations, where they developed from pyrite-containing marine sediments. These soils are a major problem because of their harmful effect on crop production, and because of the large areas affected in regions of the world where additional food production is especially important. Acidic sulfate soils are generally far more acidic than can be generated by even severe acidic precipitation problems. Acidic precipitation is unlikely to affect adversely either acidic sulfate soils or, for different reasons, potential acidic sulfate soils.

Knowledge of acidic sulfate soil chemistry, the problems experienced with such soils, and feasible management practices to mitigate some of the problems should provide information on the type and direction of chemical changes that may occur in poorly buffered soils in other regions of the world as a result of acidic precipitation.

---

*Division of Soils, Department of Agriculture, Bangkok, Thailand.
†Laboratory for Wetland Soils and Sediments, Louisiana State University, Baton Rouge, LA 70803, USA.

I. Introduction

Acidic sulfate soils form from oxidation of potential acidic sulfate soils that have accumulated considerable amounts of pyrite (FeS_2). Pyrite formation requires (1) sulfide, which is derived from the reduction of sulfate in anaerobic sediments by sulfate-reducing bacteria (that is, *Desulfovibrio desulfuricans*), (2) soluble ferrous iron from the reduction of insoluble ferric compounds in sediments, (3) organic matter, which provides energy for bacteria, and (4) an alternating aerobic-anaerobic soil or sediment environment during the pyrite formation process or at least some periods of sediment oxidation during predominantly anaerobic conditions. Upon soil drainage, oxidation of pyrite produces sulfuric acid. The soil becomes acidic if the quantity of acid produced exceeds the acid-neutralizing capacity of the soil. The excess acid often decreases the soil's pH to less than 4 and sometimes to near 2. The acid produced has a major influence on chemical and microbial processes in soils. Substances detrimental to plant growth, such as Al^{3+}, Fe^{2+}, H_2S, and CO_2, are often generated in amounts beyond the critical toxic levels for normal plant growth. Also, these soils are usually infertile because they do not contain high enough levels of essential nutrients for good crop production.

Acidic sulfate soils are generally far more acidic than can be generated by even severe acidic precipitation problems. Acidic precipitation is likely to have a minimal effect on the pH of most potential acidic sulfate soils because of redox chemistry processes associated with soil flooding that can effectively deal with acid-forming components of precipitation. Pyrite is stable in potential acidic sulfate soils because of the continuous flooding and subsequent long-term anaerobic conditions. Under these conditions, acid-forming components of acidic precipitation such as sulfate and nitrate are reduced by processes that normally do not result in acid generation. Also, other chemical processes occur in flooded soils that tend to prevent development of acidic conditions (Gambrell and Patrick, 1978). Where such soils have been manipulated (drained) to permit strong acidification to develop, the additional acid-forming tendencies of acidic precipitation will likely be too small to have a significant effect on acidic sulfate soils.

In acidic sulfate soils, the potential for very acidic soil conditions to develop results from natural processes. Often, however, cultural disturbances of these potential acidic sulfate soils, such as drainage for agricultural purposes or drainage for other reasons, cause the chemical and microbial processes to occur that result in development of severe acidic conditions. Knowledge of acidic sulfate soil chemistry, the problems experienced with such soils, and feasible management practices to mitigate some of the problems should provide information on the type and direction of chemical changes that may occur in poorly buffered soils in other regions of the world as a result of acidic precipitation.

Acidic sulfate soils are a major problem because of their harmful effect on crop production and because of the large areas affected in regions of the world where additional food production is especially important. These soils are estimated to occupy an area of 12.5 million ha (FAO/UNESCO, 1979). Acidic sulfate soils occur worldwide but are found primarily in tropical regions, although some are found in temperate regions also (Kawalec, 1973). Most acidic sulfate soils are found in coastal areas at low elevations where they developed from pyrite-containing marine sediments (Pons, 1973). However, some acidic sulfate and potential acidic sulfate soils are also found in inland areas (Poelman, 1973) where they have formed from sedimentary rocks. Exposure of these buried pyritic rocks by mining or other excavation also produces acidic sulfate soils or acidic drainage that can adversely affect plant growth or streams receiving drainage (Pons, 1973).

Large areas of acidic sulfate soils are located in areas well suited and needed for flooded rice cultivation. Thus these soils are an important potential resource in densely populated areas in developing countries where rice is the main diet, especially in the coastal areas of southeast Asia and west Africa. Thailand, for example, is a major rice-growing country and has about 1.5 million ha of acidic sulfate and potential acidic sulfate soils (Pons and van der Kevie, 1969).

The development of acidic sulfate soils from potential acidic sulfate soils is affected in a major way by management of soil drainage, that is, intentional flooding and draining cycles necessary for wetland rice cultivation. Pyrite oxidation to sulfuric acid is substantially impeded during flooding cycles for lowland rice cultivation. However, flooding soils affect the plant availability of several other soil chemical components due to changes in both pH and redox potential conditions. Reducing conditions (low redox potential) associated with flooding may transform toxic metals and other materials to more soluble forms and make worse the adverse effects of acidic sulfate soils on plant growth. Oxidized conditions (high redox potential) that develop during drainage cycles often create a deficiency of essential nutrients and again increase levels of some toxic metals, this time due to the strongly acidic conditions associated with soil drainage and oxidation. Thus, a complex interaction between production of substances adversely affecting plant growth and the alternating reduced and oxidized conditions is imposed on the acidic sulfate soils as part of the rice-cultivation system.

II. Interactions of Soil Aeration

Because pyrite formation and its oxidation require an alternating aerobic-anaerobic soil or sediment environment, we will examine the interaction of soil aeration before more specifically defining acidic sulfate soils and their genesis.

A. Well-Drained Soils

In a well-drained soil, gas-filled pore spaces interconnected with the atmosphere surround individual soil particles and aggregates and permit relatively rapid gaseous diffusion of oxygen throughout the plant-rooting depth. In some soils, gaseous oxygen content may be reduced with depth (Russell, 1961), but molecular oxygen transport across the gas-liquid interface of the soil solution is sufficient to maintain some dissolved oxygen in the solution and thus maintain the soil in an oxidized condition. This process is usually sufficient to meet the soil and root oxygen demands.

In a permeable upland soil, excess water from precipitation, irrigation, or temporary flooding rapidly drains from the upper profile through the interconnected pore spaces. After several hours of draining, most of the pore space refills with gas that is continuous with the atmosphere. Because of the hydrophilic nature of mineral and organic soil components, however, thin, oxidized moisture films of one-third bar tension or less remain around individual soil particles and aggregates. This relatively thin film of moisture provides a medium for chemical reactions affecting nutrient availability, functions as a nutrient and moisture reservoir, and supports transport of nutrients from the soil solid phase to the plant root system.

B. Flooded Soils

Flooding prevents gaseous diffusion of oxygen into the profile. The oxygen content of the soil solution immediately declines in response to the combination of prolonged flooding, continued oxygen demand for root and microbial respiration, and chemical oxidation of reduced organic and inorganic components, and the oxygen content may be depleted in several hours or a few days (Turner and Patrick, 1968). Flooding an otherwise aerobic soil may cause a surge of microbial activity that often causes strong reducing conditions to develop rapidly, especially if a substantial organic energy source is available. Most continuously flooded soils have no measurable oxygen beneath the surface. The failure of oxygen resupply to meet oxygen demand in flooded soils and sediments results in deep, reduced horizons.

Flooding and the resulting restriction of gaseous transport of oxygen into a soil or sediment does not necessarily produce a uniformly reduced profile, however. For several reasons, a small but continuous supply of dissolved oxygen is generally available at the soil- or sediment-floodwater interface. Thus, a thin, oxidized surface horizon may overlie a deep, reduced horizon as a result of dissolved oxygen from the overlying floodwater diffusing across the surface water–soil or –sediment interface (Patrick and DeLaune, 1972; Ponnamperuma, 1972). Shallow surface waters flooding soils typically contain a uniform distribution of several μg/ml of dissolved oxygen, which often approaches saturation levels. The rapid rate of oxygen transport across the atmosphere—surface water interface, the small population of oxygen-consuming organisms present, the photosynthetic oxygen production by algae within the water column, and surface water mixing by convection currents and wind action combine to produce this relatively high and uniform oxygen content. Molecular diffusion of dissolved oxygen in response to an oxygen concentration gradient provides further oxygen transport into quiescent interstitial waters.

Flooded agricultural and wetland soils and shallow sediments are usually characterized by this thin, oxidized surface layer of uniform thickness that overlies deep, anaerobic, subsurface horizons. The biological and chemical processes occurring in this oxidized surface horizon strongly influence the availability of both nutrients and toxins in flooded soils and sediment-water systems. The oxidized surface horizon is thin, ranging from a few millimeters to a few centimeters deep, because of the slow rate of oxygen transport through interstitial water and the comparatively high oxygen demand. Within this thin horizon, however, microbial metabolism and chemical transformations of nutrients and potential toxicants are similar to those in aerobic soils.

The oxidized layer's depth depends on a balance between the rate of oxygen diffusion into the surface horizon and its consumption (Mortimer, 1942). Although oxygen consumption rates have long been considered a function of microbial respiration, Howeler and Bouldin (1971) experimentally demonstrated that models including oxygen consumption for both biological respiration and chemical oxidation of both mobile and nonmobile constituents can best describe oxygen consumption rates in some flooded soils. Reduced iron and manganese ions were thought to represent the bulk of the mobile reductants that diffused upward from the reduced horizon into the oxidized zone. Precipitated ferrous iron, manganous manganese, and sulfide compounds appearing as the oxidized zone increased in thickness probably constituted much of the nonmobile constituents. Chemical reductants such as ferrous iron accounted for about 50% of the oxygen consumption in the soils and sediments studied, although Howeler (1972) noted that the ratio between biological and chemical oxygen consumption rates may vary widely and depends on the organic matter content of the soils or sediment.

The intensity of oxidation-reduction also influences the total ion content or ionic strength of soil solutions and interstitial waters. In an oxidized soil, flooding and the subsequent reduction initially causes the total ion content of the soil solution to increase. Over time the ionic strength of the reduced soil decreases and again approaches prereduction levels (Patrick and Mikkelsen, 1971). The solubilization of iron and manganese (which form from insoluble ferric and manganic oxides and hydroxides) in acidic and near-neutral soils contributes to the increase in ionic strength upon reduction. Gotoh and Patrick (1972, 1974) found substantial increases in water-soluble manganese and iron in a Crowley silt loam as the redox potential of the soil was lowered. A decrease in the reducible levels accompanying an increase in the relatively mobile forms of these metals indicated that some of the insoluble oxides and hydroxides became unstable and dissolved under reduced conditions.

III. Definition and Taxonomy of Acidic Sulfate Soils

Acidic sulfate soils have been described as having one or more of the following characteristics: (1) a pH below 4 within the top 50-cm depth due to oxidation of pyrite (FeS_2) and the subsequent formation of sulfuric acid (van Breemen, 1982), (2) a sulfuric horizon composed of either mineral or organic soil material with a pH below 3.5 and yellow jarosite mottles (USDA, 1975), and, (3) a flooded soil that contains mineral or organic materials with more than 0.75% sulfur and less than three times as much carbonate expressed as $CaCO_3$ as sulfide sulfur (USDA, 1975).

Van Breemen (1982) proposed to define sulfuric material and a sulfuric horizon as follows:

> Sulfidic material is waterlogged mineral, organic, or mixed soil material with a pH of 3.5 or higher, containing oxidizable sulfur compounds, which, if incubated as a 1-cm thick layer under moist, aerobic conditions (field capacity at room temperature), shows a drop in pH of at least 0.5 unit to a pH below 3.5 within 4 weeks. A sulfuric horizon is composed of mineral, organic, or mixed soil material, generally containing yellow jarosite mottles with a hue of 2.5Y or yellower, and a chroma of 6 or more, that has a pH <3.5 (1:1 in water) and contains at least 0.05% water soluble sulfate.

Soils that are influenced by pyrite oxidation but are not sufficiently acidic (pH is not below 4) to be classified as acidic sulfate soils are called *para* or *pseudo acid sulfate soils* (Pons, 1973). These soils may have formed where relatively small amounts of pyrite were present in the parent material, or they may have at one time been acidic sulfate soils from which most of the pyrite has oxidized and the acid leached out over time. These soils are usually classified in taxons other than those used for acid sulfate soils, frequently as Tropaquepts or Haplaquepts.

Van Breemen (1982) further defines potential and acidic sulfate soils as follows:

> Potential acid sulfate soils are either Sulfaquents (Aquents with sulfuric material within 50 cm of the mineral soil surface), Sulfic Fluvaquents (fluvaquents with sulfuric material between the 50- and 100-cm depth), or Sulfihemists (Histosols with sulfidic material within the 100-cm depth). Acid sulfate soils can be classified as: a) Sulfaquepts (Aquepts with a sulfuric horizon that has its upper boundary within 50 cm of the soil surface), b) Sulfic Tropaquepts (Tropaquepts with jarosite mottles and a pH 3.5 to 4 somewhere within the 50-cm depth, or with jarosite mottles and a pH <4 in some part between 50- to 150-cm depth), or, c) Sulfic Haplaquepts (comparable to Sulfic Tropaquept, but under a more temperate climate). Acid sulfate soils that are organic may be Sulfihemists (Histosols with a sulfuric horizon that has its upper boundary within 50 cm of the surface).

Van Breemen (1982) points out the agronomic significance of the taxonomic groupings in that Sulfaquepts are usually unsuitable for crop production without excessively expensive management practices. However, often the Sulfic subgroups can be made productive in an economically feasible manner. A proposed description of changes in time of seasonally flooded acid sulfate soils is described by Harmsen and van Breemen (1975). The soil changes in the sequence are greatly affected by drainage and leaching. In the Bangkok Plain of Thailand, for instance, acidic sulfate soil profile development exhibits three different stages. A mangrove soil in its undrained state was said to represent a Sulfaquent, and Sulfaquept and Sulfic Tropaquept refer to the older and more deeply developed acidic sulfate soils. The depth to the pyrite-containing horizon is deeper in older and more developed acidic sulfate soils and is found beneath the brown and yellow mottled (containing jarosite deposits) B horizon and a typically black A horizon. Sulfaquepts exhibit brown and yellow mottled surface soil materials and contain the gray pyritic substratum at about 50 cm beneath the surface. Soils exhibiting the least developed profile are Sulfaquents that contain very few brown mottles at the surface and the unmottled gray pyritic substratum near the surface. Van Breemen (1982) noted that as acidic sulfate soils become older, more developed, and better drained, the characteristic horizons are found at greater depths.

IV. Genesis of Acidic Sulfate Soils

The formation of acidic sulfate soils requires a physiography or favorable environment that provides the potential for pyrite formation and, subsequently, the oxidation of pyrite upon drainage as a result of natural processes or human activities. Pons and van der Kevie (1969) stated that two primary processes summarized the genesis of acidic sulfate soils, a geogenetic process and a pedogenetic process. Formation of pyrite (sulfidization or pyritization) was said to be the chief geogenetic process. Pyrite oxidation, various degrees of acidic neutralization, and formation of products as a result of pyrite oxidation were said to be the important components of the pedogenetic process.

A. Physiography and Formation of Potential Acidity

Pons and others (1982) reported that three land systems constitute environments suitable for the formation of potential acidity. These were saline and brackish swamps and marshes, saline and brackish lagoons and lakes, and sometimes poorly drained inland valleys receiving water high in sulfate.

Of these three, the most important system resulting in the development of acidic sulfate soils is brackish and saline swamps and marshes. Systems supporting herbaceous vegetation, such as mangrove swamps (*Rhizohora* sp. and *Avicennia* sp.) were said to be especially important as the highly productive vegetation serves as an organic energy source needed for pyrite formation. Sediment and dissolved sulfate are provided by tidal cycles, which also serve to remove some soluble by-products of pyrite formation.

Coastal land formation and the development of potential acidity (pyrite accumulation) in the sediment of those areas are affected by relative changes in sea level. Following the last glaciation period, the sea level rose rapidly (Blackwelder et al., 1979) to about the present level some 5,500 years BP, where it has remained fairly stable with maybe only a small drop after reaching its maximum. In regions where the rise in sea level was approximately balanced by the sediment supply, conditions favorable for pyrite formation and accumulation occurred.

Sediment accretion in coastal areas started after the late Holocene stabilization of sea level (Pons et al., 1982). Pons and others (1982) indicated that where rapid lateral coastal accretion occurred after stabilization of sea level, rapid shifting of the intertidal zone and mangrove and reed marshes happened within a relatively short time and thus limited the suitable period for pyrite formation. A rapidly changing coastline also provided an unfavorable chemical environment for pyritization.

Pons and others (1982) summarized the effect of sedimentation rate with several examples in many parts of the world. Where rapid rises in sea level and an equal accumulation of sediment and stable vegetation occurred after the last glaciation, extensive, thick, and highly pyritic sediments formed, such as interior parts of the Chao Phraya, old sea clays of Holland, the Mekong and Orinoco deltas, and parts of Sumatra. Where high rates of sedimentation and coastal accretion caused a rapid shift of the intertidal zone after stabilization of sea level, pyrite contents remained low, such as in the Irrawaddy and Mekong deltas and the Guyana coast. More recent sediments containing high pyrite contents are associated with low sedimentation rates, such as the Saigon, Niger, and Gambia Rivers, or with regions having a high density of tidal creeks. Pyritic peats have resulted in some humid regions with very low sedimentation rates, such as in the western Netherlands and the Niger delta.

B. Pyrite Formation (Geogenetic Process)

The formation of pyrite is the basis for the genesis of potential acidic sulfate soils where pyrite is commonly 2% to 10% of the mass of these soils (van Breemen, 1982). Pons and others (1982) reported that pyrite formation in sediments requires: (1) a supply of sulfate and its subsequent reduction to sulfides by sulfate-reducing bacteria using organic matter as an energy source in an anaerobic environment, (2) partial oxidation of sulfides to produce polysulfides or elemental sulfur, and either (3a) formation of FeS (from Fe-oxides, Fe-containing silicates, and soluble sulfide) and subsequent combination of FeS and S to form FeS_2 (pyrite), or (3b) direct precipitation of pyrite (FeS_2) from soluble Fe^{2+} and polysulfides. Regardless of the actual pathway, the following overall reaction was given for pyrite formation (pyritization or sulfidization) with ferric oxide in a sediment as the source of iron and organic matter (CH_2O) as an energy source.

$$Fe_2O_3 + 4SO^{2-}_{4(aq)} + 8CH_2O + \tfrac{1}{2}O_{2(aq)} \rightarrow 2FeS_2 + 8HCO^{-}_{3(aq)} + 4H_2O$$

The essential components for pyrite formation are listed below:

1. A dissolved sulfate source—usually seawater, brackish tidal water, or occasionally sulfate-rich ground waters— continuously supplied over an appreciable period (Poelman, 1973).
2. The presence of iron-containing minerals in the sediments.
3. Organic matter (CH_2O) to serve as an energy source for the sulfate-reducing bacteria, which are usually present.
4. A predominantly anaerobic environment that is generated by anaerobic and facultative anaerobic bacteria functioning in waterlogged sediments that are rich in organic matter.
5. Periods of limited aeration in the sediments, either in space or time, so that sulfide can be oxidized to disulfide.

Upon long-term submergence in the presence of the favorable conditions given above, *Desulfovibrio desulfuricans*, a sulfate-reducing bacterium, is believed to contribute to the formation of pyrite from reductions of ferric sulfates (Ivarson et al., 1982) and produce dissolved sulfide (H_2S aq. and HS^- aq.). This dissolved sulfide reacts with sedimentary iron and precipitates as an iron sulfide such as mackinawite (tetragonal FeS). In the presence of oxidants, such as oxygen and/or ferric iron, that would occur during limited periods of an oxidized environment as discussed above, part of the dissolved or solid sulfide can be oxidized to elemental sulfur. Dissolved sulfide and elemental sulfur can react to form polysulfides, and polysulfides can react with FeS to form pyrite, FeS_2. Pyrite may form directly, from polysulfides and FeS, or indirectly through griegite (cubic Fe_3S_4) that is formed as an intermediate (Goldhabor and Kaplan, 1974). The process that includes griegite as an intermediate requires atmospheric oxygen to yield framboidal pyrite. Where oxygen is not available, nonframboidal pyrite is formed (Sweeney and Kaplan, 1973).

Carbonate alkalinity, mostly HCO_3^-, which develops from oxidation of organic matter by sulfate-reducing bacteria during the formation of pyrite, may lead to supersaturation and precipitation of carbonates (Presley and Kaplan, 1968). However, this carbonate alkalinity (HCO_3^-) produced during sulfate reduction is normally not conserved in the sediment by precipitation of calcium carbonate, although groundwater in tidal marshes is commonly supersaturated with calcite. Berner (1970) indicates that dissolved organic matter may inhibit calcium carbonate precipitation, and thus the carbonate alkalinity can be removed from the system by tidal action. The fact that the potential acidity (FeS_2) is immobile and the alkalinity that forms is mobile is a primary reason for acidic sulfate soil development (van Breemen, 1973).

The process of pyrite formation in situ is not completely understood (Pons et al., 1982). Pyrite has been produced in the laboratory by several workers who indicate that pyrite forms from the reaction of acid-volatile sulfide (FeS) with excess solid elemental sulfur as shown below (Berner, 1970).

$$FeS + S \rightarrow FeS_2$$

This reaction is very slow for forming measurable amounts of pyrite (Goldhabor and Kaplan, 1974). In working with this reaction, Roberts and others (1969) noted that more than 7 days were required. However, a direct precipitation reaction between aqueous ferrous ions and polysulfide ions is much more rapid, according to Rickard's (1975) report.

$$Fe^{2+} + S_5S^{2-} + HS^- \rightarrow FeS_2 + S_4S^{2-} + H^+$$

In general, pyrite formation is enhanced under low pH conditions. For instance, Berner (1964) produced pyrite by direct precipitation in less than a day at pH 4 and room temperature. Roberts and others (1969) obtained similar results at a pH range of 4 to 6. Although these laboratory studies are of interest for learning more about formation of pyrite, the relative rates reported in laboratory studies have little significance in terms of time available for the geogenetic process.

C. Pedogenetic Process

The pedogenetic process of acidic sulfate soil formation consists of three primary steps: (1) oxidation of pyrite, (2) neutralization of acidity formed, and (3) formation of products obtained from oxidation of pyrite and neutralization of the acid formed.

Pyrite oxidation is an interaction of complex processes that include oxidation-reduction reactions, hydrolysis, complex ion formation, solubility controls, and kinetic effects (Nordstrom, 1982). Drainage is required to initiate the oxidation of pyrite and the subsequent generation of acidity. Drainage may occur naturally, as a result of a drop in relative sea level or reduced frequency of tidal flooding, by human activities to control tidal flooding and lower the groundwater table (Dent, 1986), or by impoldering (van Breemen, 1975). Also, in many of the southeast Asian mangrove areas under tidal influence, acidification occurs as buried sediment is brought to the surface by the mound-building mud lobster *Thalassina anomaly* (Andriesse et al., 1973). Appreciable aeration of potential acidic sulfate soils and subsequent acidification start only after the pyrite material remains above the water table in an oxidizing soil environment for several weeks.

The several steps to the process of oxidizing pyrite to produce acid and other products involve chemical and microbiological processes. Reactions with dissolved oxygen is one of the two ways pyrite produces ferrous iron and sulfur.

$$FeS_2 + \tfrac{1}{2}O_2 + 2H^+ \rightarrow Fe^{2+} + 2S + H_2O$$

Sulfate forms from oxidation of sulfur.

$$S + \tfrac{3}{2}O_2 + H_2O \rightarrow SO_4^{2-} + 2H^+$$

The above equations show that pyrite exposed to the atmosphere will oxidize chemically yielding Fe^{2+}, sulfate, and sulfuric acid. The summary reaction is given below.

$$FeS_2 + \tfrac{7}{2}O_2 + H_2O \rightarrow Fe^{2+} + 2SO_4^{2-} + 2H^+$$

Complete oxidation and hydrolysis of the iron to ferric oxide yields two moles of sulfuric acid per mole of pyrite oxidized (van Breemen, 1982).

$$FeS_2 + \tfrac{15}{4}O_2 + \tfrac{7}{2}H_2O \rightarrow Fe(OH)_3 + 2SO_4^{2-} + 4H^+$$

The above reaction is sulfuricization (Fanning, 1978). At a near neutral soil pH, the process is relatively slow, but it accelerates as acidity increases. In the presence of oxygen, the ferrous iron (Fe^{2+}) produced by these reactions is oxidized to ferric iron, normally a slow reaction at low pH (Singer and Stumm, 1970). However, *Thiobacillus ferrooxidans,* which functions well between pH 2.5 and 5.8 (Goldhabor and Kaplan, 1974), is effective in oxidizing reduced sulfur species and also ferrous iron at these low pH levels and thus returns ferric iron to the system.

$$Fe^{2+} + \tfrac{1}{4}O_2 + H^+ \rightarrow Fe^{3+} + \tfrac{1}{2}H_2O$$

When the pH of an oxidized system is sufficiently low for Fe^{3+} to exist in solution, Fe^{3+} may catalyze the oxidation of pyrite. The dissolved Fe^{3+} favors rapid oxidation of pyrite, especially as the pH of an oxidized system decreases below 4, according to the reaction below, where the half-time of the reaction has been reported to range between 20 and 100 minutes (Stumm and Morgan, 1970).

$$FeS_2 + 2Fe^{3+} \rightarrow 3Fe^{2+} + 2S$$

The overall oxidation reaction is given below:

$$FeS_2 + 14Fe^{3+} + 8H_2O \rightarrow 15\,Fe^{2+} + 2SO_4^{2-} + 16H^+$$

At a high pH, the oxidation of pyrite by Fe^{3+} ions is limited because Fe^{3+} is appreciably soluble only at low pH (pH ~ 4), and *Thiobacillus ferrooxidans,* on the other hand, does not function at a higher pH. Thus, in soils of high pH, ferric oxides and pyrite may be in close physical proximity, but the rate of pyrite oxidation by this process is limited by the insolubility of ferric iron.

Nordstrom (1982) summarized the oxidation of pyrite as a three-part sequence. The initiation phase producing sulfur and ferrous iron is somewhat slow. Also, the initiation phase has been reported to be pH dependent for pH levels above 4 and independent of pH below pH 3. The acid-generating phase is next, in which the soil pH begins to lower as sulfur is oxidized to sulfate and H^+ is produced. Then, if the pH of the system is brought below pH 3 by the second phase, in the catalytic phase, soluble ferric iron rapidly oxidizes pyrite. The role of *Thiobacillus ferrooxidans* is critical in providing the system with ferric iron for the catalytic phase.

D. Chemistry of Acidic Sulfate Soils

1. Response to Acidic Precipitation

Potential acidic sulfate soils and such soils that have developed extreme acidity should be little affected by acidic precipitation for different reasons. As previously discussed, potential acidic sulfate soils are those containing appreciable pyrite but are perennially wet, such that the pyrite does not oxidize with the subsequent production of acidic conditions. The pH of potential acidic sulfate soils is often near that of typical agricultural soils. Acidic sulfate soils, however, are potential acidic sulfate soils that, because of drainage, have undergone chemical changes associated with pyrite oxidation producing excessive acidity.

Because of the almost continuous flooding of pyritic zones of potential acidic sulfate soils, the pyrite is stable, pH is not adverse for wetland plant growth, and the soil is reducing. In potential acidic sulfate soils and wetland soils in general, the effect of acid-forming compounds in precipitation on soil acidity should be much less than on nearby upland aerobic soils. This effect is due to the difference in reactions of the nitrogen and sulfur oxides in wetland soils compared to upland soils, as well as due to differences in the acid-buffering capacities of wetland and upland soils.

Several conditions exist in wetlands that are likely to mitigate the impact of added nitrogen and sulfur oxides to wetland soils. If wetland soils and potential acidic sulfate soils are anaerobic, as is usually the case, the end product of nitrogen and sulfur oxides deposited with precipitation is not HNO_3 and H_2SO_4. Nitrogen oxides entering the soil undergo denitrification, which returns the nitrogen to the atmosphere as N_2 and prevents it from functioning as an acid in the soil. The reaction of sulfur oxide in an anaerobic soil is more complex, but the net effect is the same—less acidification than under aerobic conditions. Sulfur oxide entering an anaerobic soil that contains an ample energy source is reduced to sulfide instead of being oxidized to H_2SO_4, as is the case in an aerobic soil. Under most conditions, the sulfide is precipitated by ferrous iron (Fe^{2+}) and remains as a relatively inert iron sulfide precipitate as long as the soil remains anaerobic (Gambrell and Patrick, 1978).

In addition to the differences in reactions of nitrogen and sulfur oxides that lessen the acidification impact of these compounds in wetland soils as compared to upland soils, anaerobic soils usually have a larger acid-buffering capacity than aerobic soils that further reduces the impact of acid-forming substances. This increase in acid-buffering capacity in wetland soils is due almost entirely to the activity of iron and manganese compounds under anaerobic conditions. Under aerobic conditions, acid buffering in noncalcareous soils is due to the capacity of the cation-exchange capacity to adsorb aluminum and hydrogen ions associated with the nitrogen and sulfur oxides with the release of an equivalent amount of basic cations into the soil solution. Nonacid or basic soils that contain calcium and magnesium carbonate have an additional buffering capacity equivalent to the amount of acid required to neutralize the lime present in the soil.

Anaerobic soils have this same buffering capacity plus the additional acid buffering that results when ferric and manganic oxyhydroxides are microbially reduced to the ferrous and manganous forms. The soluble ferrous and manganous compounds produced in an anaerobic soil function in a similar manner to the calcium and magnesium compounds in upland soils in neutralizing acidity. In an aerobic (upland) soil, however, the iron and manganese compounds have little effect on acid buffering because of the very low solubility of the ferric and manganic forms.

Thus wetland soils and potential acidic sulfate soils are likely to interact very differently with acidic precipitation than many upland soils, even those in the same geographical area. Wetland soils in general, as well as potential acidic sulfate wetland soils, tend to be little affected by acidic precipitation. Although the probable different response of wetland soils to acidic precipitation has been recognized, at present relatively little acidic rain research has been conducted specifically addressing the fate of sulfuric acid and nitric acid components of acidic rain in wetland systems.

True acidic sulfate soils, in contrast, would be minimally affected by acidic precipitation because of the extreme acidity already present. The additional precipitation-derived acidity would be insignificant to that acidity already present.

2. Changes in pH and Redox Potential

Typical agricultural soils used for most crops contain dissolved oxygen in the soil solution and gaseous oxygen in unsaturated pore space. Rice soils differ in that the soils are drained for some of the year, but flooded for a major part of the growing season, during which time these soils become anaerobic. Upon flooding, soil reduction occurs if, simultaneously, the oxygen supply is depleted and oxygen resupply is restricted (diffusion of dissolved oxygen is *very* slow in water-saturated soils), organic matter is available as an energy source for anaerobic microorganisms, and other soil environmental conditions are suitable for microbial activity (Bouma, 1983).

Charoenchamratcheep and others (1987) measured the changes in pH and redox potential in four acidic sulfate soils of Thailand during a reduction and subsequent oxidation sequence in a laboratory study. Their results are summarized in Table 2–1. During the oxidation cycle, substantial decreases occurred in soluble iron, while soluble aluminum and sulfate-sulfur usually increased.

In a review, Dong-Qing and others (1985) have reported similar changes in redox potential and pH upon alternate submerged and drained conditions in rice soils of China.

3. Neutralization of Acidity

Acidic sulfate soils develop where the amount of sulfuric acid formed by oxidation of reduced sulfur compounds exceeds the acid-neutralizing capacity of the soil. The acid-neutralizing capacity of a soil consists of (1) carbonates, (2) exchangeable bases, and (3) easily weatherable silicates.

Calcium carbonate is a very effective neutralizing chemical component in soils. Although not present in all soils, it is present in some as a result of natural processes, and it is a common amendment for regulating the pH of other cultivated soils. If one mole of pyrite upon oxidation is equivalent to four moles of H^+ as previously shown, the acidity from the oxidation of 1% by mass of pyrite sulfur is approximately balanced by 3% of $CaCO_3$ (Pons et al., 1982). In most marine sediments of the humid tropics, the calcium carbonate content is low; however, appreciable amounts may be found in sediments of arid and humid temperate

Table 2–1. Redox potential (Eh in millivolts) and pH changes in four Thailand soils over a reduction and subsequent oxidation cycle.

	Beginning of reduction cycle		End of reduction cycle		End of oxidation cycle	
Soil	pH	Eh	pH	Eh	pH	Eh
Rs[a]	3.9[b]	+ 400	6.0	− 175	4.1	+ 425
Rs-a	3.5	+ 325	4.1	+ 100	3.5	+ 450
Bg	2.8	+ 475	4.5	+ 200	2.0	+ 800
Ma	4.6	+ 450	6.4	− 175	3.8	+350

Reproduced from Soil Science Society of America Journal, Volume 51, No. 3, May–June 1987, pages 630–634, by permission of the Soil Science Society of America, Inc.

[a] Soil types include Rs = Rangsit; Rs-a = Rangsit, very acid; Bg = Bang Pakong; Ma = Maha Phot.

[b] Approximate values taken from published figures.

regions (Pons et al., 1982). Pons and others (1982) state that if all the dissolved sulfate in seawater trapped in sediment is reduced to sulfide, the HCO_3^- levels produced should lead to supersaturation with calcium carbonate. Citing Berner (1970), Pons and others point out that calcium carbonate rarely precipitates, presumably due to an inhibiting effect of dissolved organic matter. Van Breemen (1973) noted that the alkalinity of interstitial waters in nonalkaline soils rarely exceeds 10 mEq L^{-1}, and, at moisture contents up to 100%, the dissolved alkalinity was calculated to contribute, at a maximum, to the neutralization of 1 mEq of acid per 100 g of soil. Thus Van Breemen (1973) stated that seawater, which contains a lower alkalinity, cannot be considered an effective acid-neutralizing agent, even if large quantities are applied for leaching purposes. An exception was noted in the eastern part of the Central Plain of Thailand, where seasonal flooding with moderately alkaline water (2 to 5 mmol HCO_3^- L^{-1}) may have increased the pH of the upper horizons of acidic sulfate soils to near neutrality; however, this process occurred over a period of centuries.

Dong-Qing and others (1985) also reported that the buffering capacity of soil humus is 10 to 100 times greater than the mineral part of the soil (excluding carbonates) and thus accounts for many rice soils being better buffered in the surface horizon than in deeper horizons.

4. Exchangeable Bases

Because of large specific surface areas and the negative charge associated with clay minerals, the soil solid phase can be important in immobilizing excess acidity. In soils with both high organic matter and/or clay contents, the sorption of H^+ associated with the formation of nonexchangeable acidity may substantially neutralize strong acid under neutral to slightly acidic soil conditions. Van Breemen (1973) presented data showing between 5 and 10 mEq of acid per 100 g of soil was immobilized by the exchange complex of typical acidic sulfate soils in Thailand when the pH drops from 7.5 or 7 to about 5. Van Breemen (1973) also noted that at a lower pH, more H^+ can be immobilized because exchangeable Al^{3+} enters the soil solution. The total amount of acid immobilized by the exchange complex of an acidic soil is approximately equal to the difference between the cation exchange capacity (C.E.C.) at pH 7 and the amount of exchangeable bases at the soil pH. Van Breemen, citing the work of Sombatpanit (1970), indicated these amounts have been shown to be 10 to 30 mEq per 100 g of acidic sulfate soils (pH 3.5 to 4) from Thailand.

Most heavy clay soils derived from marine sediments have considerable amounts of smectite clay minerals, and their cation exchange capacity, when fully saturated with bases, is capable of immobilizing most of the acidity released by the oxidation of up to 0.5% pyrite-S, such that the pH will not drop below 4.0 (Pons et al., 1982). Pons goes on to say that if the clay fraction is predominantly kaolinitic (less cation-exchange capacity than smectites) or if clay contents are low, oxidation of less than 0.5% pyrite-S can make the soil strongly acidic.

5. Weatherable Minerals

At pH levels less than 4, some clay minerals are transformed to simpler silicate clays, and this transformation consumes H^+ in the process. The rate of this clay transformation is very slow such that this acid-immobilizing process should have a negligible impact on the development of acidic sulfate conditions in most situations. However, the intensity of acidity development would be reduced to some degree. As an example, the transformation of Mg montmorillonite to kaolinite is shown to consume some acidity (van Breemen, 1973).

$$\text{Mg-montmorillonite} + 2H^+ + 23H_2O \rightarrow Mg^{2+} + 7\ \text{kaolinite} + 8H_4SiO_4$$

Kaolinite is the end product of most clay mineral weathering processes, especially under acidic conditions, and, as expected, it occurs widely in older soils and acidic sulfate soils.

6. Products from Oxidation and Neutralization

Some of the major products from pyrite oxidation and neutralization, in addition to sulfuric acid, include jarosite, iron oxides, and sulfates. Most of the sulfate is leached out of the soil, but some remains as jarosite.

Jarosite is a mixture of basic sulfates with the general formula $AB_3(SO_4)_2(OH)_6$ in which *A* is K, Na, H_3O, ½Pb, NH_4, or Ag, and *B* represents Fe (III) (jarosite) or Al (alunites) (van Breemen, 1973). The most important members of the jarosite group are jarosite, natrojarosite, and hydronium jarosite. Jarosite—particularly the colors associated with its presence in soils—is a conspicuous, characteristic indicator of acidic sulfate soils. Jarosite usually precipitates as pale yellow deposits (2.5–5Y, 8/3–8/6), as fillings in biopores such as root channels, or as efflorescences on ped faces and pore walls. Where individual particles form, they are often smaller than 1 μm, with a diameter usually less than 5 μm (Andriesse et al., 1973). Jarosite forms only under oxidizing, acid ($pH < 4$) conditions. In acidic sulfate soils, jarosite is metastable and ultimately is hydrolyzed to goethite (van Breemen, 1982).

Iron oxides (ferric oxides and hydroxides) precipitate directly by oxidation of dissolved Fe^{2+} as the pH of the soil remains above 4. Although this iron may become less available with time and subsequent increasing crystallization, it is the source of "active" iron that may be dissolved by acids and reducing or complexing agents (Xie-Mind, 1985) and is the source of most excess soluble iron if the soil again becomes reducing. Under oxidizing conditions, fine-grained goethite (FeOOH) may form, either directly as soluble ferrous sulfate is released during pyrite oxidation or more slowly by hydrolysis of jarosite. The reactions are part of the sulfuricization process under acidic soil conditions releasing H^+ (van Breemen, 1982). During oxidation of pyrite in drainage water, goethite is the most commonly identified iron oxide, although it may be transformed slowly to hematite (Fe_2O_3) under some conditions (van Breemen, 1982). In developed acidic sulfate soils, part of the ferric oxides in the B horizon may exist as hematite, resulting in the presence of red mottles.

Most of the iron in pyrite remains in the soil profile as acidic sulfate soils develop. Only a small portion of the sulfate released remains in the profiles, as drainage tends to leach most of the soluble sulfur forms (primarily sulfate) from the oxidized soil profile. At a greater depth, however, oxidized sulfur forms leaching through the profile may again be reduced to sulfide (Dent, 1986). The residual forms of sulfur that may persist for long periods are jarosite and gypsum ($CaSO_4$). Gypsum is relatively soluble and will not persist where substantial leaching is occurring, although it is sometimes found in coastal soils, especially in dryer soils or those with some calcium carbonate.

E. Acidic Sulfate Soils and Rice Production

Agricultural crops or natural plant populations grown on soils impacted by acidic precipitation do not usually experience the severity of stress from soil chemical conditions as do plants growing on acidic sulfate soils. However, many of the plant growth stress factors associated with soil chemistry are the same except for intensity. Thus the brief discussion below on acidic sulfate soils and rice growth indicates the nature of the problems of soil acidification and the direction of soil chemical changes from acidic precipitation that may affect plant growth.

Acidic sulfate soils generally show characteristics unfavorable for agricultural production. In most cases, the important crop in regions where acidic sulfate soils predominate is rice. The suitability of these soils for rice production depends on the potential or actual acidity conditions and the availability of affordable and feasible management practices to overcome the soil chemical problems. These problems are briefly discussed in the following sequence:

1. Soil acidity
2. Sulfide toxicity
3. Iron toxicity
4. Aluminum toxicity
5. Salinity
6. Toxicity of organic acids and carbon dioxide
7. Nutrient deficiency

1. Soil Acidity

The direct adverse effect of H^+ on plants has been observed at an acidity stronger than pH 3.5 to 4.0 (Ponnamperuma et al., 1973). The evidence was obtained mostly from plants grown in solution culture media in order to rule out toxicity from other elements that may be more soluble at low pH. The probability of soil acidity contributing directly to plant growth problems on some acidic sulfate soils has been reported (Brinkman and Pons, 1973).

Van Breemen and Pons (1978) have reported pH levels of 3 and lower in young acid sulfate soils (Sulfaquepts) and rapidly oxidized potential acidic sulfate soils (Sulfaquents). Occasionally, pH values of approximately 1 to 2 have been reported in oxidized horizons of acidic sulfate soils (Tanaka and Yoshida, 1970), and the authors have seen pH levels of less than 2 on acidic sulfate soils from west Africa.

2. Sulfide Toxicity

Sulfate is reduced under strongly anaerobic conditions in submerged soils or sediments; as discussed previously, this step is the first in pyrite formation and acidic sulfate soils development. Sulfate may be reduced to sulfites, hydrogen sulfide, elemental sulfur, and to other reduction products. Sulfate formed from pyrite oxidation during the drainage cycle in rice soils is subject to reduction again upon flooding the soil for rice cultivation.

Sulfide inhibits respiration in rice and reduces the oxidizing power of rice roots and hence retards the uptake of various nutrients and causes poor growth (Vamos, 1967). Rice seedlings have been reported to die in some circumstances from hydrogen sulfide toxicity when an acidic sulfate soil is allowed to be reduced too long before rice is sown (Brinkman and Pons, 1973).

The rate of sulfate reduction in flooded soil depends on soil properties. Ponnamperuma (1972) reported that in the neutral and alkali soils studied, concentrations of sulfate as high as 1,500 mg kg^{-1} may be reduced to zero within 6 weeks of submergence. Due to inhibited microbial activity in strongly acidic soils, the rate of sulfate reduction to sulfide is very much less in acidic soils than in near-neutral soils. Connell and Patrick (1968, 1969) reported that a redox potential of −150 mv in a pH range of 6.5 to 8.5 was very favorable for sulfide formation. The optimum pH for sulfide formation has been reported to be 6.7 in one study (Jakobsen et al., 1981). Liming and/or prolonged flooding to raise the soil pH above 5 tends to increase sulfide formation and the possibility of plant toxicity, according to van Breemen (1975) and Dent (1986). Although liming, if economically feasible, certainly reduces the severity of many of the problems with acidic sulfate soils, Komes (1973) also mentioned precautions are necessary when liming the acidic sulfate soils of Thailand to avoid accelerating sulfate reduction to sulfide at a higher pH. Thus the message is clear that overliming can cause problems.

Peak concentrations of hydrogen sulfide levels in soils have been found to vary with soil pH, organic matter content, and the availability of reactive reduced Fe and Mn (IRRI, 1973) such that hydrogen sulfide levels tend to be greater in soils low in iron or high in organic matter. High concentrations of ferrous iron found in many reduced soils tend to precipitate dissolved sulfide by the formation of FeS, which is a major process controlling dissolved sulfide levels. Patrick and Reddy (1978) state that this mechanism should somewhat protect plants from the toxic effects of dissolved sulfide. Adding chemical oxidants such as nitrate or manganic oxides to retard sulfide formation has been suggested as a management practice to lessen the problem of sulfide formation (Ponnamperuma, 1965; Engler and Patrick, 1973). This process was believed to function by stabilizing the redox potential above the critical potential for sulfate reduction or possibly by competing with organic acid fermentation for molecular hydrogen.

3. Iron Toxicity

In acidic sulfate soils, iron toxicity is an important factor affecting plant growth (Nhung and Ponnamperuma, 1966). A low pH and/or a reducing soil environment favors elevated levels of reduced, soluble iron. Ponnamperuma (1955) indicates iron toxicity can result from the presence of excess iron alone; Howeler (1973) and Benckiser and others (1984) indicate it can be due to an unbalanced availability of iron relative to low available levels of Ca, Mg, K, and P. Rice is relatively tolerant of the elevated soluble iron concentrations typical of anaerobic, flooded soils compared to upland plants, which normally are exposed to very low levels of available soil iron because of their oxidized rooting zones. Although other mechanisms may exist (Yoshida and Tadano, 1978), oxidation and thus precipitation of excess iron in the rhizosphere is one important control mechanism for rice. Unlike upland plants, rice and other wetland plants exhibit a high rate of oxygen diffusion into the roots. Air enters rice plants through aboveground tissue, primarily leaves, and moves to the roots through air passages known as *aerenchyma tissue*. Oxygen thus transported to the roots is used primarily for respiration. After root respiration requirements are met, excess oxygen may diffuse out from the roots and create a thin oxidized zone around the roots (Armstrong, 1971) that allows oxidation and immobilization of several potentially toxic ions, including iron, at or near the root surface.

Although the mechanism described above mitigates iron toxicity to some degree, depending on local soil conditions, iron toxicity does occur in rice. Tian-Ren (1985) reports that rice farmers in China anticipate reduced rice production "if rusting colloidal materials exist on the surface of standing water in the field, or if rusting waters are used for irrigation." This condition was attributed to the presence of excess iron in the water, and that liming would reduce this adverse effect on rice was noted.

Prasittikhet (1987) used laboratory microcosms of acidic sulfate (Sulfic Tropaquept) and nonacidic sulfate (Typic Tropaquept) soil material from Thailand to investigate the effects of controlled pH and redox potential conditions on transformations of several nutrients and their effects on the growth of rice. He incubated some microcosms at selected controlled redox potential conditions (500, 250, 50, and −150 mV) with no pH control. Others were incubated at various combinations of controlled pH and redox potential levels (pH levels of 3.5, 4.5, and 5.5 and redox potential levels ≥ 500, 250, and 50 mV, respectively). These soil suspensions were planted with rice seedlings of varieties tolerant to acidic sulfate soils (IR 46) and sensitive to acidic sulfate soils (IR 26). Three weeks after rice transplanting, the soil solution and plant tissue samples were processed and analyzed for various soil and plant properties. Metal activities in the soil solution were determined with a computer program.

His results suggested that both redox potential and pH significantly affected the transformations of various fractions of iron in both soil types. Water-soluble iron and exchangeable iron were inversely related to both pH and redox potential, and reducible iron was positively related. His exerimental data indicated that amorphous $Fe(OH)_3$ probably controlled solubility of ferric iron at pe + pH around 12.95 and above, and that goethite (FeOOH) may regulate ferric iron solubility at pe + pH of about 11.95 and below.

Prasittikhet (1987) found that increases in both redox potential and pH negatively affected iron uptake in rice. Iron uptake was strongly related to water-soluble iron, ferric iron activity, and E'-Fe. His results indicated that the critical levels of the minimum pFe^{2+} and E'-Fe (divalent charge fraction in soil solution due to Fe) that trigger excessive uptake of iron by the rice plant could probably be defined at pFe^{2+} of 3 and E'-Fe of 0.45, respectively. The average iron content in the shoot tissue of the senstivie variety (IR 26) was greater than that of the tolerant variety (IR 46). Growth of both rice varieties was negatively related to the ratio of iron to manganese in the shoot tissue. He noted that a ratio of iron to manganese in the shoot tissue of 4.5 and higher is probably harmful to rice growth.

Prasittikhet (1987) suggested that the negative relationship he found between manganese uptake and $p(Mn^{2+}:Fe^{2+})$ activity ratio indicated that iron may have an antagonistic effect on manganese uptake. The ratio between iron and manganese in shoot tissue was the most important variable negatively associated with weight gain; this suggests an antagonistic effect of the two metals on rice growth. Under controlled pH and redox potential conditions, the ratio between iron and manganese in shoot tissue was also the most important variable negatively associated with the rice weight gain in his experiments.

4. Aluminum Toxicity

Aluminum toxicity to various crop plant species has received much research attention (Foy et al., 1965). In most cereal crops such as rice, the roots are often the first tissue to show symptoms of aluminum injury. Prasittikhet (1987) reviewed the mechanism of aluminum toxicity. A number of processes, including precipitation of aluminum phosphate in the rhizosphere or impairing phosphate utilization with the plant, may be involved.

Aluminum toxicity in acidic sulfate soils may occur at pH levels below 5 for seedlings, but older plants do not show symptoms until the pH drops to near 4 and below. However, other interactions may be involved. For example, the low availability of phosphorus in acidic sulfate soils may enhance aluminum toxicity compared to other soils (van Breemen and Pons, 1978).

Aluminum toxicity in acidic sulfate soils decreases where soil pH increases upon flooding and anaerobic conditions develop, a common occurrence as soils become more reducing (Gambrell and Patrick, 1978).

Prasittikhet (1987), in his experiments with rice seedlings in acidic sulfate and nonacidic sulfate soils, found that pH strongly influenced the solubility of aluminum in acidic sulfate soils. Water-soluble aluminum and the percentage aluminum saturation of the cation-exchange capacity were negatively related to pH in both the acidic and nonacidic sulfate soils under controlled pH and redox potential conditions but negatively related to pH only in the acidic sulfate soil under controlled redox potential conditions. Aluminum activity in the soil solution was the only aluminum parameter negatively associated with pH in both soil types and over all controlled conditions. The activity of Al^{3+} consistently adversely affected the weight gain of both rice varieties tested in acidic sulfate soils.

5. Salinity

Excess salinity is the presence of too much soluble salt in the soil solution. The common major ions of these salts are sodium, calcium, magnesium, chloride, and sulfate, where sodium and chloride usually predominate. The proximity of many acidic sulfate soils to saltwater tidal areas is a factor in most salinity problems in acidic sulfate soils. Elevated salinity is believed to make worse other toxicity problems both by weakening the plants and by increasing the concentration of other toxic ions such as iron and aluminum in solution (Pasricha and Ponnamperuma, 1976).

6. Toxicity of Organic Acids, Phenols, and Carbon Dioxide

Carboxylic acids such as formic, acetic, propionic, and butyric acids are known to occur in flooded, anaerobic soils where acetic acid has been reported to be the most abundant (Motomura, 1962). These organic acids have been reported to be toxic to rice seedlings at concentrations of 10^{-2} mol L^{-1} and less (Yoshida and Tadano, 1978). There is little if any indication they are toxic under typical upland soil conditions; in flooded soils, problems have been noted to be more likely with increasing time of submergence.

Possible mechanisms of growth inhibition include direct toxicity in the undissociated state, thus the greater potential for injury in acidic sulfate soils. At higher pH levels, carboxylic acid functional groups tend to be ionized, where they may compete for certain nutrients or act as carriers that increase total soluble concentrations, as has been reported for iron (Motomura, 1961).

Another class of potentially toxic organics to rice is the phenolic compounds. Van Breemen and Moormann (1978) suggest these compounds may be more important in terms of toxicity than the more familiar alcohols, aldehydes, carboxylic acids, and organic sulfur compounds that generally occur in small concentrations in anaerobic soils. Inhibition of rice growth at levels greater than 0.1 to 10 mmol L^{-1} may occur.

The toxicity of carbon dioxide to rice in flooded acidic sulfate soils has been reported (Ponnamperuma, 1972), as acidic soils may accumulate high concentrations of carbon dioxide (IRRI, 1965).

7. Nutrient Deficiencies

In acidic sulfate soils, phosphorus is tightly bound in forms unavailable to plants, such as highly insoluble iron and aluminum phosphates (Moormann, 1963) or phosphate fixed on clay surfaces. Under anaerobic conditions, some of the iron and aluminum phosphate will become available (Patrick, 1964; Patrick et al., 1985). However, Dent (1986) believes increasing crystallization of the oxides with time gradually reduces the degree of mobilization by flooding. Although a number of interacting processes affect phosphorus availability, conditions in acidic sulfate soils are such that phosphorus is very often a major limitation for plant growth, and it is even difficult to improve the situation with phosphorus fertilizer applications.

Patrick and Mahapatra (1968) found little or inconsistent response by lowland rice to phosphorus fertilization, although upland crops grown on the same soils respond markedly to applied phosphorus. The phosphorus requirements of rice are about the same as those of other crops, so soil factors regulating transformations affecting the availability of phosphorus to plants must be strongly influenced by the reducing conditions that develop with waterlogging.

Ponnamperuma (1972) suggested that soluble phosphorus increased in flooding acidic soils because of hydrolysis of ferric and aluminum phosphates, release of phosphorus adsorbed to clays and hydrous oxides by anion-exchange processes, and reduction of ferric compounds containing phosphate to the more soluble ferrous compounds. He noted that an increase in pH with soil reduction might facilitate the first two of these processes. The calcium phosphate compounds that predominate in alkaline soils become more soluble as pH approaches 7 upon reduction.

A few studies have reported soil test results that have enabled researchers to predict adequately the response of lowland rice to phosphate fertilizers. Most soil test methods commonly used to estimate available phosphate for upland crops, however, have not been satisfactory indicators of soil phosphorus availability to rice, as Patrick and Mahapatra (1968) noted. Because these tests have usually been applied to air-dry soil samples, they do not take into account the increase in phosphate availability to lowland rice caused by soil submergence and subsequent reduction. Mahapatra and Patrick (1971) later reported extractable iron, aluminum, and calcium phosphates of air-dry soils to be highly correlated with the Bray-extractable phosphorus of reduced soils in Louisiana. They suggested that identifying and measuring phosphorus forms in air-dry soils, which became available upon reduction, would allow phosphate fertility to be evaluated in flooded soils by chemical testing of air-dry samples.

Khalid and others (1976) observed that phosphorus adsorbed from 0.1 and 0.2 mg/ml phosphorus solutions under reduced conditions was highly correlated with rice yields. The currently used procedure of testing soil in air-dry samples was less useful for predicting fertilizer phosphorus requirements for lowland rice in this study. They concluded that modifying soil-testing procedures to account for the influence of soil reduction processes on phosphorus availability might improve the correlations between soil test results for phosphorus and rice yields.

Nitrogen is also frequently a limiting factor for good crop growth in acidic sulfate soils (Moormann, 1963). In fact, nitrogen is the nutrient most limiting to agricultural production in both well-drained and flooded agricultural soils throughout the world. In flooded rice soils in particular, nitrogen deficiencies are common despite fertilizer applications because soil and sediment reduction in submerged environments favors physical, chemical, and biological processes that remove available nitrogen. Oxidation-reduction intensity strongly influences these processes, as does the presence of a well-oxidized surface water or soil layer overlying a deep, reduced horizon, because reduced fertilizer nitrogen may diffuse to the oxidized zone to be nitrified, and the nitrate thus formed can then diffuse back into the anaerobic zone where denitrification removes it from the system.

The plant availability of soil nitrogen is often restricted by the slow mineralization of organic matter to available mineral forms and by unfavorable soil conditions for the process of nitrogen fixation (Dent, 1986). *Mineralization* is the biological transformation of organically combined nitrogen to ammonium nitrogen during the degradation of organic matter. Because organic matter degrades considerably faster in oxidized than in reduced soils, mineralization is also slower in reduced soils, although the nitrogen requirements for anaerobic metabolism have been shown to be lower than for aerobic metabolism (Patrick and Wyatt, 1964; Waring and Bremner, 1964). Some acidic sulfate soils of Thailand have been reported to mineralize less than 5% of their organic nitrogen, compared to 10% to 20% for other tropical soils (Kawaguchi and Kyuma, 1969).

Flooded soils also lose substantial amounts of nitrogen through the nitrification-denitrification sequence, as described above (Tusneem and Patrick, 1971; Broadbent and Tusneem, 1971). In fact, some studies have revealed that more ammonium was lost than was initially present in the surface oxidized layer of wetland soils (Patrick and Reddy, 1976a, 1976b). Abichandani and Patnaik (1955) estimated that 20% to 40% of the ammonium fertilizers applied to rice was lost; in Japan, Mitsui (1954) found losses of applied nitrogen ranging from 30% to 50%. Laboratory and field tracer studies conducted on flooded soils have confirmed reports of substantial losses of nitrogen from these soils. In a field study, Patrick and Reddy (1976b) applied labeled ammonium sulfate to rice by deep placement and determined that the rice recovered 49% of the applied nitrogen, 26% remained in the soil (in roots or as soil organic nitrogen), and 25% was lost, presumably through the nitrification-denitrification sequence.

Acidic sulfate soils usually have much smaller amounts of important nutrient cations such as Ca, Mg, and K compared to other soils because the acidic conditions displace these cations from exchange sites where they are removed from the profile by leaching. Then, aluminum becomes abundant on the exchange complex and causes further problems for normal plant growth.

F. Reclamation and Management of Acidic Sulfate Soils to Improve Rice Yields

All agricultural soils are managed to varying degrees to enhance productivity. Despite the rather severe constraints of acidic sulfate soils, different reasonably successful methods have been applied to make them economically productive. In part, this effort is motivated by the pressing need for bringing these soils under cultivation where food production is a primary concern. These successful reclamation methods must take advantage of the particular differences in soil properties, topography, hydrology, climate, and other factors associated with different acidic regions where acidic sulfate soils are found. Also important are the resources available to the local community to make the soils productive. Cost versus the anticipated economic benefits is a major consideration in attempting to cultivate acidic sulfate soils.

Several approaches have been applied to improve acidic sulfate soils. Prasittikhet (1987) has reviewed development of the following reclamation and management practices that have been applied to improve rice yield in acidic sulfate soils:

1. Rice cultivation in tidal marshes
2. Intensive shallow drainage
3. Drainage and leaching
4. Liming and fertilization

Three of the four approaches take advantage of managing the soil chemistry of toxic substances by flooding and/or drainage cycles that affect redox potential conditions, and subsequently the chemistry of the elements adversely affecting rice growth. In addition to managing flooding and drainage cycles, one of these involves the additional component of reducing toxic levels of substances by seawater leaching and exchange of more toxic components with cations predominating in seawater. The final approach involves chemical amendments to reduce levels of toxic substances by pH control and adding nutrients that are limiting. These management approaches are briefly discussed below.

1. Rice Cultivation in Tidal Marshes

Potential acidic sulfate soils do not acidify where frequent tidal flooding occurs, as this prevents the extended period of aeration required for pyrite oxidation. In some areas, rice can be grown only along riverbanks in the wet season where the normal tidal cycle backs up fresh water (van Breemen and Pons, 1978). In some places, farmers transplant rice one or more times so that the final transplanting is in potential acidic sulfate soils kept reduced by frequent tidal flooding. Transplanting is required so that the plants are sufficiently large to survive the tidal flooding cycles. The success of this method lies in the fact that the potential acidic sulfate soil is not allowed to oxidize and thus produce strong acidic conditions.

2. Intensive Shallow Drainage

In flooded rice-growing areas of the Mekong Delta, where the soils are predominantly young acidic sulfate soils, rice performs well on raised beds (about 9 m wide and 36 m long). These beds are drained by a network of shallow ditches (about 1 m wide and 0.3 to 0.6 m deep). Oxidation of pyrite and iron and subsequent acidification occur during the dry season, and then leaching of toxic oxidation products and excessive acidity into the drainage ditches occur during the first of the wet season. When the floodwater reaches the surface of the beds, the drainage network is opened to the river at low tide. This cycle is repeated a few times. In describing this system, Dent (1986) reports rice yields are increased around an order of magnitude. One problem with this practice is the increased acidity of floodwaters reaching crops in adjacent areas.

3. Drainage and Seawater Leaching

In areas where better methods are not economically feasible, seawater leaching has been used with some success. Presumably, cations in seawater displace some of the aluminum that is causing plant toxicity (Bloomfield and Coulter, 1973). However, to be successful, the soil permeability and local groundwater hydrology (perhaps with managed drainage) must allow leaching, and fresh water must be available for a final leaching to reduce salinity to acceptable levels.

4. Liming and Fertilization

Because nitrogen and phosphorus are normally limiting on acidic sulfate soils, fertilization offers the potential for enhancing yields. However, the situation with phosphorus is complex because of the high capacity of acidic sulfate soils for immobilizing this nutrient in unavailable forms. Although this area requires additional research to maximize fertilizer use efficiency, fertilization has been shown to enhance yields.

Liming is an effective chemical amendment for raising soil pH and thus reducing levels of toxic aluminum and iron available to crops (Ponnamperuma et al., 1973).

Often, however, because of the expense of the large quantities required, fertilization and liming are not always economically feasible or affordable for some farmers. Sometimes regional transportation and satisfactory application methods for wet soils cause obstacles. For example, 5 to 10 tons of lime per hectare might be necessary to make some soils productive if liming were the only management practiced. Although the expense of liming usually prevents application of too much lime, remember from the previous discussion in this reivew that overliming can favor increased sulfide production during the flooding cycle for rice production and cause toxicity problems. Prasittikhet (1987) reviewed other problems associated with overliming related to nutrient chemistry and increased microbial activity.

V. Summary

Acidic sulfate soils are generally far more acidic than can be generated by even severe acidic precipitation problems. Acidic precipitation is unlikely to affect adversely either acidic sulfate soils or, for different reasons, potential acidic sulfate soils. However, knowledge of acidic sulfate soil chemistry, the problems experienced with such soils, and feasible management practices to mitigate some of the problems should provide information on the type and direction of chemical changes that may occur in poorly buffered soils in other regions of the world as a result of acidic precipitation. The potential for very acidic conditions to develop in these soils results from natural processes. Often, however, cultural disturbances of these potential acidic sulfate soils, usually for agricultural purposes, cause the chemical and microbial processes that produce severe acidic soil conditions.

Pyrite (FeS_2) is the distinguishing chemical feature of potential acidic sulfate soils, and it forms by natural processes primarily in coastal sediments. Upon soil drainage, pyrite oxidizes producing sulfuric acid. The soil becomes acidic if the quantity of acid produced exceeds the acid-neutralizing capacity of the soil. The excess acid often decreases the soil's pH to less than 4, and sometimes to near 2. The acid produced has a major influence on chemical and microbial processes in soils. Substances detrimental to plant growth, such as Al^{3+}, Fe^{2+}, H_2S, and CO_2, are often generated in amounts beyond the critical toxic levels for normal plant growth. Also, these soils are usually infertile, containing levels of essential nutrients below what is required for good crop production.

Acidic sulfate soils are a major problem because of their harmful effect on crop production and because of the large areas affected in regions of the world where additional food production is especially important. These soils are estimated to occupy an area of 12.5 million ha. Acidic sulfate soils occur worldwide in most climate zones but are found primarily in tropical regions. Most acidic sulfate soils are found in coastal areas at low elevations, where they developed from pyrite-containing marine sediments. However, some acidic sulfate and potential acidic sulfate soils are also found in inland areas, where they have formed from sedimentary rocks. Exposure of these buried pyritic rocks by mining or other excavation also produces acidic sulfate soils or acidic drainage that can adversely affect plant growth or streams receiving drainage.

Large areas of acidic sulfate soils are located in areas well suited and needed for flooded rice cultivation. Thus these soils are an important potential resource in densely populated areas in developing countries where rice is the main diet, especially in the coastal areas of southeast Asia and west Africa. Thailand, for example, is a major rice-growing country and has about 1.5 million ha of acidic sulfate and potential acidic sulfate soils.

The development of acidic sulfate soils from potential acidic sulfate soils is affected in a major way by management of soil drainage, such as the intentional flooding and draining cycles necessary for wetland rice cultivation. Flooded rice cultivation offers the benefits of retarding additional pyrite oxidation during the flooding cycle. However, flooding soils affects the plant availability of several soil components due to changes in both pH and redox potential conditions. Reducing conditions (low redox potential) associated with flooding may render toxic metals and other materials more soluble in the soil solution and thereby aggravate the adverse effects of acidic sulfate soils on plant growth. Oxidized conditions (high redox potential) during drainage cycles, however, often create a deficiency of essential nutrients and again increase levels of some toxic metals, this time due to the strongly acidic conditions associated with soil drainage. Thus, there is a complex interaction between production of substances adversely affecting plant growth and the alternating reduced and oxidized conditions imposed on the soil as part of the rice cultivation system. Knowledge of the chemistry of these soils under flooding and drainage cycles is leading to management practices that in some cases are making a substantial increase in rice yields economically feasible. However, additional research is needed to continue progress in this area.

References

Abichandani, C. T., and S. Patnaik. 1955. Internat. Rice Comm. Newsletter 13:11.

Andriesse, J. P., N. van Breemen, and W. A. Blokhuis. 1973. *In* H. Dost, ed. *Acid sulfate soils,* 11–39. Proc. Int. Symp. ILRI. Pub. 18, vol II, Wageningen, The Netherlands.

Armstrong, W. 1971. Plant Physiol 25:192–197.
Benckiser, G., S. Santiago, H. U. Neue, I. Watanabe, and J. C. G. Ottow. 1984. Plant and Soil 79:305–316.
Berner, R. A. 1964. J Geol 72:293–306.
Berner, R. A. 1970. Am J Sci 268:1–23.
Blackwelder, B. W., O. H. Pilkey, and J. D. Howard. 1979. Science 204:518–520.
Bloomfield, D., and J. K. Coulter. 1973. Adv Agron 25:256–326.
Bouma, J. 1983. *In* L. P. Wilding, N. E. Smeck, and G. F. Hall, eds. *Pedogenesis and soil taxonomy. I. Concepts and interactions,* 253–281. Elsevier Science Publishers B.V., Amsterdam.
Brinkman, R., and L. J. Pons. 1973. *In* H. Dost, ed. *Acidic sulfate soils,* 169–203. Proc. Int. Symp. ILRI Pub. 18, Vol. I. Wageningen, The Netherlands.
Broadbent, F. E., and M. E. Tusneem. 1971. Soil Sci Am Proc 35:922.
Charoenchamratcheep, C., C. J. Smith, S. Satawathananont, and W. H. Patrick, Jr. 1987. Soil Sci Soc Am J 51:630–634.
Connell, W. E., and W. H. Patrick, Jr. 1968. Science 159:86–87.
Connell, W. E., and W. H. Patrick, Jr. 1969. Soil Sci Soc Am Proc 33:711–715.
Dent, D. L., and R. W. Raiswell. 1982. *In* H. Dost and N. van Breemen, eds. *Proc. Bangkok symp. on acid sulfate soils,* 73–79. ILRI Pub. 31. Wageningen, The Netherlands.
Dong-Qing, C., W. Jing-Hua, and Z. Xiao-Nian. 1985. *In* Y. Tian-ren, ed. *Physical chemistry of paddy soils,* 131–156. Science Press, Beijing.
Engler, R. M., and W. H. Patrick, Jr. 1973. Soil Sci Soc Am Proc 37:685–688.
Fanning, D. S. 1978. *Soil morphology, genesis, classification, and geography*. Dept. of Agron., Univ. of Maryland, College Park.
FAO/UNESCO. 1979. *Soil map of the world, scale 1:5,000,000*. Vols. 1–X. UNESCO, Paris.
Foy, C. D., W. H. Armiger, L. W. Briggle, and D. A. Reid. 1965. Agron J 57:413–417.
Foy, C. D., H. N. Lafever, J. W. Schwartz, and A. L. Fleming. 1974. Agron J 66:751–758.
Gambrell, R. P., and W. H. Patrick, Jr. 1978. *In* D. D. Hook and R. M. M. Crawford, eds. *Plant life in anaerobic environments,* 375–423. Ann Arbor Science Publishers, Ann Arbor, MI.
Goldhabor, M. B., and I. R. Kaplan. 1974. *In* E. D. Goldberg, ed. *The sea,* vol. 5, *Marine chemistry,* 527–655. Wiley Interscience, New York.
Gotoh, S., and W. H. Patrick, Jr. 1972. Soil Sci Soc Am Proc 36:738.
Gotoh, S., and W. H. Patrick, Jr. 1974. Soil Sci Soc Am Proc 38:66.
Harmsen, K., and N. van Breemen. 1975. Soil Sci Soc Am Proc 39:1148–1153.
Howeler, R. H. 1972. J Environ Qual 1:366.
Howeler, R. H., and D. R. Bouldin. 1971. Soil Sci Am Proc 35:202.
IRRI (International Rice Research Institute). 1965. Annual Report 1965. Los Banos, Philippines. 335 pp.
IRRI (International Rice Research Institute). 1973. "Research Highlights" in Annual Report 1973. Los Banos, Philippines.
Ivarson, K. C., G. J. Ross, and N. M. Miles. 1982. *In* J. A. Kittrick, D. S. Fanning, and L. R. Hossner, eds. *Acid sulfate weathering,* 57–76. Spec. Pub. 10. Soil Science Society of America, Madison, WI.
Jakobsen, P., W. H. Patrick, Jr., and B. G. Williams. 1981. Soil Sci 132:279–287.

Kawaguchi, K., and K. Kyuma. 1969. *Lowland rice soils in Thailand*. Nat. Sci. Ser. N-4. Center for Southeast Asian Studies, Kyoto University, Japan. 270 pp.

Kawalec, A. 1973. *In* H. Dost, ed. *Acid sulphate soils,* 292–295. Proc. Int. Symp., ILRI Pub. 18, vol. I, Wageningen, The Netherlands.

Khalid, R. A., W. H. Patrick, Jr., and F. J. Peterson. 1976. *In Proc. 16th rice tech.,* 103–104. Working group, Lake Charles, LA.

Komes, A. 1973. Thai J Agric Sci 6:127–143.

Mahapatra, I. C., and W. H. Patrick, Jr. 1971. *In Proc. internat. symp. soil fert. eval., New Delhi* 1:53.

Mitsui, S. 1954. *Inorganic nutrition, fertilization, and amelioration for lowland rice.* Yokendo, Ltd., Tokyo, Japan.

Moormann, F. R. 1963. Soil Sci 95:271–275.

Mortimer, C. H. 1942. J Ecol 30:147.

Motomura, S. 1961. Soil Sci Plant Nutr 7:54–60.

Motomura, S. 1962. Soil Sci Plant Nutr 8:20–29.

Nhung, M. M., and F. N. Ponnamperuma. 1966. Soil Sci 102:29–41.

Nordstrom, D. K. 1982. *In* J. A. Kittrick, D. S. Fanning, and L. R. Hossner, ed. *Acid sulfate weathering,* 37–56. Spec. Pub. 10, Soil Science Society of America, Madison, WI.

Pasricha, N. S., and F. N. Ponnamperuma. 1976. Soil Sci Soc Am J 40:374–380.

Patrick, W. H., Jr. 1964. *In Soil fertility and plant nutrition,* 605–608. Trans. 8th. Int. Congr. Soil Sci., vol. IV. Publishing House of the Academy of the Socialist Republic of Romania, Bucharest.

Patrick, W. H., Jr., and R. D. DeLaune. 1972. Soil Sci Soc Am Proc 36:573.

Patrick, W. H., Jr., and I. C. Mahapatra. 1968. Adv Agron 20:323.

Patrick, W. H., Jr., and D. S. Mikkelsen. 1971. *In* R. D. Dinauer, ed. *Fertilizer technology and use,* 2d ed., 187. Soil Science Society of America, Madison, WI.

Patrick, W. H., Jr., D. S. Mikkelsen, and B. R. Wells. 1985. *In Fertilizer technology and use,* 3d ed., 197–228. Soil Science Society of American, Madison, WI.

Patrick, W. H., Jr., and K. R. Reddy. 1976a. J Environ Qual 5:469.

Patrick, W. H., Jr., and K. R. Reddy. 1976b. Soil Sci Soc Am J 40:678.

Patrick, W. H., Jr., and K. R. Reddy. 1978. *In* F. N. Ponnamperuma, ed. *Soils and rice,* 361–380. International Rice Research Institute, Los Banos, Philippines.

Patrick, W. H., Jr., and R. Wyatt. 1964. Soil Sci Am Proc 28:647.

Poelman, J. N. B. 1973. *In* H. Dost, ed. *Acid sulphate soils,* 197–207. Proc. Int. Symp. ILRI Pub. 18, vol. II. Wageningen, The Netherlands.

Ponnamperuma, F. N. 1955. The chemistry of submerged soils in relation to the growth and yield of rice. Ph.D. dissertation. Cornell University, Ithaca, NY.

Ponnamperuma, F. N. 1965. *In The mineral nutrition of the rice plant,* 295–328. Johns Hopkins University Press, Baltimore.

Ponnamperuma, F. N. 1972. Adv Agron 24:29–95.

Ponnamperuma, F. N., T. Attanandana, and G. Beye. 1973. *In* H. Dost, ed. *Acid sulphate soils,* 391–406. Proc. Int. Symp. ILRI Pub. 18, vol. II. Wageningen, The Netherlands.

Pons, L. J. 1973. *In* H. Dost, ed. *Acid sulphate soils,* 3–17. Proc. Int. Symp. ILRI Pub. 18, vol. I. Wageningen, The Netherlands.

Pons, L. J., N. van Breemen, and P. M. Driessen. 1982. *In* J. A. Kittrick, D. S. Fanning, and L. R. Hossner, ed. *Acid sulfate weathering,* 1–18. Spec. Pub. 10. Soil Science Society of America, Madison, WI.

Pons, L. J., and W. van der Kevie. 1969. *Acid sulphate soils in Thailand*. Soil Survey Rept., Land Dev. Dept., Bangkok, SSR-81. 69 pp.
Prasittikhet, J. 1987. Metal availability and rice growth under controlled redox potential and pH conditions in acid sulfate soils of Thailand. Ph.D. dissertation, Louisiana State University, Baton Rouge.
Presley, B. J., and I. R. Kaplan. 1968. Geochim Cosmochim Acta 32:1037–1048.
Rickard, D. T. 1975. Am J Sci 275:636–652.
Roberts, W. M. B., A. L. Walker, and A. S. Buchanan. 1969. Mineral Deposita 4:18–29.
Russell, J. E. 1961. *Soil conditions and plant growth*. John Wiley and Sons, New York.
Singer, P. C., and W. Stumm. 1970. Science 167:1121–1123.
Sombatpanit, S. 1970. Acid Sulfate Soils: Their nature and properties. Unpublished thesis, Royal Agricultural College, Uppsala, India.
Sombatpanit, S., and L. Wangpaiboon. 1973. Study on the lime requirement of an aluminum-rich acid sulfate soil. Paper presented at the 12th Conf. Agric. Biol. Sci., Kasetsart University, Bangkok.
Stumm, W., and J. J. Morgan. 1970. *Aquatic chemistry: An introduction emphasizing chemical equilibria in natural waters*. Wiley Interscience, New York. 583 pp.
Sweeney, R. E., and I. R. Kaplan. 1973. Econ Geol 68:618–634.
Tanaka, A., and S. Yoshida. 1970. *Nutritional disorders of the rice plant in Asia*. IRRI Tech. Bull. 10, Los Banos, Philippines. 51 pp.
Tian-Ren, Y. 1985. *In* Y. Tian-Ren, ed. *Physical chemistry of paddy soils,* 197–217. Science Press, Beijing.
Turner, F. T., and W. H. Patrick, Jr. 1968. *Trans. 9th Internat. Cong. Soil Sci.* 4:53.
Tusneem, M. E., and W. H. Patrick, Jr. 1971. *Nitrogen transformations in waterlogged soil*. Louisiana Agric. Exp. Sta. Bull. 657, Baton Rouge.
United States Department of Agriculture. 1975. *Soil taxonomy: A basic system of soil classification for making and interpreting soil surveys*. USDA Agric. Handbook 436, U.S. Government Printing Office, Washington, DC.
Vamos, R. 1967. Soil Fert Abstr 30:2438.
van Breemen, N. 1973. *In* H. Dost, ed. *Acid sulfate soils,* 66–130. ILRI Pub. 18, vol. I. Wageningen, The Netherlands.
van Breemen, N. 1975. Soil Sci Soc Am Proc 39:1153–1157.
van Breemen, N. 1982. *In* J. A. Kittrick D. S. Fanning, and L. R. Hossner, eds. *Acid sulfate weathering,* 95–108. Spec. Pub. 10. Soil Science Society of America, Madison, WI.
van Breemen, N., and F. R. Moormann. 1978. *In* F. N. Ponnamperuma, ed. *Soils and rice,* 781–800. International Rice Research Institute, Los Banos, Philippines.
van Breemen, N., and L. J. Pons. 1978. *In Soils and rice,* 739–761. International Rice Research Institute, Los Banos, Philippines.
Waring, S. A., and J. M. Bremner. 1964. Nature 201:951.
Xie-Ming, B. 1985. *In* Y. Tian-ren, ed. *Physical chemistry of paddy soils,* 69–91. Science Press, Beijing.
Yoshida, S., and T. Tadano. 1978. *In* G. A. Jung, ed. *Crop tolerance to subtropical land conditions,* 233–255. Spec. Pub. 32. American Society of Agronomy, Madison, WI.

Aluminum Speciation: Methodology and Applications

Paul M. Bertsch*

Abstract

There exists ample evidence to suggest that chemical speciation regulates Al mobility in the environment and its bioavailability and toxicity to many aquatic organisms and plant species. Several Al speciation methods are useful for providing *estimates* of organically and inorganically complexed Al and adequately differentiating soluble from colloidal Al components. Additionally, comprehensive geochemical speciation models can be used effectively to delimit the boundary conditions describing Al solubility and solution species distribution. Neither of these approaches, however, can provide reliable information regarding specific molecular distributions of Al. Little work has focused on Al species transformation in natural systems, yet such reactions may be the primary regulators of Al mobility in natural systems, particularly during episodic events.

I. Introduction

Ever since the hypothesis was proposed that Al remobilization in soils and subsequent transport to aquatic systems may be accelerated by atmospheric inputs of strong acids (Cronan and Schofield, 1979), the biogeochemistry of Al in natural systems has been studied extensively (e.g., Johnson et al., 1981; Driscoll et al., 1983, 1985; David and Driscoll, 1984; James and Riha, 1984; Turner et al., 1985; Lawrence et al., 1986, 1988; Cozzarelli et al., 1987). The toxicity of Al to aquatic organisms has also been widely studied (Baker and Schofield, 1980, 1982; Driscoll et al., 1980; Hall et al., 1985). Additionally, a renewed interest in Al toxicity to plants has ensued, resulting partly from the suggestion that Al may be toxic to forest vegetation (Ulrich et al., 1980; Vogelman, 1982) and also as a result of new analytical techniques and approaches that have stimulated numerous recent

*Division of Biogeochemistry, University of Georgia, Savannah River Ecology Laboratory, P.O. Drawer E, Aiken, SC 29801, USA.

research efforts in the long-studied toxicity of Al to crop species (e.g., Foy et al., 1978; Pavan and Bingham, 1982; Blamey et al., 1983; Alva et al., 1986; Parker et al., 1988).

Central to all of these studies is the concept of *aluminum speciation,* the distribution of Al among its various physicochemical forms. It has been difficult to elucidate specific mechanisms operative in Al remobilization in soil and sediments, its transport through terrestrial and aquatic ecosystems, and its bioavailability, biocycling, and toxicity to aquatic and terrestrial organisms. Incomplete understanding of Al species distribution in complex multicomponent systems and the general lack of information regarding the processes and rates involved in Al species transformations are largely responsible for the inability to adequately define these mechanisms. Notwithstanding, it is widely accepted that speciation regulates Al mobility, transportability, bioavailability, and ultimately its toxicity to terrestrial and aquatic organisms.

The literature dealing with Al speciation methods and their applications has become voluminous. It is not my intent to review all of these studies in detail, but rather to present the salient features of Al speciation methods, approaches, and their applications. It is hoped that this approach will more clearly demonstrate the importance of speciation in defining the environmental biogeochemistry of Al and exemplify the many limitations inherent in Al speciation methodology, as well as potential consequences of misuse of these methods and approaches.

II. The Evolution of a Species

Long before Garrels popularized the concept of metal speciation with respect to mineral solubility (Garrels and Thompson, 1962), the speciation of Al hydrolytic products had been widely studied and debated (e.g., Hayden and Rubin, 1976; Bertsch, 1989). Thus, the concept of free ion activity utilized in the calculation of hydrolytic Al species distribution and Al solubility had been well established (e.g., Frink and Peech, 1962, 1963). Among the polynuclear Al species proposed in the literature are the $[Al_2(OH)_2]^{4+}$, $[Al_3(OH)_4]^{5+}$, $[Al_4(OH)_{16}]^{2+}$, $[Al_6(OH)_{15}]^{3+}$ through $Al_{54}(OH)_{144}(H_2O)_{36}^{18+}$ series (hexameric or gibbsite fragment model), $[Al_8(OH)_{20}]^{4+}$, $[Al_9(OH)_n]^{(27-n)+}$, $[AlO_4Al_{12}(OH)_{24+n}(OH)_{12-n}]^{(7-n)+}$ (Al_{13}), $[Al_{13}(OH)_{32}]^{7+}$, $[Al_{14}(OH)_{34}]^{8+}$, and the $[Al_{15}(OH)_{36}]^{9+}$ complexes. However, only the gibbsite fragment model and the $[Al_2(OH)_2]^{4+}$ and Al_{13} species have been suggested with due consideration to structure, and of these, only the dinuclear and Al_{13} polynuclear species have convincing direct spectroscopic evidence for their existence (Bertsch, 1989). The mononuclear hydrolytic Al species proposed in the literature are the $Al(OH)(H_2O)_5^{2+}$, $Al(OH)_2(H_2O)_4^{+}$, $Al(OH)_3(H_2O)_3^{\circ}$, $Al(OH)_4^{-}$, and $Al(OH)_5^{2-}$ species, but only the mono, di, and tetrahydroxo mononuclears are widely accepted. Controversy continues to surround the existence of the trihydroxo Al complex, and there is little supportive evidence for the $Al(OH)_5^{2-}$ species (Bertsch, 1989).

It was largely the work of Adams and coworkers (e.g., Adams and Lund, 1966; Richburg and Adams, 1970; Adams, 1974) that first considered the importance of free ion activity of the hexaaqua Al^{3+} species after correcting for various inorganically complexed forms in more complex soil solutions. Important inorganic complexes of Al include the sulfate, fluoride, phosphate, and silica complexed species. Of these, the sulfato- and fluoro-Al complexes have been best defined thermodynamically (Nordstrom and May, 1989). There is strong evidence for the existence of both mono- and polynuclear phosphato- and silicato-complexed Al forms (White et al., 1976; Wada and Wada, 1980), although they have not been properly defined thermodynamically. The organic Al complexed species have been defined thermodynamically for only a few organic acids, whereas the humic-bound Al forms remain poorly described. Other complexes that have been ignored entirely are the mixed organic and organic-inorganic species. Spectroscopic evidence suggests that formation of such complexes does occur (Akitt et al., 1971; Bertsch, 1987a). Problems resulting from our lack of knowledge regarding Al complexed species will be discussed in greater detail subsequently.

During the 1960s and 1970s, most investigators studying Al chemistry in soils and aquatic systems and Al toxicity continued to report total Al values, generally failing to consider the important role of speciation in their results. This produced an extensive database that was, and unfortunately is still, of limited usefulness. In addition to the need for speciating the total dissolved Al fraction, it became increasingly evident that distinguishing between physicochemical forms of Al—that is, dissolved, polynuclear colloidal, and crystalline phases—was complicated by evidence that portions of these solid phases could pass through even 0.1 μm pore-size filters (Kennedy et al., 1974; Jones et al., 1974; Barnes, 1975). By the early 1980s it was well established that Al species distribution was the primary regulator of Al mobility in soils and aquatic systems and of Al toxicity to plants and fish (Driscoll, 1980; Driscoll et al., 1980; Johnson et al., 1981; Baker and Schofield, 1982; Pavan and Bingham, 1982; Blamey et al., 1983), although there is much disagreement concerning the relative toxicity of many of the species.

III. Methods and Approaches

Three general approaches to Al speciation can be identified. The first involves the analytical separation of various Al fractions based on differential reaction kinetics with complexation agents and/or the physicochemical separation of Al fractions based on size or charge, often followed by operationally defined categorization of these distinguishable fractions as distinct Al species. The second approach involves the computational differentiation of Al species from an analytically determined "total" Al fraction utilizing a thermodynamically based geochemical speciation model, usually employing mass balance constraints. The third general approach involves some combination of one or more analytical separation techniques along with a geochemical speciation model.

A. Analytical Methods

1. Timed Spectrophotometric Methods

The original Al speciation methods were devised to differentiate mononuclear and polynuclear hydroxo Al complexes. Okura and others (1962), expanding on an earlier study, observed differential reaction of 8-hydroxyquinoline with Al and suggested that the Al fraction interacting instantaneously was hexaaqua- and hydroxo-mononuclear species, whereas the fraction interacting more slowly was polynuclear Al forms. This method was further refined and calibrated by Turner (1969, 1971, 1976; Turner and Sulaiman, 1971) for a number of partially neutralized Al solutions, where it was reported that three distinct forms of Al could be differentiated based on reaction kinetics. Consistent with Okura and others (1962), Turner suggested that the Al fraction reacting instantaneously with 8-hydroxyquinoline was mononuclear Al (Al_a) including the hexaaqua $Al(H_2O)_6^{3+}$, mono- and dihydroxo $Al(OH)(H_2O)_5^{2+}$, $Al(OH)_2(H_2O)_4^{+}$ mononuclears, as well as the aluminate $Al(OH)_4^-$ species present in higher pH solutions. He identified a more slowly reacting Al fraction, which conformed to pseudo-first-order kinetics, and assigned this to polynuclear complexes (Al_b), and finally an Al fraction nonreactive with 8-hydroxyquinoline, which he assigned to one or more forms of colloidal solid phases (Al_c). A similar method for differentiating mono- and polynuclear Al complexes was proposed by Smith (1971) and Smith and Hem (1972), where differential reaction kinetics of Al was ferron (8-hydroxy-7-iodo-5-quinoline-sulfonic acid) were utilized. These studies produced results very similar to the investigations of Turner (1969, 1971; Turner and Sulaiman, 1971), although the two had divergent interpretations concerning the nature of Al polynuclears present in solution (Bertsch, 1989).

These two timed spectrophotometric methods are, in general, the basis for the many different speciation methods that have been developed for a variety of applications. Experimentally these methods are quite simple: The ferron or 8-hydroxyquinoline is added to the solution to be speciated and the absorbance monitored as a function of time (Figure 3–1A). (Note: The 8-hydroxyquinoline method requires extraction into an organic phase to quench further reaction and extract the 8-hydroxyquinoline complexed Al.) The data are generally transformed into first-order rate plots to provide a more accurate estimate of the mononuclear and polynuclear Al components existing in solution (Figure 3–1B). Recently it has been suggested that a more accurate estimate of the Al_a or mononuclear Al fraction can be obtained by assuming two simultaneous irreversible first-order reactions:

$$Al_a + F \xrightarrow{ka} Al\text{-}F \qquad (1)$$

$$Al_b + F \xrightarrow{kb} Al\text{-}F \qquad (2)$$

where Al*a* = mononuclear Al species, Al*b* = reactive polynuclear Al species, *ka* is the mononuclear-ferron rate coefficient, k_b is the polynuclear-ferron rate coefficient, and Al-F is the common ferron-Al complex (Jardine and Zelazny, 1986). More recently it was demonstrated that ferron interactions with hydroxo-

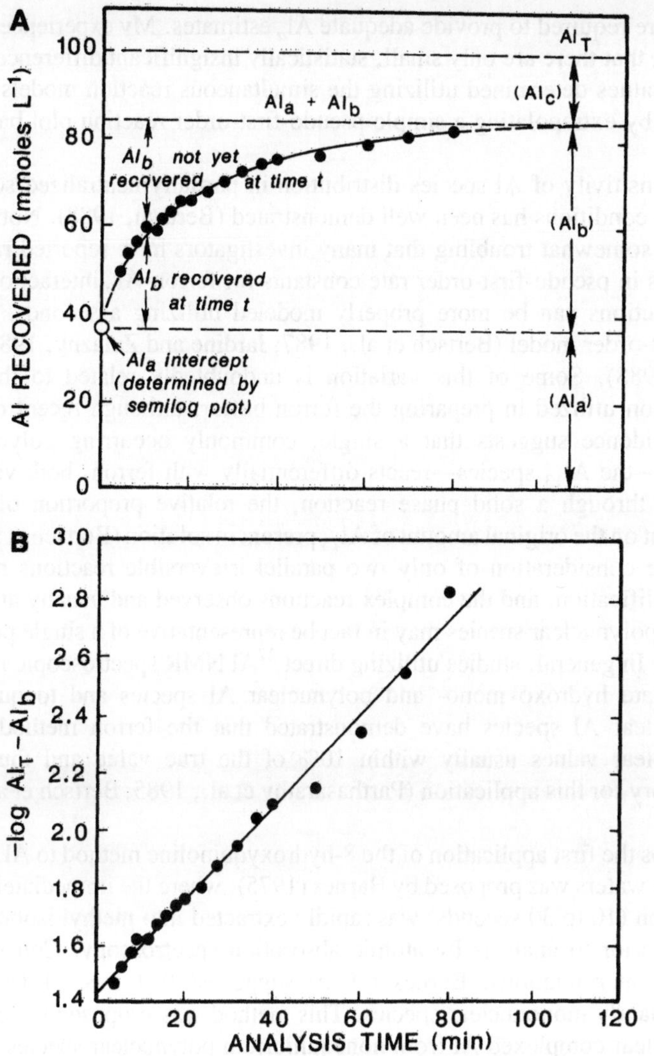

Figure 3–1. General representation of (A) absorbance versus time plot for an Al solution by the ferron method and (B) the pseudo-first-order rate plot used to extrapolate the Al_b fraction to zero time.

mono- and polynuclear Al species could be more accurately modeled by utilizing a binary species general order reaction function where two simultaneous second-order reactions provided the best fit (Jardine and Zelazny, 1987a, 1987b). These studies also demonstrated that differentiation of hydroxo-mono- and -polynuclear species in solutions containing inorganic and organic ligands could also be made by utilizing the simultaneous second-order reaction model, although for strongly complexing ligands (e.g., $H_2PO_4^-$, F^-, citrate, and oxalate) low ligand/Al molar ratios were required to provide adequate Al_a estimates. My experience indicates, however, that there are only small, statistically insignificant differences between the Al_a values determined utilizing the simultaneous reaction models and those obtained by extrapolating a simple pseudo-first-order reaction plot back to zero time.

The sensitivity of Al species distribution in partially neutralized solutions to synthesis conditions has been well demonstrated (Bertsch, 1988). Notwithstanding, it is somewhat troubling that many investigators have reported rather wide variations in pseudo-first-order rate constants for ferron-Al_b interactions, or that these reactions can be more properly modeled utilizing a second-, or even a fractional-order model (Bertsch et al., 1987; Jardine and Zelazny, 1987a; Parker et al., 1988). Some of this variation is undoubtedly related to the specific formulation utilized in preparing the ferron buffer, although recent direct ^{27}Al NMR evidence suggests that a single, commonly occurring polynuclear Al complex—the Al_{13} species—reacts differentially with ferron, both via solution and also through a solid phase reaction, the relative proportion of which is dependent on the original amount of Al_{13} present in solution (Bertsch et al., 1989). Thus, the consideration of only two parallel irreversible reactions may be an oversimplification, and the complex reactions observed and usually attributed to multiple polynuclear species may in fact be representative of a single polynuclear complex. In general, studies utilizing direct ^{27}Al NMR spectroscopic methods to differentiate hydroxo mono- and polynuclear Al species and to quantify the mononuclear Al species have demonstrated that the ferron method produces mononuclear values usually within 10% of the true value and thus is quite satisfactory for this application (Parthasarathy et al., 1985; Bertsch et al., 1986a, 1986b).

Perhaps the first application of the 8-hydroxyquinoline method to Al speciation in natural waters was proposed by Barnes (1975), where the immediately reacting Al fraction (10 to 30 seconds) was rapidly extracted into methyl isobutyl ketone (MIBK) prior to analysis by atomic absorption spectroscopy. Consistent with previous interpretations, Barnes (1975) suggested that this Al fraction was predominantly mononuclear species. This method was proposed to separate the mononuclear complexed Al from nonequilibrium polynuclear species and finely divided colloidal precipitates that passed a 0.1 μm pore-sized filter. Prior to complexation and extraction, the samples were adjusted to $\simeq$pH 8.3 to minimize possible interferences from fluoride. No references to potential problems with organically complexed Al species were made in this study or in the subsequent improvement of the method (May et al., 1979). In another modification of the 8-hydroxyquinoline method (pH = 5.0), however, Bloom and others (1978) noted significant time dependence of contact with the complexing reagent prior to extraction with butyl acetate and increased Al recovery, and suggested that this resulted from organically complexed Al forms. Bloom and others (1978) also indicated that a minimal contact time with reagents before extraction would probably provide a better estimate of the quantity of uncomplexed monomeric Al than other commonly used methods. The method presented, however, utilized a 15-minute reaction time to allow for the complexation of Fe by phenanthroline in order to eliminate Fe interference. In a later study on acid-washed peats, Bloom and others (1979) performed "immediate" extraction with butyl acetate to minimize decomplexation of organic-Al complexes by the 8-hydroxyquinoline. In another investigation utilizing the pH 8.3 version of this method on low-ionic-strength surface waters, it was concluded that Al-organic complexes were totally dissociated following the 15-second reaction time at pH 8.3 (LaZerte, 1984). This finding is inconsistent with the results of Lalande and Hendershot (1986), who found that only 66% of the Al associated with lower-molecular-weight organics was recovered by the 15-second hydroxyquinoline extraction at pH 8.3, the remainder being recovered in the fraction assigned to polynuclear or colloidal Al.

The 8-hydroxyquinoline method was further modified by James and others (1983), who employed a 15-second reaction with 8-hydroxyquinoline (HQ) buffered at pH 5.2, followed by extraction into butyl acetate. The method employed a correction for Fe by measuring the absorbance at a second wavelength (600 nm) and subtracting the corresponding contribution of the Fe absorbance at 395 nm. In samples containing only OH^- ligands, the method was demonstrated to predominantly measure the hexaaqua-$Al(H_2O)_6^{3+}$ and hydroxo-monomers $Al(OH)^{2+}$, $Al(OH)_2^+$ after 15 seconds of reaction, consistent with previous findings. The presence of other complexing ligands such as F^-, citrate, and Si, however, produced variable results that were dependent on ligand:Al mole ratios and pH, with generally greater Al being recovered following the 15-second reaction than the predicted thermodynamic distribution of only the aquated and hydroxo-complexed forms. This variability with model ligands led James and others (1983) to define operationally the 15-second reactive Al as the "labile" Al fraction because it included a kinetically labile Al fraction that differed with the specific chemical conditions of the solutions studied.

Another timed spectrophotometric method for differentiating mono- and polynuclear Al species was introduced by Blamey and others (1983). This method utilized the aluminon (aurine tricarboxylic acid) assay of Hsu (1963), in which it was assumed that the Al fraction reacting after 30 minutes was the Al_a fraction. The method was calibrated utilizing a single, partially neutralized Al solution [50 μmol L^{-1} with respect to total Al and having a OH/Al mole ratio (ñ) of 2.0], where it was demonstrated that adequate differentiation of the Al_a and Al_b fractions could be made. The method was used without further calibration in several phytotoxicity studies employing nutrient solutions containing various complexing ligands, such as phosphate, fluoride, citrate, and sulfate. Recent studies have demonstrated, however, that strong complexing ligands do interfere with the Al_a-Al_b differentiation (Jardine and Zelazny, 1987a, 1987b; Noble et al., 1988a; Alva et al., 1989). Such use of the aluminon method to estimate the Al_a fraction in complex solutions demonstrates the importance of properly calibrating a newly introduced Al speciation method with at least a minimum of model ligands before its indiscriminate use. Any spectrophotometric method involving a complexing organic requiring colloidal "lake" formation, such as aluminon, should generally be avoided for speciation methods because the reaction pathways and stoichiometries are so poorly defined.

Perhaps the most promising timed spectrophotometric method has recently been introduced by Bartlett and others (1987). Utilizing the pyrocatechol violet method of Dougan and Wilson (1974), these investigators (Bartlett et al., 1987) referenced the reaction rate of Al in solutions containing added complexing ligands or in natural soil and lysimeter solutions against the reaction rate of Al standards by regressing the absorbances at 60-second increments against log time in 60-second units. From this analysis three fractions were operationally defined: the rapidly reacting Al (RRAl) fraction, the moderately rapid Al (MRAl) fraction, and the total reactive Al (TRAl) fraction. It was demonstrated that the RRAl fraction by this method represented a smaller fraction of the total Al and less organically complexed Al than the "labile" Al fraction determined by the 15-second HQ method of James and others (1983), whereas the MRAl fraction tended to produce values more similar to the 15-second HQ "labile" Al fraction. The major advantages of this method include its simplicity, its ability to produce a range of kinetically reactive Al values that can be subsequently related to some observed effect (e.g., increase of mobility or toxicity), its need for only a small sample volume, and its relative precision and sensitivity compared to similar methods. An additional advantage of this method is the ability to correct for Fe interference by absorbance measurements at a second wavelength, rather than by adding reducing or complexation agents to the samples. This method is and probably will continue to be used in many field studies requiring rapid Al speciation methods.

In general, one must exercise caution in direct comparisons of data utilizing different methods or versions of a given spectrophotometric method because significant variability in the specifics of the methods may exist. For example, even when identical pH buffers are used, there may be differences in the concentration of the complexation agent or aging times of the reagents used between investigators, resulting in variability that may easily be overlooked (Jardine and Zelazny, 1986).

2. Physicochemical Separation Methods

The most basic of the physicochemical separation methods probably is simple filtration. The well-established delineation between "dissolved" and "total" Al has been based on filtration through a 0.45 μm filter. It has long been recognized that this operationally defined "dissolved" Al fraction or even filtrate passing a 0.1-μm filter can include significant contributions from finely divided colloidal solid phases containing Al either within the matrix or as an adsorbed species (Jones et al., 1974; Kennedy et al., 1974; Barnes, 1975). Again, the presence or absence of these finely divided phases is highly system specific; thus, studies demonstrating that only a small fraction of Al passing a 0.45-μm filter is colloidal in nature can also be readily identified (e.g., Royset et al., 1987).

As indicated, the original Al speciation methods attempted to differentiate between "truly dissolved" and colloidal Al phases. Many of the speciation methods are quite reliable for this differentiation in samples that do not contain significant quantities of strongly (e.g., organically) complexed Al. However, for many samples of natural origin there are problems in clearly differentiating the dissolved and colloidal Al fractions. The choice of filter materials and the proper pretreatment of these materials can be quite important in terms of sample contamination, loss of dissolved Al, or dissolved organic carbon (Campbell et al., 1983; Jardine et al., 1986; Hodges, 1987; Royset et al., 1987). Thus, filtering introduces an initial operationally defined fractionation that must be carefully considered. Another complicating factor regarding filtration can arise when water chemistry information is coupled with a geochemical model for Al speciation. Often the Al in the filtrate is entered into a model as the total dissolved Al component even when the filtrate may contain one or more Al-containing solid phases.

Several Al fractionation methods differentiate various complexed Al species based on their size and/or charge. The most popular of these speciation methods utilizes a strongly acidic sulfonic resin to separate "organically" and "inorganically" complexed Al forms. The method originally proposed by Driscoll (1980) utilized a sulfonic resin column and the ferron assay which Driscoll (1984) changed to the 8-hydroxyquinoline assay using methyl isobutyl ketone for the organic solvent extraction, following the methods of Barnes (1975). The method measures three Al fractions that have been operationally defined: (1) a total reactive Al fraction determined following acidification of the sample to pH = 1 for 1 hour (Al_r), (2) a "total" monomeric Al fraction (Al_m) determined on a nontreated sample by the method of Barnes (1975), and (3) the "nonlabile" monomeric Al fraction determined on a sample passed through the sulfonic resin column (Al_o) (Figure 3–2). The Al*o* fraction has been ascribed to organically complexed Al, which is assumed to pass the ion exchange column without interacting. It was further assumed that all inorganically complexed mononuclear Al species would completely dissociate and be retained within the ion exchange column. Thus, the Al_m less the Al*o* fraction ($Al_m - Al_o = Al_l$) was assigned to a labile monomeric fraction assumed to be predominantly mononuclear, inorganically complexed Al(Al_l), for example, hydroxo mononuclear Al species [$Al(OH)^{2+}$, $Al(OH)_2^+$, and $Al(OH)_3^\circ$, and the sulfato-, fluoro-, and phosphato-Al complexes]. Driscoll (1984) pointed out the importance of matching as closely as possible the ionic strength and pH within the column environment to that of the original samples and discussed proper resin pretreatment methods for this purpose. Other sources of significant variation in this method are column size, flow rate, resin bead size, or any factor influencing the sample resonance time within the column.

Driscoll (1984) compared the organic Al values estimated by the resin column method (Al_o) against those determined by the fluoride electrode technique (to be discussed subsequently) and found adequate agreement only at low Al_o values (Figure 3–3). Significant deviations were observed at less than 4 μmol L^{-1} Al_o, and these deviations were for the column method to consistently *underestimate* Al_o, suggesting that some dissociation of Al-organic complexes may be occurring.

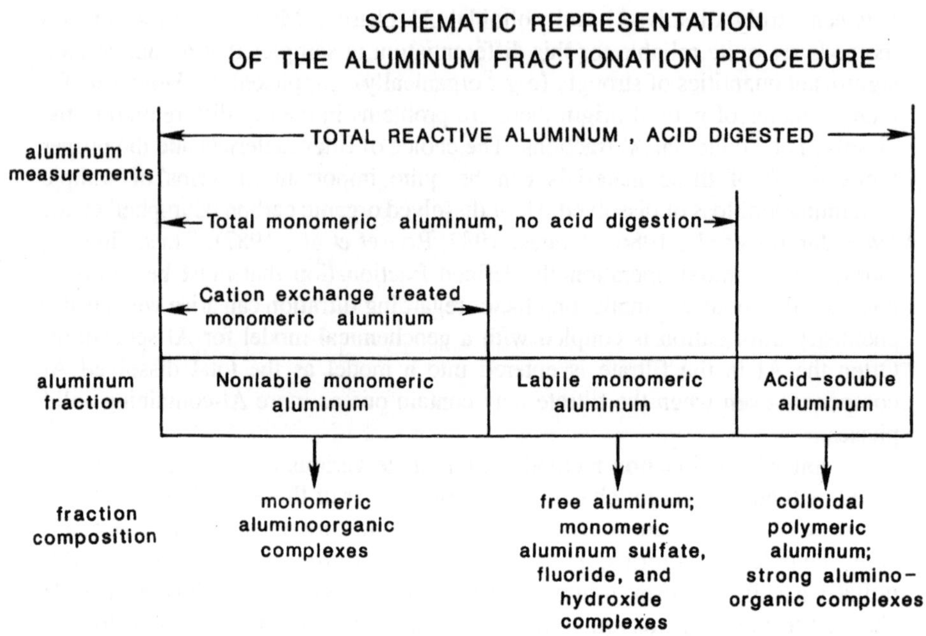

Figure 3–2. Schematic representation of the steps used in the strongly acidic ion exchange column method to derive the nonlabile (predominantly organically bound), labile (predominantly inorganically complexed), and the acid-soluble Al fractions (polynuclear, colloidal, and strong Al-organic complexes). (After Driscoll, 1980, with permission.)

Hodges (1987) also found evidence for some decomplexation of organic Al complexes when using the sulfonic resin column method. However, he also reported that strongly chelating organic ligands such as citrate could desorb inorganic Al separated from previously processed samples, producing anomalously high Al_o estimates. Thus, the number of samples run through a given column must also be carefully regulated.

LaZerte and others (1988) compared a modified sulfonic ion exchange column method with their previously published dialysis method (LaZerte, 1984) for estimating the Al_o and Al_I fractions. The comparison demonstrated remarkable agreement for the 267 dilute natural water samples they investigated, with only a 6% deviation observed (lower Al_I for the dialysis method) (Figure 3–4). An almost identical comparison by Backes and Tipping (1987) for isolated naturally occurring humic materials indicated a much larger deviation; up to 25%, and the discrepancy increased with the relative amount of Al associated with the organic fraction. My experience has been that, in general, the ion exchange column technique performs best for surface waters having moderate concentrations of dissolved organic carbon (DOC). The method performs quite variably, however, for complex soil solutions or higher ionic strength surface or groundwaters representing a more complex mixture of organic components. It has been

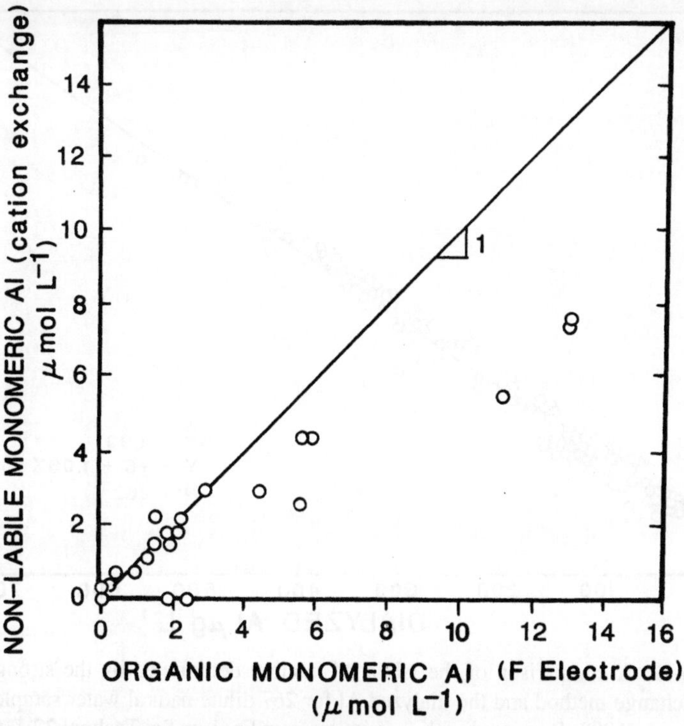

Figure 3–3. A comparison of the nonlabile Al fraction determined by the strongly acidic cation exchange column method and the organically complexed Al estimated by the F^- ion specific electrode technique for surface waters from the Adirondack Mountains. (Driscoll, 1984, with permission.)

suggested that low ionic strength surface waters may represent an optimal situation for the DOC to be composed primarily of a higher molecular weight fraction (Thurman, 1989, personal communication). This may explain the observations of LaZerte and others (1988) and invalidate indiscriminate extrapolation of such results to waters having diverse chemistries or origins. Also, for solutions high in F^- relative to Al, some leakage to fluoro-Al species can be expected. Several variations of the sulfonic ion exchange resin column technique have appeared in the literature, including fully automated versions (Seip et al., 1984; Rogeberg and Henricksen, 1985; Hodges, 1987). The method is simple, fast, and easy to employ in field studies and can provide information concerning the relative distribution of organically and inorganically complexed Al that can be quite useful in describing Al mobility in natural systems.

Batch equilibration of solutions with an iminodiacetate chelating cation exchange resin (Chelex-100) has also been used to speciate solution Al (Campbell et al., 1983; Dufresne and Hendershot, 1986; Litaor, 1987; Hodges, 1987). Campbell and others (1983) utilized the general methods of Figura and McDuffie

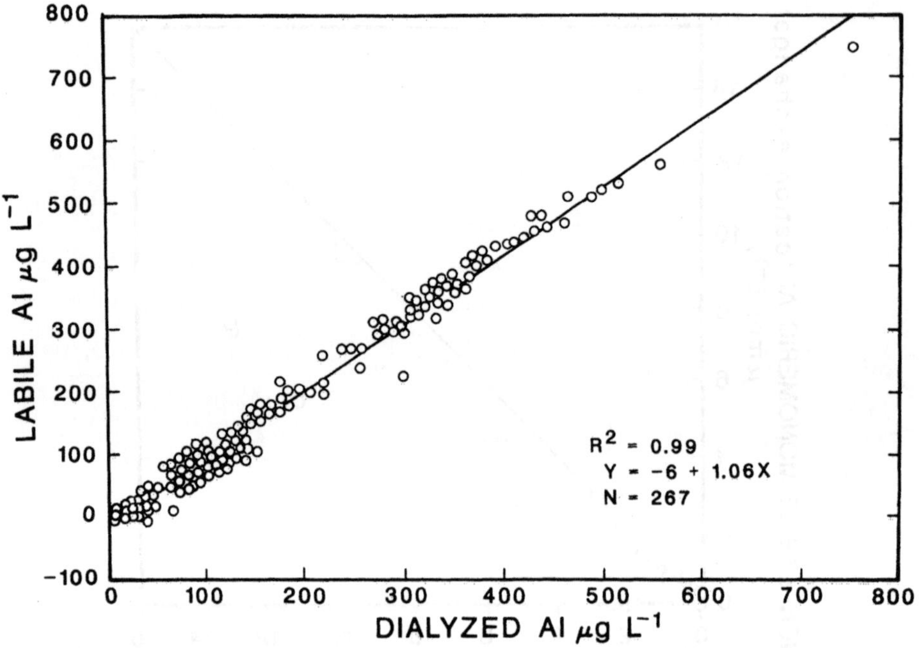

Figure 3–4. A comparison of the labile Al fraction determined by the strongly acidic cation exchange method and the dialyzed Al for 267 dilute natural water samples. (From LaZerte et al., 1988. Reprinted with permission from Environ Sci Technol 22:1106–1108. Copyright 1988, American Chemical Society.)

(1979, 1980) in an attempt to categorize the "readily" exchangeable and "nonexchangeable" forms. The Chelex-100 resin was fractionally loaded with Ca and Mg such that the equilibrium Ca and Mg concentrations were similar to those of the natural waters they were studying (1.0 and 0.5 mg L^{-1} Ca and Mg, respectively). For synthetic solutions, mono- and polynuclear hydroxo-complexes and fluoroaluminum species (F:Al = 0, 0.84, 1.7) exchanged readily (>85%) after a 30-minute reaction period. Aluminum solutions to which peat-extracted fulvic and humic acids had been added were found to exchange much more slowly (<5% after 30 minutes at a DOC:Al of 155:1). For streamwater samples, the Al exchange rate was found to be intermediate. Ultraviolet irradiation of natural samples completely eliminated the nonexchangeable fraction, leading Campbell and others (1983) to conclude that the nonexchangeable Al fraction was predominantly an organically complexed form. Because the method is kinetically based and because the rate of the exchange reaction is highly dependent on the specific ligands and Al:ligand ratio of the solutions investigated (Hodges, 1987), this method also provides only an operationally defined differentiation of readily exchangeable and nonexchangeable Al fractions. It has also been demonstrated that only the higher molecular weight organic complexes are included in the nonexchangeable fraction, because lower molecular weight humics and organic acids are readily retained by the Chelex-100 resin (Hodges, 1987). The method is not as easily adaptable for field use as the ion exchange column method previously described and requires a larger sample volume.

LaZerte (1984) employed an equilibrium dialysis technique coupled with the 8-hydroxyquinoline method of Barnes (1975) to differentiate between inorganically and organically complexed Al, as well as polynuclear and noncrystalline forms. It was reported that the 15-second or "fast reactive" Al fraction determined by a slightly modified version of the 8-hydroxyquinoline method of Barnes (1975) represented both mononuclear inorganic and organically complexed Al. The "fast reactive" Al fraction determined on the dialyzate is a measure of the inorganic mononuclear Al species, with the organically bound Al determined by difference. A "total reactive" Al fraction is also determined by a long (3 to 6 hour) equilibration of the sample with 8-hydroxyquinoline prior to extraction into MIBK.

Several potential problems with the dialysis method were discussed by LaZerte (1984) and by Bloom and Erich (1989). Perhaps the most significant is the presence of lower molecular weight organics that can also enter the dialysis tube [1,000 nominal molecular weight cutoff (NMWCO)]. LaZerte (1984) suggested that only 6% to 12% of the DOC that absorbs at 330 nm was found to penetrate the 1,000-NMWCO membrane. However, this would be system specific and might be quite different for certain stream or lysimeter waters containing a more complex assemblage of DOC. Another potential problem with the method is the assumption that all of the organically complexed Al reacts rapidly with 8-hydroxyquinoline, a factor that, as discussed previously, has not been demonstrated universally. The method also requires rather large sample volumes and is time consuming, easily subject to contamination, and quite cumbersome for field processing of samples.

Ultrafiltration has also been used to separate organically complexed and polynuclear Al from "inorganic" mononuclear Al forms (Aveston, 1965; Parthasarathy et al., 1985; Bloom and Erich, 1989; Parker et al., 1989). The method suffers from the potential problem of Al and organics binding to the membrane filters, similar to those discussed for the dialysis methods (Bloom and Erich, 1989). The method has proved useful, however, in suggesting a size range of the predominant polynuclear Al species in hydrolyzed Al solutions (Parthasarathy et al., 1985; Parker et al., 1989).

Various chromatographic techniques have been used to speciate solution Al. Gardiner and others (1987) utilized a size-exclusion chromatographic technique with a Superose 12 column. They found that the Al in surface soil solution extracts was present in three major fractions corresponding to distinct elution time ranges (see Figure 3–8). The first region was characterized by Al that coeluted with a UV-absorbing (254 nm) organic fraction that referenced against calibration standards having a nominal molecular weight of >10,000. The second region generally corresponded to a lower molecular weight UV-absorbing organic fraction. The Fe present in these filtered samples also eluted within these two regions (Figure 3–5). The third region was characterized by a low molecular weight Al fraction that did not correspond to a UV-absorbing fraction but did

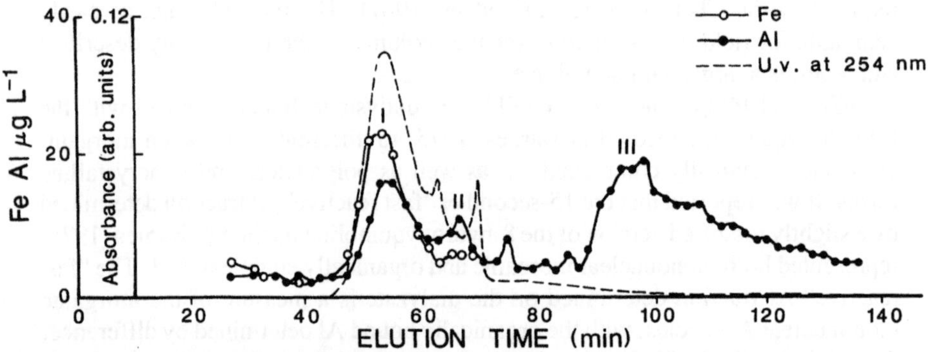

Figure 3–5. Distribution of Fe and Al in a surface soil solution as compared to UV absorbance at 254 nm following size-exclusion chromatography. (From Gardiner et al., 1987. Reprinted with permission of Kluwer Academic Publishers.)

correspond to the predominant Al fraction from an acidic subsoil sample and also to an inorganic Al standard solution, suggesting that this fraction represents Al primarily in the inorganic form. This method contains a number of problems, including organic and inorganic interactions with the column material, but it can provide some useful gross information regarding organic Al distributions and has the potential to be combined with other speciation methods. Size-exclusion methods have also been used to characterize hydrolyzed Al and Fe solutions (Rengasamy and Oades, 1979; Akitt et al., 1981).

An ion chromatographic method in which the predominantly inner-sphere mononuclear aquahydroxo-, fluoro-, oxalato-, and citrato-Al complexes were speciated has recently been advanced (Bertsch and Anderson, 1989). The method utilizes a low-charge sulfonic exchange resin to separate the various Al species and post column derivitization with Tiron (4, 5-dihydroxy-m-benzenedisulfonic acid) for quantitation of the separated Al complexes. Excellent quantitative agreement between predicted species concentrations via the thermodynamic speciation model GEOCHEM and the analytically determined concentrations was obtained for samples prepared in the eluent matrix. The predominantly outer-sphere sulfato-acetato-, propinato- and benzoato-Al complexes were not observed to elute as reduced charged species, but rather exhibited retention times indistinguishable from the $Al(H_2O)_6^{3+}$ species, suggesting complete degradation of the complexes within the eluent-column environment (Figure 3–6). The technique appears useful for discriminating between inner- and outer-sphere Al complexes. The method was employed on partially neutralized Al solutions and was found to provide values for mononuclear aqua-hydroxo-Al species within 2% of the values derived from a quantitative ^{27}Al NMR method (Bertsch et al., 1986a, 1986c). Complex redistribution resulting from differences between sample and eluent ionic strength and pH was influenced by the specific ligand and ligand: Al mole ratios, thus limiting the application of this method for natural samples. The method has been

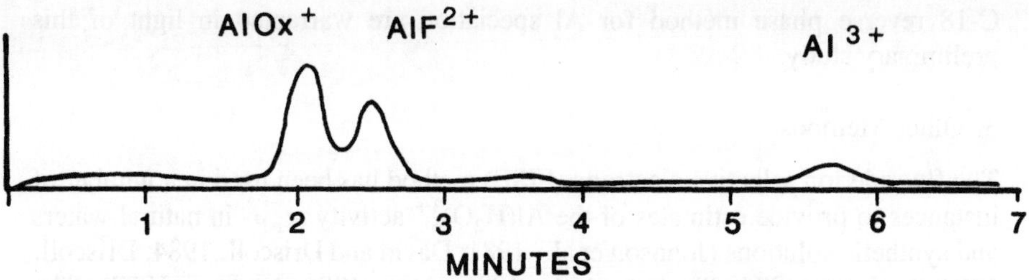

Figure 3–6. The ion chromatogram for a solution containing 18.5 μmol L^{-1} Al, 9.2 μmol L^{-1} F, and 18.0 μmol L^{-1} oxalate, demonstrating that the lower-charged complexes elute much more rapidly than the $Al(H_2O)_6^{3+}$ species. (Adapted from Anderson and Bertsch, 1988. Soil Sci Soc Am J 52:1597–1602 with permission of the Soil Society of America, Inc.)

demonstrated useful, however, in characterizing well-defined Al solutions used in kinetic and toxicological investigations (Anderson and Bertsch, 1988).

Another method that, to my knowledge, has not been fully developed for Al speciation is reverse phase chromatography, which has been more commonly employed for the study of copper-organic complexes (Mills and Quinn, 1981; Brown et al., 1983). Preliminary studies in my laboratory suggest that utilizing a C-18 reverse phase column may be quite useful for separating Al associated with larger humic materials from inorganically complexed forms and that Al fraction associated with organic acids and lower molecular weight humic components. Applications of this method to surface and lysimeter waters of the Eastern Brook Lake watershed in the eastern Sierra Nevada was made in conjunction with the 8-hydroxyquinoline method of James and others (1983), where labile, nonlabile, and total Al were measured on filtered (0.4 μm) and on filtered plus C-18 processed samples (Bertsch, 1988). The preliminary results from this watershed suggest that the "labile" (15-second) Al fraction is predominantly associated with larger dissolved humic components that are retained on the C-18 column, whereas much of the nonlabile Al passing the C-18 column appeared to be associated with lower molecular weight humics and organic acids (Table 3–1). There also appears to be a greater fraction of Al associated with larger dissolved humic components at the higher elevations within the watershed. More in-depth investigations of the

Table 3-1. Labile (15-second hydroxyquinoline) Al_ℓ ($\%Al_T$) before and after C-18 reverse-phase processing for seep and surface waters at the Eastern Brook Lakes Watershed, Eastern Sierra Nevada, Bishop, CA.

		Labile Al (% Al_T)		
Sample	Elevation	Not processed	C-18 processed	Al_ℓ lost (%)
South meadow	3,205	24.1	N.D.	100
East meadow	3,171	30.8	7.3	76
Main meadow	3,153	46.3	25.0	46
South inlet	3,153	38.4	15.1	61

C-18 reverse phase method for Al speciation are warranted in light of this preliminary study.

3. Other Methods

The fluoride ion selective electrode (FISE) method has been used in a number of instances to provide estimates of the $Al(H_2O)_6^{3+}$ activity $a_{Al^{3+}}$ in natural waters and synthetic solutions (Johnson et al., 1981; David and Driscoll, 1984; Driscoll, 1984; LaZerte, 1984; Plankey et al., 1986; Ares, 1986; Hodges, 1987). The method utilizes an appropriate decomplexation buffer to determine total F^- and direct a_{F^-} measurements, with the assumption that Ca, Mg, Si, and Fe fluoride complexes are negligible and that the system is at thermodynamic equilibrium such that:

$$[F^-]_{Total} = [F^-] + [Al\text{-}F] + [H\text{-}F] \tag{3}$$

Thus Al-F is calculated from pH, F^-_{Total}, and a_{F^-} by simple rearrangement of Equation 3.

Then $a_{Al^{3+}}$ is calculated by:

$$\alpha_{Al^{3+}} = Al\text{-}F \,/\, \frac{\alpha_{F^-} K_{1,2}}{\gamma_2} + \frac{2\alpha_{F^{2-}} K_{1,2}}{\gamma_1} + 3\alpha_{F^-} + \frac{4\alpha_{F^{4-}} K_{1,4}}{\gamma_1} + \frac{5\alpha_{F^{5-}} K_{1,5}}{\gamma_2} + \frac{6\alpha_{F^{6-}} K_{1,6}}{\gamma_3} \tag{4}$$

where γ_y is the single ion activity coefficient for the fluoro-Al complex of valence y, and $K_{1,x}$ is the equilibrium constant for the formation of the appropriate fluoro-Al complex according to:

$$Al^{3+} + xF^- \longrightarrow Al\,F_x^{(3-x)}\; K_{1,x} \tag{5}$$

There are several potential problems associated with the application of this method, especially to natural samples. Among these are the assumption that the system is in equilibrium, that no other important fluoro-complexes are present, and the potential interferences to electrode response caused by dissolved organic matter. Additionally, LaZerte (1984) pointed out that since $AlF_x^{(3-x)}$ complexes make up only a small portion of F_{Total} at higher pH values, small errors in a_{F^-} and F_{Total} determinations result in large deviations in computed $a_{Al^{3+}}$ values. At pH 5.5, LaZerte (1984) demonstrated that a 10% error in either a_{F^-} or F^-_{Total} measurements would result in a 69% change in the computed $a_{Al^{3+}}$ value. Based on this sensitivity analysis, it was suggested that the method be used only for waters with pH values less than 5.5, and only with extreme caution between pH 5.0 and 5.5. Other studies have used the FISE technique as a point of comparison for other speciation methods, often with the implication that the FISE method provides some direct measurement of $a_{Al^{3+}}$. It is also often assumed that, after accounting for all other important inorganic Al complexes by calculation, a measure of the organically complexed Al can be obtained by difference. Another problem with the use of the FISE technique for Al speciation is the choice of thermodynamic constants for the various AlF_x complexes, which differ enough to cause substantial variations in the $a_{Al^{3+}}$ estimates. In light of the discussion above, interpretations of Al speciation data in this manner are very tenuous.

Another approach using the FISE technique was proposed by Ares (1986), where solutions of interest were titrated with NaF and the kinetics of fluoro-Al complexation were monitored. The observed reaction paths were then related to the equilibrium composition of the original solutions to provide estimates of the $a_{Al^{3+}}$ and hydroxo-complexed Al concentrations. In addition to the problems discussed above, differential complexation kinetics of F with hydroxo- and organic Al species could provide a complicated reaction path for Al:F complexation in multiligand systems. Recently, the influence of oxalate and citrate (Anderson and Bertsch, 1988) and fulvate (Plankey and Patterson, 1988) on enhancing the rate of Al:F complexation has been reported. Results from these studies indicate that these ligands differentially enhance the complexation of fluoro-Al species, suggesting that the application of the FISE technique to complex natural systems would be a difficult endeavor, providing semiquantitative speciation information at best.

Fluorescence methods have recently been advanced for the study of organic Al complexes (Blaser and Sposito, 1987; Plankey and Patterson, 1987; Shotyk and Sposito, 1988). Plankey and Patterson (1987) used the increased fluorescence intensity of soil derived fulvic acid (FA) in the presence of Al to assess Al-FA complexation. They argued that, as there was no evidence for coagulation or precipitation in their solutions (based on light scattering at 350 nm), the increased fluorescence intensity was directly proportional to Al-FA complexation, as has been demonstrated previously for fluorescence quenching induced by paramagnetics (Saar and Weber, 1980; Ryan and Weber, 1982a, 1982b). An assumption made in the utilization of fluorescence spectroscopy for differentiating organically complexed from uncomplexed Al is that the fluorophores present in the organic material of interest form similar complexes and have similar stabilities to more populous complexation sites that are nonfluorophores (Ryan and Weber, 1982a). This assumption has been demonstrated reasonable for Cu complexation with FA (Saar and Weber, 1980; Ryan and Weber, 1982a). Furthermore, Plankey and Patterson (1987) indicated that the fluorescence technique could be used satisfactorily to reproduce kinetics data for the reaction of salicylic acid with Al, thus justifying its use for FA-Al complexation studies.

Blaser and Sposito (1987) and Shotyk and Sposito (1988) utilized fluorescence quenching to estimate the mole fraction of Al complexed with organic ligands derived from a leaf litter water extract. From these data they calculated an overall stability coefficient for Al complexation by the organics as a function of pH. Although the technique provides some useful information and may be applicable for more uniform organics, it is difficult to envision a suitable application for these coefficients because water extracts of leaf litter contain a complex admixture of organic ligands, thus violating the aforementioned assumption concerning similar stabilities (Tukey, 1966). Tukey (1966) reported more than 8 carbohydrates, 22 amino acids, and 15 organic acids, including ascorbic, citric, fumaric, glutaric, lactic, pyruvic, succinic, and acid glycoside, in the water extracts of plant foliage that he examined. Thus, it appears that coefficients generated in such a mixture would be too gross to be of specific quantitative value.

B. Computational Approaches

Prior to the appearance of computer models to speciate soluble complexes in soil-water systems, computational speciation methods were used by relatively few investigators, primarily because of the tediousness involved in the calculations (Adams, 1974). Beginning in the late 1960s, publications began to appear in which computer models were utilized to calculate speciation in natural systems (Helgeson et al., 1969, 1970), and by the late 1970s, several comprehensive geochemical speciation models were available (Nordstrom et al., 1979). By the mid-1980s, many of these modified geochemical speciation models became widely available and were regularly used for applications involving Al speciation.

The determination of equilibrium speciation of aqueous solutions is generally accomplished using either a free energy minimization method or the equilibrium constant approach. The latter approach is more commonly employed for the more popular speciation models and is often preferred for comprehensive models used for natural systems, as the database is normally composed of measured equilibrium constants that tend to be more reliable than a tabulated free energy database (Nordstrom and Ball, 1984). The equilibrium constant approach formulates the equilibrium problem as a set of mass balance equations for each component and a set of mass action expressions for each individual species. The models then solve a series of nonlinear mass action expressions with the linear mass balance equations. Thus, an analytical total Al determination, in combination with the analytical determined values for all other important dissolved components, is entered into the models and the equilibrium speciation of Al, and all other dissolved constituents for which thermodynamic data exist are then calculated.

Limitations to the use of thermodynamic models for speciating aqueous Al in natural systems include the arbitrary physical separation of "dissolved" Al by filtration, the fact that many of the equilibrium constants for Al complexes are suspect or unavailable, and the often tenuous assumption that the system is in thermodynamic equilibrium. Estimates of the free ion activity of $Al(H_2O)_6^{3+}$ generated with geochemical speciation models have been demonstrated to relate better to an observed biological effect or chemical mobility than simply a total Al value, yet how accurate these estimates are in predicting the actual microscopic molecular distribution of dissolved Al complexes in natural samples is a matter of conjecture.

The difficulty in identifying the truly dissolved Al fraction in many natural water samples containing finely divided colloidal precipitates was discussed previously. The significance of this to geochemical speciation models is in defining the total dissolved Al concentration (Al_T) that must be input into the model. Most comprehensive thermodynamic speciation models allow solid phases to form when the ion activity product (IAP) exceeds the solubility product (K_{SP}), although this is often an optional operation. If one assumes that the Al fraction passing a 0.45 or 0.1 μm filter represents Al_T, then the saturation index (SI):

$$SI = \log \frac{IAP}{K_{SP}}$$

should not exceed 1 (supersaturation) if the system is in thermodynamic equilibrium. In instances where the SI > 1, most investigators have suggested that the system is supersaturated with respect to a given Al solid phase (usually gibbsite or a noncrystalline analogue). The acceptance of this interpretation inherently suggests that the system of interest is in thermodynamic disequilibrium and the problem is kinetically controlled. Thus, in the absence of data suggesting how rapidly the system is proceeding toward equilibrium, the speciation of Al_T into its various components for a single sample is of limited usefulness, and the decision to allow precipitation (or not) must be made.

An alternate interpretation is that finely divided solid-phase components are passing the membrane filter. If light-scattering data are available to support this conclusion (as is sometimes the case), one may, in fact, choose to allow precipitation to proceed and consider the speciation of the thermodynamically predicted Al_T. The problem in this instance is that the solid phases to be considered in the problem must be defined, and the accuracy of their associated constants confirmed. Additionally, the possibility of authentic supersaturation, which may actually result if the system is not in thermodynamic equilibrium (perhaps the rule rather than the exception) or if some other factor (e.g., dissolved organics or Al adsorbed to colloidal phases, "pseudocolloidal") is not properly defined in the problem, must be dismissed. If a comprehensive geochemical model is being utilized (e.g., MINTEQ, Felmy et al., 1983), the user must often have a good knowledge of mineralogy in order to assess properly the allowed solids database. My experiences have indicated that some inexperienced users do not modify such databases, and thus environmentally irrelevant solid phases (e.g., diaspore) are often left to control Al solubility in the equilibrium problem.

Problems surrounding the accuracy of thermodynamic databases are not trivial. Schecher and Driscoll (1987) utilized the speciation model ALCHEMI and a Monte Carlo method to evaluate the uncertainty associated with Al equilibrium calculations. They considered the uncertainty of equilibrium constants associated with the experimental generation of the constant (experimental error), and the uncertainty associated with the choice of thermodynamic constants from the literature (literature error). Overall, the uncertainty resulting from the experimental error associated with the determination of complexation constants was small (Figure 3–7). In the region of minimum Al solubility (circumneutral pH range), where hydroxo- and fluoro-Al complexes are important, the imprecision in equilibrium constants was greatest (0.16 log units). When experimental uncertainty associated with solubility constants was also considered, the variation in model output was much greater (Figure 3–8), and most of this uncertainty resulted from the errors in solubility constant determinations. The most striking example of

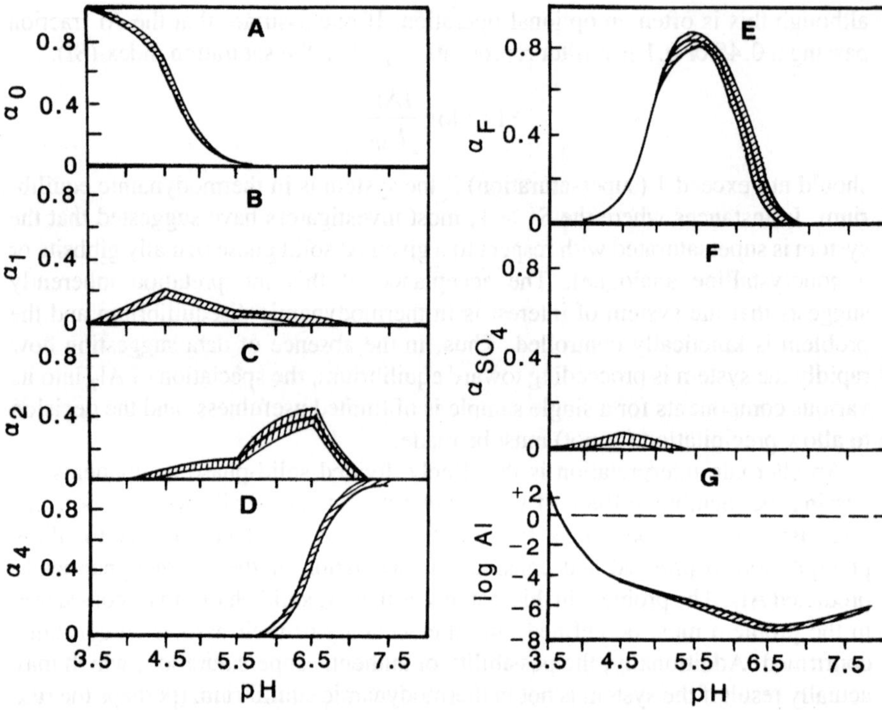

Figure 3–7. Variation in mean model output within 98% confidence limits for experimental errors associated with the determination of complexation constants. Ionization fractions for (A) $Al^{3+}(\alpha_o)$, (B) $Al(OH)^{2+}$ (α_1), (C) $Al(OH)_2^+$ (α_2), (D) $Al(OH)_4^-$ (α_4), (E) Al-F(α_F), (F) Al = SO_4 (αSO_4), and (G) Al_T. (Schecher and Driscoll, 1987, with permission.)

variation in model output, however, was in the analysis of "literature error," where the 98% confidence limits of calculated Al species distribution was extremely broad over the whole pH range, with the uncertainty in Al_T in the circumneutral pH range extending more than eight orders-of-magnitude (Figure 3–9). These calculations demonstrate the importance of the thermodynamic database to Al speciation and accentuate the often repeated warnings concerning "critical evaluation" of thermodynamic constants prior to their utilization in geochemical models. No standard methods for such a "critical evaluation" exist, and our experience has been that some "critically evaluated" constants within many databases are inappropriate or incorrect. Thus, the evaluation of the database is entirely the responsibility of the user. Unfortunately, an evaluation of the databases within the larger models is a formidable challenge and frequently an impractical exercise. A recent, very thoughtful compilation of thermodynamic constants for Al complexes has been published (Nordstrom and May, 1989) and will not be repeated herein; however, users of geochemical models are encouraged to check their Al database against the values presented therein.

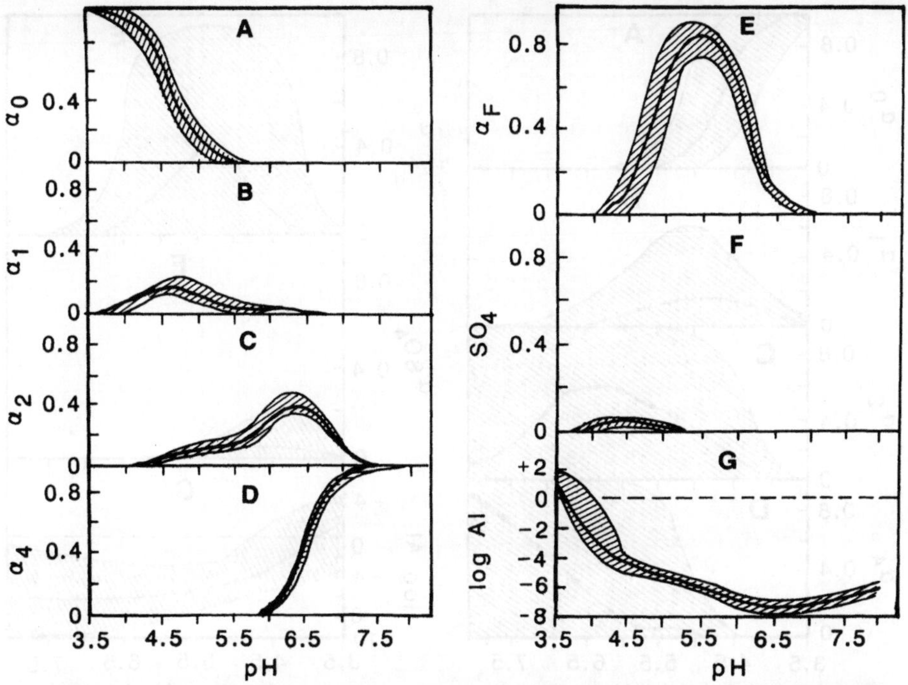

Figure 3–8. Variation in model output (98% confidence) due to experimental uncertainty in aqueous complexation constants and solubility constants. Panels (A) to (F) are the same as in Figure 3–7. (Schecher and Driscoll, 1987, with permission.)

The previous discussion was not meant to discourage the use of geochemical speciation models, as they remain the most powerful tool available for handling large sets of chemical data from natural systems. The model predictions can provide a significant amount of information about a complex system, establish boundary conditions, and often expose potential problems. (Schecher and Driscoll, 1988). For example, large cation-anion imbalances in a data set may suggest that an important component is not being measured. Additionally, significant supersaturation with respect to an important solid phase that can form in an environmentally significant time frame may suggest that important complexation reactions that are either poorly defined or not defined in the database (e.g., organic complexes, silicato-Al complexes, or mixed ligand complexes) may be important in a system, or that the system is in complete disequilibrium because of some environmental or physicochemical anomaly. Furthermore, as will be discussed subsequently, the free ion activity of Al^{3+} predicted by most geochemical models is an extremely useful correlative parameter with respect to bioavailability, toxicity, adsorption, or mobility. Thus, the discussion concerning the problematic areas of geochemical modeling are meant to place the predictions derived from this very useful tool in proper perspective.

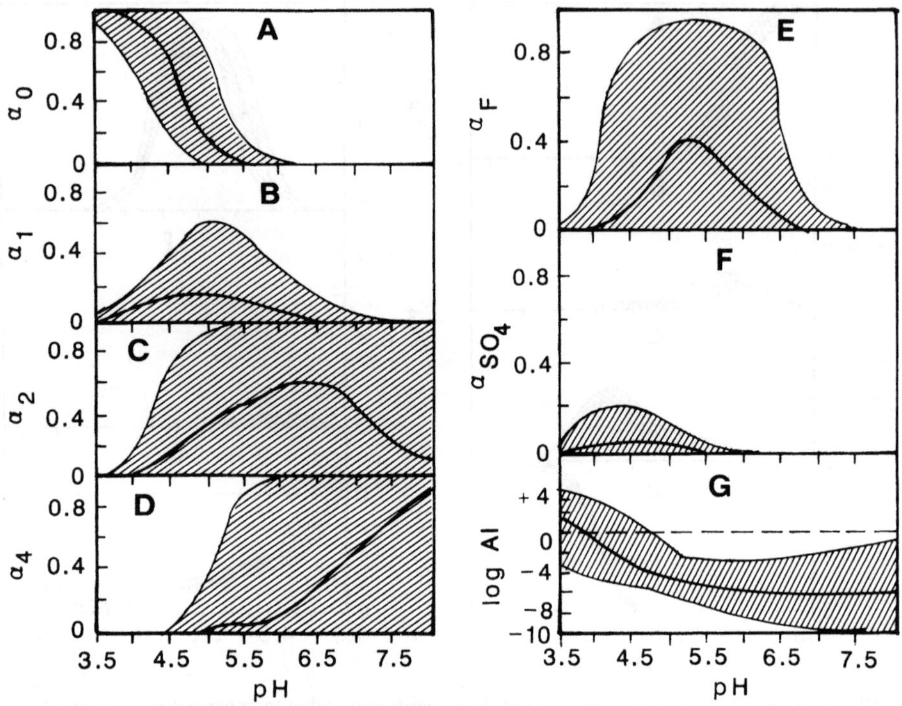

Figure 3–9. Variation in model output (98% confidence) due to variations in thermodynamic constants obtained from the literature. Panels (A) to (F) are the same as in Figure 3–7. (Schecher and Driscoll, 1987, with permission.)

C. The Analytical and Computational Approach

Perhaps one of the more widely used and potentially useful approaches to Al speciation is a hybrid method using both analytical and computational Al speciation methods together. The potential problems of colloidal Al phases are minimized because most analytical speciation methods are adequate at separating truly dissolved Al from colloidal solid-phase components. Most useful are methods that adequately separate inorganically and organically complexed Al species. The inorganic mononuclear Al can then be used as the Al_T value only if inorganics are considered as ligands in the formulation of the problem. The model can then be used to provide predictions for the hydroxo-, fluoro-, sulfato-, and silicato-Al species (if the appropriate thermodynamic data are available). One problem, however, is when a method that does not clearly differentiate between inorganically and organically complexed forms is used as the input parameter for a geochemical model. The 8-hydroxyquinoline (15-second) and aluminon (30-minute) labile Al fractions (Blamey et al., 1983; Alva et al., 1986; Wright et al., 1987) have been used in such a manner, even though many studies have indicated that some, and in certain instances, a significant fraction of the labile Al determined by these methods is organically bound and may also include variable amounts of fluoro-complexed species (James et al., 1983; Adams and Hathcock, 1984; Noble et al., 1988a). I have also noted instances where a labile Al fraction measured by timed spectrophotometric methods is input as Al_T along with organic acids as ligands, even though these labile fractions may exclude a significant portion of the organically complexed Al species. Thus, extreme caution must be exercied in the utilization of analytical Al speciation data with a thermodynamic speciation model so that the input value for Al_T (and ligands) makes chemical sense and so that ligands are not considered twice or that they are not mutually exclusive. Additionally, analytically determined Al fractions are often compared to thermodynamic equilibrium calculations, resulting in conclusions that will be dependent on the specific thermodynamic data employed in the calculations and subject to the variations therein.

IV. Chemical Speciation and Al Toxicity

A. Aquatic Organisms

It has been well documented that Al is toxic to a number of terrestrial and aquatic organisms, although the mechanisms involved in this toxicity are poorly understood. Schofield and Trojnar (1980) concluded that Al concentration was the primary factor influencing brook trout (*Salvelinus fontinalis*) stocking success in the 53 Adirondack lakes they investigated. Bioassays have indicated significant increases in mortality at Al concentrations of $\geqslant 0.2$ mg L^{-1} for both brook trout and brown trout (*Salmo trutta*), with gill damage observed in the former at 0.1 mg Al L^{-1} (Cronan and Schofield, 1979; Muniz and Leivestad, 1980). Baker and Schofield (1982) found that Al toxicity to white suckers (*Castostamus commersoni*) and brook trout depended both on the solution pH and life history stage. Solution Al concentrations of 0.1 and 0.2 mg L^{-1} were found to reduce the survival and growth rates of white sucker and brown trout larvae and post larvae, respectively. Toxicity was reported to be greatest in solutions containing hydroxo-Al complexes, that is, solutions having pH values of 5.2 to 5.4. Brown (1983) suggested that Al toxicity in brown trout was highly dependent on Ca^{2+} concentration, being more severe at Ca^{2+} concentrations of 0.5 mg L^{-1} or less, and less severe at 1.0 mg L^{-1} or more solution Ca^{2+}. This observation led to the conclusion that the physiological effect of aluminum is mediated via the ion regulatory system. In agreement with the findings of Schofield and Trojnar (1980) and Baker and Schofield (1982), Brown (1983) reported greater toxicity of Al to brown trout with increasing pH in the 4.5 to 5.4 range. Aluminum has also been demonstrated to be toxic to a number of other aquatic organisms, including mayfly nymphs, chironomids, dixid midges, stone flies, isopods, and caddis larvae (Hall et al., 1985, 1987; Ormerod et al., 1987; Raddum and Fjellheim, 1987). It has been suggested that increases in inorganic mononuclear Al associated with episodic decreases in pH during snowmelt resulted in a fish kill of Atlantic salmon (*Salmo solar*) in Norway (Henriksen et al., 1984). Additionally, attenuation of Al toxicity to blackfly larvae and brown trout in the presence of elevated dissolved humics has been reported (Karlsson-Norrgren et al., 1986; Petersen et al., 1986); however, few studies have clearly focused on the influence of Al speciation on Al toxicity to aquatic organisms. Driscoll and others (1980) reported that fluoride and citrate significantly reduced the toxicity of Al to brook trout and that the mortality of white sucker fry appeared to be related to be labile inorganic Al fraction as determined by the ion exchange column method.

The observed increase in Al toxicity with increased pH has led to the suggestion that the hydroxo-complexed Al species may be more toxic and even to the speculation that the mitigation of Al toxicity to fish by complexing ligands may result from restricted Al hydrolysis (Driscoll, 1984). These hypotheses require more rigorous testing, and specific hydroxo-Al complexes (mononuclear versus polynuclear) need to be evaluated for their relative toxicities, as liming acidic surface waters is an ever-increasing mitigation procedure that would be expected to enhance Al hydrolysis and polymerization reactions.

B. Plants

The toxicity of Al to plants has been a topic of considerable interest since the early part of this century, as it has been identified as one of the most significant growth-limiting factors on acidic soils (e.g., Adams and Pearson, 1967; Foy et al., 1978; Foy, 1984; Foy et al., 1987). Additionally, increased soluble Al resulting from acidic deposition has been proposed as a possible stress factor contributing to forest decline in some areas of North America and Europe (Ulrich et al., 1980). As with aquatic organisms, the mechanisms controlling Al toxicity to plants are not well elucidated, and considerable controversy surrounds the various hypotheses. Unlike the studies on aquatic organisms, however, a significant body of literature exists and Al speciation has been a major focus of phytotoxicity studies (Pavan and Bingham, 1982; Pavan et al., 1982; Blamey et al., 1983; Wagatsuma and Ezoe, 1985; Alva et al., 1986; Suhayda and Haug, 1986; Cameron et al., 1986; Hue et al., 1986; Tanaka et al., 1987; Kerven and Edwards, 1987; Kinraide and Parker, 1987a; 1987b; Wagatsuma and Kaneko, 1987; Wright et al., 1987; Noble et al., 1988a, 1988b, 1988c; Parker et al., 1988, 1989).

In early studies, the toxicity of Al to a variety of crop species was found to be best related to the activity of the hexaaquaaluminum species in solution (calculated from a total "dissolved" Al measurement, usually not accounting for complexed Al species) and only variably related to other measurements of "available" Al, such as total soluble Al, exchangeable Al, and percent Al saturation (Adams and Lund, 1966; Brenes and Pearson, 1973). Another early study clearly demonstrated the ability of citrate to detoxify Al (Bartlett and Riego, 1972a), an observation that has been well supported by subsequent investigations (Adams and Hathcock, 1984; Hue et al., 1986; Suhayda and Haug, 1986). Hue and others (1986) conducted a thorough investigation on the effects of organic acids on Al toxicity to cotton roots (*Gossypium hirsutum* L.). The results of this study demonstrated that organic acids could be divided into three groups based on their ability to reduce Al toxicity: (1) strong detoxifiers (citric, oxalic, tartaric); (2) moderate detoxifiers (malic, malonic, salicylic); and (3) weak detoxifiers (succinic, lactic, formic, acetic, phthalic). The ability of these organic acids to detoxify Al was positively correlated with the relative position of the OH/COOH groups on their main chain, when such positioning favored the formation of stable 5- or 6-bond ring structures with Al. Many investigators have also demonstrated that Al toxicity is generally reduced during short-term bioassays in the presence of a number of inorganic ligands (e.g., fluoride, sulfate, and phosphate), although the relative toxicities of the various inorganically complexed Al species remain an area of debate. Pavan and Bingham (1982) utilized the comprehensive thermodynamic speciation model GEOCHEM to calculate the free ion activities of the aqua-, hydroxo-, and sulfato-complexed Al species in their hydroponic bioassays where coffee (*Coffea aralico* L.) seedlings were used as the test species. They (Pavan and Bingham, 1982) reported that the ion activity of the $Al(H_2O)_6^{3+}$ species was the best predictor of Al toxicity compared to either total solution Al or the total Al^{3+} concentration, producing a continuous function. Although they did not state what constants they used for phosphato-Al complexes (or which specific complexes were considered) and apparently did not explicitly account for the possible Al complexes formed with the EDDHA (added as Fe EDDHA at EDDHA:Al mole ratios of $\simeq$2.4 and 0.6), their study clearly demonstrated the utility of the free ion activity of Al^{3+} as calculated via a comprehensive thermodynamic speciation model as a useful correlative parameter to describe bioavailability and toxicity.

Since that time, however, many investigators have attempted to relate the predicted activities of the hexaaqua-, hydroxo-, sulfato-, and fluoro-complexed Al to observed toxicities and have correlated an observed biological response (usually root length) to one or more specific Al complexes. The results of these studies have been confusing, some suggesting that the activities of the hydroxo mononuclear complexed Al may be relatively phytotoxic or even more toxic than $Al(H_2O)_6^{3+}$ (Kerridge, 1969; Moore, 1974; Adams and Moore, 1983; Blamey et al., 1983; Alva et al., 1986; Noble et al., 1988b), with others suggesting the opposite (Bartlett and Riego, 1972b; Pavan et al., 1982; Cameron et al., 1986; Parker et al., 1988, 1989; Shann and Bertsch, 1989). Likewise, a number of studies have implicated sulfato Al species as relatively toxic (Blamey et al., 1983; Alva et al., 1986a, 1986b), whereas others have suggested that these complexes are much less toxic or nontoxic to plants (Cameron et al., 1986; Tanaka et al., 1987; Kinraide and Parker, 1987a; Noble et al., 1988b, 1988c). Generally, most studies have indicated that fluoride greatly reduces the toxicity of Al to plants and that the fluoro Al complexes are either much less toxic than $Al(H_2O)_6^{3+}$ or nontoxic (Cameron et al., 1986; Tanaka et al., 1987; Noble et al., 1988c).

Many of these discrepancies can be explained by considering one or more of the following: (1) The different investigators used nutrient solutions of varying complexity, often containing organic chelators that were not explicitly accounted for; (2) the various studies employed speciation models differing in their degree of complexity and often having different thermodynamic databases; (3) highly significant correlations between specific complexes and toxicity may have resulted from the collinearity between the activities of the various complexes, including the $Al(H_2O)_6^{3+}$ species; (4) many of the studies used different approaches to estimate the total mononuclear Al concentration that was input into the model calculations (some have used timed spectrophotometric methods, and others have simply employed a total dissolved Al measurement or assumed that the total Al added in a treatment remained as the total mononuclear fraction); and (5) the toxicity may have been influenced by some other factor not considered in the analysis, such as cation amelioration mechanisms (Kinraide and Parker, 1987b) or the presence of polynuclear Al species (Parker et al., 1988, 1989; Shann and Bertsch, 1989).

Differences in thermodynamic databases could lead to significant differences in the relative toxicity of a given complex as deduced from correlations with model-generated Al species distributions. This has recently been considered by Kinraide and Parker (1989) in relation to the relative toxicities of the hydroxo-complexed Al species. They used the data of Alva and others (1986b) to evaluate the relative root length versus species activity of the hexaaqua- and hydroxo-Al complexes, utilizing three sets of formation constants (Table 3–2). This analysis demonstrated that conclusions regarding the relative toxicity of the various Al complexes are model dependent. In light of the uncertainty analysis of Schecher and Driscoll (1987), it is evident that any number of conclusions could be made regarding the relative toxicities of a given complex, depending on the selection of thermodynamic constants. This, coupled with (1) the situations where organic ligands are sometimes present in the nutrient solutions, yet not considered in the speciation calculations; (2) the well-known fact that conditions within the rhizosphere are constantly changing (e.g., pH, organic exudates, etc.); and (3) the often overlooked fact that single ion activities have no thermodynamic significance (Sposito, 1984), would all suggest that such model calculations provide merely an approximation of Al species distribution. In the absence of other confirmatory evidence, these approximations have no specific microscopic molecular significance, although they may be quite useful correlative parameters to describe a system or explain an observed biological response or physicochemical behavior. Thus, in very simple solutions containing a single ligand, the calculations can be interpreted with some certainty (although complications of collinearity with the hydroxo-mononuclear species and with disequilibrium remain), whereas for multiple-ligand systems, or especially those supersaturated with

Table 3-2. Coefficients of determination (r^2) for the exponential regression of relative root length (RRL) versus hexaaqua- and hydroxo-Al species activities.

	RRL versus		
Source of constants	Al^{3+}	$A\ell(OH)^{2+}$	$A\ell(OH)_2^+$
Nordstrom and May (1989)	0.888	0.804	0.162
Lindsay (1979)	0.793	0.942	0.371
Parks (1972)	0.666	0.933	0.691

Calculations from Kinraide and Parker (1989) using data from Alva et al. (1986).

respect to an Al hydroxide solid phase, and in the absence of other evidence, the speciation calculations provide only general information that may be used simply to delimit the thermodynamic boundary conditions of the system.

Considerable effort has been expended in relating the labile Al fraction as determined by a number of spectrophotometric assays to observed Al phytotoxicity (Adams and Moore, 1983; Blamey et al., 1983; Adams and Hathcock, 1984; Alva et al., 1986; Wright et al., 1987; Kerven and Edwards, 1987; Noble et al., 1988b, 1988c). There are many inconsistencies in the approach used among different investigators and often significant errors in their interpretations. As discussed previously, the rapidly reacting Al fraction or labile Al has often been input into thermodynamic speciation models, often without regard to the method dependence of this measurement. For example, some methods, although demonstrated to include a portion of the organically complexed Al species and perhaps not all of the fluoro-Al complexes within the labile estimate, are input as the total inorganic mononuclear fraction. Clearly, any estimate of "toxic" Al that is substantially less than the total dissolved Al concentration will better relate to observed biological responses or physicochemical behavior in most complex solutions, regardless of the inability of the method or analysis to predict accurately Al species distribution. Such improved correlations are not proof of molecular reality.

A widespread misconception is that the timed spectrophotometric assays provide an estimate of the hexaaqua- and hydroxo-mononuclear Al species. Several of the methods were demonstrated to produce such an estimate in partially neutralized solutions containing *only* these mononuclear and hydroxo-polynuclear Al species. Few claims have been made by method developers that this estimate is valid in multiligand environments. Rather, investigators who have adopted these methods for phytotoxicity studies have sometimes made this claim, usually citing the original methods paper for support. An example is the revised 8-hydroxyquinoline method of James and others (1983). The original paper states, "These comparisons of theoretical and experimental results indicated that labile (15s. reactive) Al comprised Al^{3+} and $Al(OH)^{2+}$ in systems containing *no other complexing ligands*" (James et al., 1983). Systems containing complexing ligands provided variable amounts of fluoro- and citrato-complexed Al within the labile Al fraction, being highly dependent on ligand:Al mole ratio, total Al concentration, and pH. Similar results have been reported for the ferron method (Jardine and Zelazny, 1987a, 1987b). This situation is what led James and others (1983) to define the 15-second fraction as labile Al because the exact molecular significance was ill-defined and would be system specific. Regardless of this, or that the labile Al fraction determined by this method failed to distinguish between phytotoxic and nontoxic Al in some soil solutions (Adams and Hathcock, 1984), several investigators have incorrectly cited the James and others (1983) modification of the 8-hydroxyquinoline method, suggesting that the labile Al fraction reportedly measures only the Al^{3+}, $Al(OH)^{2+}$, and $Al(OH)_2^+$ species in multiple ligand environments (Shotyk and Sposito, 1988; Noble et al., 1988a). There have been enough comparisons published to demonstrate the method dependence and system specificity of the identified labile Al fraction (Hodges, 1987; Wright et al., 1987; Kerven and Edwards, 1987; Parker et al., 1988; Noble et al., 1988a; Alva et al., 1989). The different methods provide only operationally defined fractions, which will vary with solution composition (including type of ligand, ligand:Al mole ratio, and pH), buffer preparation, time of reaction, and even the choice of thermodynamic constants if model calculations are used as a point of reference. These differences do not diminish the usefulness of the methods for providing a reasonable estimate of phytotoxic Al (e.g., Wright et al., 1987; Noble et al., 1988b, 1988c) but suggest that as the model predictions, they are useful only as a correlative tool. Therefore, further detailed method comparisons will provide little additional information and should be generally discouraged.

An additional area of controversy surrounds the relative toxicity of the hydroxo-mononuclear Al species. It is difficult to draw definitive conclusions regarding the observation that Al toxicity may increase with pH in the 4.0 to 5.0 range (Kerridge, 1969; Moore, 1974; Blamey et al., 1983; Alva et al., 1986; Wagatsuma and Kaneko, 1987; Parker et al., 1988, 1989; Shann and Bertsch, 1989). Many investigators have related this to the greater toxicity of the mononuclear hydroxo-Al species compared to $Al(H_2O)_6^{3+}$ (Moore, 1974; Blamey et al., 1983; Alva et al., 1986b). However, as discussed previously, differences in available thermodynamic constants for the various hydroxo-complexes, and collinearity between these and the hexaaqua-Al cation would weaken these arguments when based solely on the calculated thermodynamic species distributions. Furthermore, the results of one study in which nutrient solutions containing Al were partially neutralized, suggested that polynuclear Al species were nonphytotoxic (Blamey et al., 1983). This study has been widely cited (Alva et al., 1986a, 1986b; Cameron et al., 1984; Hodges, 1987; Noble et al., 1988a, 1988b), even though the 30-minute aluminon method used by Blamey and others (1983) poorly discriminates between mononuclear, small polynuclear, and colloidal Al forms.

In contrast, other studies have indicated that polynuclear Al may be more phytotoxic than the hexaaqua-Al cation (Bartlett and Riego, 1972b; Wagatsuma and Kaneko, 1987; Parker et al., 1988, 1989; Shann and Bertsch, 1989). The studies of Parker and others (1988, 1989) are particularly noteworthy, because these investigators grew their indicator plants in solutions containing only Al and Ca. Furthermore, they provided adequate differentiation between the mononuclear, polynuclear, and colloidal Al fractions by using the ferron timed spectrophotometric assay, and they provided independent direct ^{27}Al NMR spectroscopic data. In the absence of hydroxo-polynuclear Al species, increased pH resulted in reduced Al toxicity, suggesting that the hydroxo-mononuclear Al species were *less* toxic than the hexaaqua-Al cation (Parker et al., 1988). Solutions containing polynuclear Al species demonstrated a much greater phytotoxicity even when mononuclear Al concentrations were below levels generally considered toxic ($\leqslant 3$ μM) (Parker et al., 1989). They also reported that the two wheat (*Triticum aestivicum* L.) cultivars exhibited a differential response to Al^{3+} but not to hydroxo-polynuclear Al, indicating that there may be different physiological mechanisms controlling the toxicity. ^{27}Al NMR spectroscopic investigations provided evidence that the Al_a fraction determined by the ferron assay was closely related to the true mononuclear Al concentration [Al^{3+}, $Al(OH)^{2+}$, $Al(OH)_2^+$] and that the polynuclear Al fraction was dominated by the $AlO_4Al_{12}(OH)_{24}(H_2O)_{12}^{7+}$ (Al_{13}) polynuclear species (Parker et al., 1989). Another recent study has provided even more support for the high toxicity of the Al_{13} polymer to six wheat cultivars, all exhibiting differential tolerance to Al^{3+} toxicity (Shann and Bertsch, 1989). These studies (Parker et al., 1988, 1989; Shann and Bertsch, 1989) raise questions regarding the explanation that increased toxicity at higher pH values is attributable to the hydroxo-mononuclear species. Although the environmental significance of the Al_{13} polynuclear has yet to be demonstrated, the ease with which it forms even in dilute, partially neutralized solutions (Bertsch, 1987), coupled with its very high relative toxicity, would raise doubt concerning any phytotoxicity study involving pH adjustment where the formation or absence of this species has not been explicitly verified. Furthermore, the toxicity of other possible Al polynuclear also requires further investigation. Likewise, because the relative distribution between mononuclear, Al_{13}, and colloidal hydroxo Al (Al_c) is so highly dependent on specific synthesis conditions (Bertsch et al., 1986a, 1986b; Bertsch, 1987b), studies that have utilized partial neutralization without adequate differentiation of polynuclear from colloidal Al (Al_c) (e.g., aluminon method) would also be suspect, as Al_c has been demonstrated to be rather innocuous in short-term phytotoxicity studies (Parker et al., 1989; Shann and Bertsch, 1989). Thus, caution should be exercised in interpreting Al toxicity as influenced by pH until more conclusive studies regarding the relative toxicity of hydroxo-mono- and polynuclear Al species are available.

Aluminum is toxic to a number of tree species, although there exists a considerable range of reported Al concentrations required to produce toxicity, thus causing considerable disagreement concerning the potential role of Al stress as a factor in forest decline attributed to acidic inputs (e.g., Rost-Siebert, 1983; Cumming et al., 1985, 1986; Neitzke and Runge, 1985; Van Praag and Weisson, 1985; Thornton et al., 1986; Truman et al., 1986; Arp and Ouimet, 1986; Ryan et al., 1986a, 1986b; Entry et al., 1987; Berggren and Fiskesjo, 1987; Bergkvist, 1987; Zhao et al., 1987; Brassard et al., 1988; Bengtsson et al., 1988; Asp et al., 1988; Ohno et al., 1988; Shortle and Smith, 1988). Although some of this variability probably results from genetically controlled tolerance mechanisms, much of it may be attributed to artifacts resulting from experimental conditions leading to significant differences in the chemical speciation of Al (Brassard et al., 1988). Even several recent studies evaluating Al toxicity to various tree species have failed to consider Al solubility limits within the nutrient solutions employed. Other studies have carefully avoided exceeding solubility limits, yet they have not considered Al complexation reactions, or in some instances have considered only the mononuclear hydroxo-Al species, even in complex nutrient solutions or buffers containing several good complexing ligands. Thus, the results from many of these studies should be carefully reviewed prior to the acceptance of critical toxicity levels or even ranges for a given tree species.

Specific studies considering the influence of chemical speciation on Al availability or toxicity to forest species are less common than those for crop species. Van Praag and Weissen (1985) grew spruce [*Picea abies* (Karst)] and beech (*Fagus sylvatica* L.) seedlings on plots of undisturbed humus O layers and on disturbed plots with the organic horizons removed (bare mineral soil). Both P and Al concentrations were generally higher in the stems, needles, and roots of 2-year-old spruce seedlings growing on the undisturbed plots compared to those growing on the disturbed plots (mineral soil). Van Praag and Weissen (1985) also reported that "total" and Al^{3+} values were greater in the disturbed compared to the undisturbed plots, leading the authors to suggest that soluble organic Al complexes in the undisturbed organic horizons were more readily absorbed and translocated within the seedlings. In a subsequent study (Van Praag et al., 1985), both EDTA and citrate reduced Al injury (as inferred from enhanced Ca and Mg absorption and translocation); however, there was no clear evidence that these chelators enhanced Al uptake or translocation. Arp and Ouimet (1986) studied the influence of oxalate complexation on Al uptake by black spruce [*Picea mariana* (Mill)], where it was found that more Al was accumulated in the roots in the presence of oxalate. The significance of this finding is unclear, as there was no evidence that this Al fraction was either toxic or nontoxic to the seedlings and because a similar trend in root Ca levels was noted. Conversely, Berggren and Fiskesjo (1987) indicated that Rost-Siebert (1983) reported a decrease in Al toxicity to spruce and beech seedlings grown in EDTA-amended solution cultures. Thus, the influence of low molecular weight organic acids on Al toxicity to trees is less clear than for crop species, although a general trend for detoxification in their presence is indicated. It should be noted that the assays for the tree species are usually much longer in duration, and this could produce some differences. Furthermore, these data illustrate that chemical speciation influences on bioavailability do not necessarily provide information regarding Al toxicity, this being consistent with the observation that many plants growing naturally on acidic soils tend to accumulate significant quantities of Al while demonstrating very high Al tolerance (Matsumoto et al., 1976; Foy et al., 1978, 1987; Truman et al., 1986).

Bengtsson and others (1988) grew beech seedlings in soil solutions collected from two depths (15 and 50 cm) of a soil profile in a beech forest in southern Sweden. The total and labile (by 15-second 8-hydroxyquinoline) Al fractions were determined to be 184 and 17 μM, and 430 and 410 μM for the soil solutions taken from the surface and subsurface horizons, respectively. Although the Al_T for the soil solutions collected from the surface horizons exceeded the Al_T demonstrated toxic (as inferred from Ca^{2+} and P accumulation) in hydroponic solutions, no toxicity was observed. Conversely, the beech seedlings grown in the soil solution obtained from the subsurface horizon demonstrated drastically reduced Ca and P uptake. Bengtsson and others (1988) concluded that dissolved humics were probably responsible for reducing the labile Al fraction of the soil solution obtained from the surface soil horizon, and thus the toxicity. Berggren and Fiskesjo (1987) also utilized soil solutions collected from beech, birch, and spruce forest sites in southern Sweden to conduct Al toxicity studies using onion (*Allium cepa* L.) as the test species. Utilizing the method of Driscoll (1984), they found very good agreement (negative correlation) between the labile mononuclear Al fraction (inorganic) and the root length of the plants grown in the soil solutions. Additionally, the observed root inhibition followed a similar relationship to that found for plants grown in nutrient solutions containing only inorganic Al. Thus, the Al speciation method of Driscoll (1984) was useful in discriminating between toxic and nontoxic Al species, generally representing inorganic and organically bound Al. Further studies will be required to demonstrate the influences of inorganic ligands on Al toxicity to forest species.

V. Chemical Speciation of Al in the Environment

A. Aqueous Al

The body of literature on Al speciation in aquatic and terrestrial systems has become voluminous in recent years. As with other applications of Al speciation discussed herein, there exists considerable variation in methods and approaches used by various investigators. Notwithstanding, significant information on the processes regulating Al mobility in the environment has been generated, as data on Al speciation are far more informative than total dissolved Al measurements, albeit the speciated fractions are method dependent. However, differences between methods are typically <25% (Hodges, 1987), thus minimizing the influence on interpretations extrapolated to whole watersheds, particularly if the methods adequately differentiate inorganic Al from organically bound and colloidal Al forms (e.g., Campbell et al., 1983; LaZerte, 1984; Driscoll, 1984; Hodges, 1987).

The investigation of Johnson and others (1981) was among the first comprehensive watershed studies in which Al speciation was carefully evaluated. Employing an earlier modification of the strongly acidic ion exchange column method (Driscoll, 1980), these investigators reported spatial variations in Al speciation within a headwater stream located in the Hubbard Brook Experimental Forest, New Hampshire, where substantial changes in the amount and type of dissolved Al along an elevation gradient were noted. Decreases in elevation from 812 to 445 m were accompanied by an observed increase in stream pH (4.75 to 5.11) and decrease in total dissolved Al. The Al speciation data, however, revealed a marked decrease and increase in inorganic and organically complexed forms, respectively, with decreasing elevation (Johnson et al., 1981). Similar trends have been reported for other watersheds, although some differences in the relative distribution between inorganic and organic Al were observed (e.g., Lawrence et al., 1986). Another important finding in many of these investigations is that fluoro-Al complexes are an important component of the inorganic Al fraction and also that the operationally defined organic Al fraction is strongly correlated to dissolved organic carbon concentrations (e.g., Johnson et al., 1981; Campbell et al., 1983; LaZerte 1984; Driscoll et al., 1984, 1987, 1988; Driscoll and Newton, 1985; Lawrence et al., 1986; David, 1986; Sullivan et al., 1986).

A recent comparison between an acidic watershed in British Columbia (Jameison Creek) and Hubbard Brook suggested that the source of acidic inputs has a strong influence on Al speciation (Driscoll et al., 1988). The Jameison Creek watershed reportedly receives low inputs of sulfuric acid; thus, its acidity was attributed to natural sources, that is, organic acids. The Hubbard Brook watershed, conversely, received significant sulfuric acid inputs, and these reportedly are largely responsible for the acidic conditions observed within the watershed. The Hubbard Brook streams had a much higher total dissolved Al ($\sim$23 μmol L^{-1} vs. $\sim$4 μmol L^{-1}) than the Jameison Creek system. Perhaps even more striking was the much greater fraction of Al_T present as labile (inorganically bound) Al in the Hubbard Brook waters ($\sim$45% to 90% vs. $\sim$25% to 30%). This finding led Driscoll and others (1988) to hypothesize that inputs of strong inorganic acids largely from acidic deposition have resulted in elevated Al concentrations in impacted surface waters, in addition to a shift in speciation from nonlabile (predominantly organic) to the presumably more toxic labile, predominantly inorganic Al forms. Some evidence for this was provided recently when sulfuric acid added to two streams in southern Norway resulted in large increases in Al_T, with a significant fraction of this as labile inorganic Al (Henriksen et al., 1988). The observation that the mobilization of organic Al complexes may also be enhanced by decrease in pH (Driscoll et al., 1987; McAvoy, 1989) might complicate the generalization of this observation, however, as the relative distribution between inorganic and organically complexed Al as a fraction of the total mobilized may be system specific, being temporally and spatially variable (McAvoy, 1989). Furthermore, James and Riha (1984) reported a decrease in DOC with decreasing pH in soil solutions, corresponding to an observed decrease in the nonlabile Al fraction. The total and labile Al fractions increased with decreasing pH, however, suggesting that Al was being ion exchanged from the soluble organic constituents before their removal or that Al from other solid phase sources was being mobilized, primarily in an inorganically complexed form (James and Riha, 1984). The labile and nonlabile fractions were determined by the method of James and others (1983), and thus an alternate explanation might be that more of the organically complexed Al was included in the 15-second labile Al fraction as pH decreased. Regardless, the independent DOC measurements clearly indicate an opposite trend than that reported by Driscoll and others (1987) and McAvoy (1989) for stream waters, indicating the potential system specificity of the process and the importance of having independent DOC measurements to complement the operationally defined Al speciation data.

Many investigations have also reported important temporal variations in Al speciation within surface waters (e.g., Driscoll et al., 1984; Henriksen, et al., 1984; LaZerte, 1984; Siep et al., 1984; Hooper and Shoemaker, 1985; Bull and Hall, 1986; Hendershot et al., 1986; Cozzarelli et al., 1987; Henriksen et al., 1988; Lawrence and Driscoll, 1988; Lawrence et al., 1988; McAvoy, 1988, 1989; Driscoll et al., 1989). Several investigators have reported elevated Al concentrations during periods of higher flow rates (e.g., LaZerte, 1984; Hooper and Shoemaker, 1985; Bull and Hall, 1986; Hendershot et al., 1986; Sullivan et al.,

1986; Henriksen et al., 1988; Lawrence et al., 1988; McAvoy 1988, 1989), although the distribution between the various Al fractions has been somewhat variable between studies. LaZerte (1984) observed an increase in total and labile Al for several Ontario streams during spring snow melt, a trend that closely followed increased flow and decreased pH. The fluoro-complexed Al species were important inorganic forms of Al mobilized during these periods, and even a highly colored stream had a significantly increased inorganic Al fraction (LaZerte, 1984). Lawrence and others (1988) reported a similar situation for an acidic headwater stream at the Hubbard Brook Experimental Forest but only at lower elevations (544 m). At the high elevation site (732 m), increased flow was associated with an increased pH and a decrease in the labile, inorganically complexed Al fraction. This difference led Lawrence and others (1988) to suggest that the relationship between surface water chemistry and hydrologic flow paths was more complicated than previously thought and that both spatial and temporal factors needed to be considered in watershed-modeling efforts.

Seasonal and spatial variation in Al speciation for a second-order Massachusetts stream was recently reported by McAvoy (1988). Unlike the results of Lawrence and others (1988), McAvoy (1988) reported large increases in labile inorganic Al during high flow conditions at the higher-elevation sites. This labile fraction was as high as 90% of the total, in contrast to low flow conditions, where it comprised only ~30% of the total Al fraction. The downstream site also had elevated inorganic Al at high flow (~70% Al_T), whereas during low flow conditions it was only ~10% of the Al_T. The Al chemistry at an intermediate wetland site was predictably influenced by fluctuations in DOC, where the nonlabile organically bound Al fraction comprised approximately 30% and 80% of the Al_T during high and low flow conditions, respectively. The inorganic mononuclear Al concentrations were lower in the wetland stream compared to the mountain stream during high flow yet were higher during low flow conditions, reflecting the lower pH values maintained during the summer months in the wetland stream. Similar trends were evident during episodic rainfall events at these three sites, suggesting that conclusions about Al mobilization based on overall seasonal trends related to flow need to be modified to include the potential for significant Al mobilization related to episodic events.

The influence of other perturbations, such as whole tree harvesting, have also been demonstrated to influence Al speciation of stream waters (Lawrence et al., 1987; Lawrence and Driscoll, 1988). Stream pH decreased following the harvest, apparently as a result of stimulated nitrification, and this was accompanied by an increase in inorganic mononuclear Al. Additionally, changes in the calculated speciation of the inorganic mononuclear Al fraction were noted, with an observed shift from predominantly fluoro- to aqua- and hydroxo-complexed Al species, as the relative increase in Al_T was much greater than the increase in dissolved F. This trend was observed at all sampling locations, suggesting that a larger fraction of inorganic Al was also being transported as aqua- and hydroxo-complexed forms (Lawrence and Driscoll, 1988).

There also have been a number of investigations where Al speciation in soil solutions was estimated (e.g., Driscoll et al., 1983; David and Driscoll, 1984; Driscoll et al., 1985; Cozzarelli et al., 1987), with several others providing a great deal of speculation concerning soil solutions and solid phase distributions based on stream water chemistry (Lawrence et al., 1986; Lawrence et al., 1988; McAvoy, 1988, 1989). Driscoll and others (1983) reported greater total mononuclear Al levels in soil solutions from a higher elevation (760 m) site, which was observed to decrease at an intermediate (640 m) and a lower-elevation site (520 m). This observation was consistent with the streamwater chemistry reported at the same sites (Johnson et al., 1981). The Al_T at all sites was predominated by nonlabile or organically complexed forms (as determined by the ion exchange column method of Driscoll, 1984), although there was greater labile, inorganically complexed Al at the highest elevation, consistent with the more acidic conditions (Driscoll et al., 1983). David and Driscoll (1984) compared soil solution chemistries in hardwood and conifer stands located at the Huntington Forest in the central Adirondack Mountains of New York. The Al concentrations were about two times higher at the conifer site than at the hardwood site for all soil horizons, and this corresponded to generally lower pH and higher DOC for the soil solutions collected at the conifer site. However, the H^+ concentration was not found to correlate well to Al_T at either location, as the organically complexed Al was a significant fraction of Al_T (~82% and 93% from O-horizon leachates of the conifer and hardwood sites, respectively). The organic- to inorganic-bound Al decreased with depth at both sites, with only 67% and 58% of Al_T being organically complexed in B horizon leachates for the conifer and hardwood sites. As has been reported for many stream and surface waters, the fluoro-Al complexes were an important component of the inorganically complexed Al, ranging from 58% to 92% of the total inorganic Al fraction. The results from these and other studies (e.g. Driscoll et al., 1985; Driscoll et al., 1989) have indicated that much of the mobile Al being exported out of surface soil horizons in forested watersheds is organically complexed. Furthermore, many of these studies have suggested that much of the mobile organic Al is being deposited in subsurface horizons, consistent with the more traditional view of spodosol formation (Driscoll et al., 1983; David and Driscoll, 1984; Driscoll et al., 1985; Driscoll et al., 1989). These observations counter the more recently proposed model of podzolization, where soluble silicato-Al species are thought to be the primary mobile Al species being deposited in subsurface horizons as the poorly ordered mineral phases imogolite or allophane (Farmer et al., 1979; Farmer et al., 1980; Farmer and Fraser, 1982; Anderson et al., 1982; Childs et al., 1983). Clearly, more data are required to differentiate between these mechanisms and to delineate the conditions under which one or both may be operative.

Cozzarelli and others (1987) reported Al speciation in soil solutions collected from the White Oak Run watershed located in the Blue Ridge Mountains of central Viginia. Zero-tension and tension lysimeters were employed within soil profiles to differentiate between gravitational macropore flow and slower capillary micropore flow. Soil waters collected from the O horizon at two locations (low slope and midslope) during a May storm event were different for the two types of lysimeters, with a much higher Al_T in waters collected from the zero-tension lysimeters

(593 ± 140 vs. 39 ± 14 μg L^{-1}). Conversely, the micropore and macropore waters collected from C horizons were more similar, although the macropore Al_T exceeded that for the micropore water (124 ± 21 vs. 75 ± 25 μg L^{-1}). The Al speciation data revealed that a large fraction of Al_T from both macropore and micropore soil waters was in the acid-soluble form. Inorganic mononuclear Al fractions were observed to increase from 5% in the O horizon to 59% in the C horizon for the macropore waters, whereas this fraction was much more constant at depth for the micropore waters. Cozzarelli and others (1987) found that fluoro-Al complexes predominated the inorganic mononuclear Al fraction, which is consistent with studies conducted in northeastern United States. The observation that the inorganic Al fraction was greater at depth for the macropore water was suggested as proof that the water was moving rapidly through the soils with minimal interaction. It was also suggested that macropore flow could be a significant contributor to streamwater Al during storm events (Cozzarelli et al., 1987). Similar discussions concerning residence time and soil horizon source have been advanced to explain variations in streamwater chemistry within watersheds (e.g., Lawrence et al., 1988; McAvoy, 1988, 1989). Further investigations where Al speciation is estimated in lysimeter waters will undoubtedly provide greater delineation of the mechanisms and processes regulating Al mobility in soils and surface waters.

The various studies reporting speciation of aqueous Al in surface, ground, and soil waters, in addition to using slightly different methodologies, have also differed in the time the Al speciation was performed. Several investigators perform the Al speciation in the field, whereas others transport the samples to a laboratory prior to conducting the speciation analysis. Still other investigators do not specify when the Al speciation analysis was conducted in relation to the collection time. Campbell and others (1986) reported no significant redistribution of Al between its physicochemical forms within 24 hours of collection, although significant temporal variations were observed for surfacewater samples collected throughout the year. For one sample collected during low flow, Al speciation using the method of LaZerte (1984) revealed decreases in the nonlabile and increases in labile (8-hydroxyquinoline extraction) Al fractions, and it was suggested that some unstable polynuclear Al was converting to a mononuclear form on aging. The conclusion of this study was that the distribution of Al between its various physicochemical forms is established within the stream. As shown in Figure 3–10, significant changes in the rapidly reacting Al fraction (by the pyrochatechol violet method of Bartlett, 1988) were observed for lysimeter waters collected from soils in the Whiteface Mountain region within the first 40 hours of collection (Huntington and Miller, 1988, unpublished). The initial decrease in the rapidly reacting Al fraction (comparable to the labile Al fraction by the 8-hydroxyquinoline method) within the first 18 hours can probably be explained by CO_2 degassing, although it is currently unclear what reactions are involved in the subsequent increase and leveling off of this fraction between 20 and 60 hours. These two reports indicate that surfacewater samples may be less sensitive to the time of speciation than soil water samples. More intensive studies evaluating the

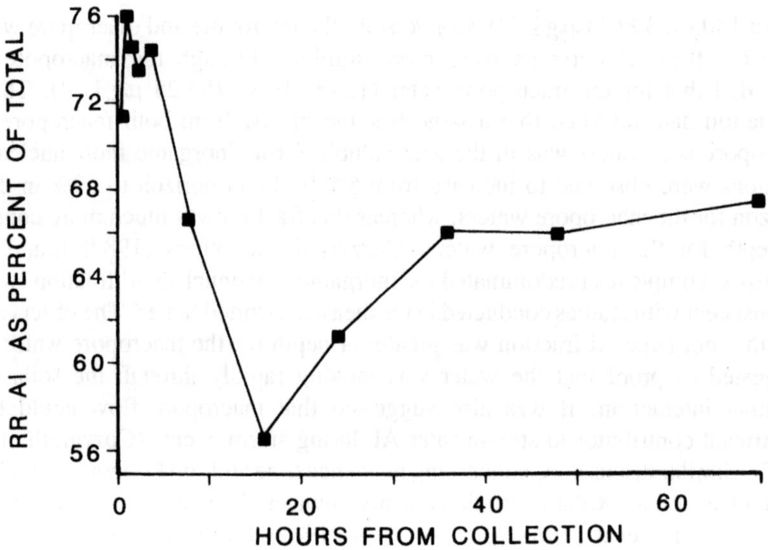

Figure 3–10. The rapidly reacting Al fraction (% Al_T) of a soil water sample determined by the pyrocatechol violet method of Bartlett and others (1987) as a function of time following collection. (Huntington and Miller, unpublished, with permission.)

time and sample handling dependence on Al speciation measurements in field studies are required before generalization can be made, yet it would appear that the sooner the measurements are made following collection, the better they will reflect the distribution of Al species within the samples.

B. Solid Phase Al

Data on aqueous Al chemistry and speciation in surface, ground, and soil waters are often discussed in relation to a solid phase Al component that presumably controls the Al solubility within the system (e.g. Johnson et al., 1981; Driscoll et al., 1983, 1984; David and Driscoll, 1984; Hooper and Shoemaker, 1985; Cronan et al., 1986; Cozzarelli et al., 1987; Litaor, 1987; Lawrence et al., 1988; McAvoy, 1989). There have been many suggestions concerning the appropriate solid-phase control on Al solubility, including the trihydroxide Al mineral phase gibbsite as well as a number of poorly crystalline analogues, kaolinite, and other poorly crystalline 1:1 aluminosilicate solid phases (including imogolite and allophane), hydroxysulfate Al phases such as alunite or jurbanite, various 2:1 clay minerals, and exchange-phase Al. Many studies have reported that Al solution concentrations have closely followed the theoretical solubility of gibbsite (e.g., Johnson et al., 1981; Driscoll et al., 1984). Most of these investigations have not provided evidence for the existence of this mineral phase in the soils or sediments within the watersheds considered. Furthermore, Al hydroxy-interlayered 2:1 clay minerals and exchange-phase Al have been demonstrated to describe the solubility of Al better or equally as well (Cronan et al., 1986; Bartlett, 1988; Dahlgren et al., 1989). There has been much discussion recently concerning the ability to predict Al concentrations in surface waters by assuming a specific solid phase control of solubility as many reports of under- or oversaturation, particularly during episodic events, are accumulating (e.g., Hooper and Shoemaker, 1985; Cronan et al., 1986; Lawrence et al., 1988; McAvoy, 1988; Dahlgren et al., 1989; McAvoy, 1989). Many of these acidic soils high in organically complexed Al are typically undersaturated with respect to gibbsite, and this has been attributed to a solubility control by Al adsorbed to solid-phase organics (Bloom et al., 1979; Driscoll et al., 1985; Cronan et al., 1986). In most acidic heterogeneous soil systems where many complex Al-containing solid phases coexist, it would be very surprising if the *mineral* gibbsite actually controlled the solubility of Al. The adherence of a solution data set to one solid phase or another is not proof that the solid phase either exists in the soil or sediment of interest or is actually controlling Al solubility. Instances where the solubility of one solid phase or another effectively predicts solution Al concentration as a function of pH, for example, may be fortuitous, and thus the utilization of solubility models for such predictions is purely empirical, albeit potentially useful.

Because the release of Al during acidification events is primarily kinetically controlled, it is useful to establish the relative distribution of Al between various labile solid-phase components. Only a few studies have considered solid-phase Al components in relation to the potential mobility of Al in soils (Driscoll et al., 1983; David and Driscoll, 1984; Dahlgren et al., 1989; Driscoll et al., 1989). Most of these investigations have partitioned solid-phase Al into exchangeable (KCl extractable), pyrophosphate extractable (organic bound), and NH_4-oxalate extractable (organic and poorly crystalline) Al. Driscoll and others (1983, 1985) reported exchange-phase, pyrophosphate, and oxalate-extractable Al values in soils that were generally consistent with soil solution Al speciation data and the traditional model of podzolization; that is, the pyrophosphate Al levels were high in the O_2 horizon, were minimum in the E horizon, and peaked in the Bhs_1 horizon, before decreasing with further depth. Similar trends have also been reported by Driscoll and others (1989). There have not, to my knowledge, been studies that have specifically related these various phases to an easily mobilizable Al fraction. There is a good chance that one or more of these operationally defined soil pools regulate Al remobilization. Further method development and refinement to elucidate better solid-phase Al associations and greater utilization of such methods will be required before the mechanisms and processes regulating Al release and deposition in acidic systems are better understood.

VI. Summary and Conclusions

Only recently has it been demonstrated that the chemical speciation of Al regulates its mobility, bioavailability, and toxicity. However, there continues to be conflicting evidence in the literature concerning the relative mobility or toxicity of a specific Al fraction or complex. Most of these discrepancies can be attributed to differences in speciation techniques, the failure to consider all significant ligands, the utilization of different thermodynamic constants, the methods by which samples were prepared or handled, and/or the misuse of a method or an approach. It is difficult to recommend a specific speciation method because the choice depends largely on the specific application. In order to elucidate general trends, for example, a method that generally separates inorganically and organically bound Al and colloidal solid-phase Al components seems to be adequate for many investigations. Although the Al associated with specific organic components appears to be very important in describing the complex behavior of Al in watersheds, most currently used methods do not discriminate between different classes of organic Al complexes. Likewise, toxicological investigations continue to be concerned with differentiating the toxicity of specific inorganic and organic Al complexes, and currently available methods and approaches generally are inadequate for providing such detailed microscopic molecular information. There has been significant, although not quite convergent, progress made in understanding the differential toxicity of Al to both terrestrial and aquatic organisms and in describing the complex behavior of Al in acidified watersheds by the utilization of existing Al speciation techniques. It is increasingly clear, however, that the specific elucidation of the underlying mechanisms regulating toxicity or of the complex processes involved in the biogeochemical cycling of Al in forested ecosystems will require greater refinement of Al speciation methods, particularly with respect to the organic Al fractions. Furthermore, it is becoming increasingly established that important reactions involving the mobilization and transport of Al occur during episodic events, where nonequilibrium processes may be quite important. Thus, the rates and mechanisms controlling Al release and Al species transformations are very important yet poorly understood processes. Further development and refinement of Al speciation techniques will also be required if these important controls on the biogeochemical cycling of Al as influenced by acidic deposition are to be better defined.

Acknowledgments

Gratitude is expressed to Carl Strojan, Domy Adriano, and Stephen Mitz for reviewing the manuscript and providing useful criticism and suggestions, and to Jean Coleman for drafting the figures. The author was partially supported during the preparation of this manuscript by a grant from Southern California Edison and by Contract DE-AC09-765R00-819 between the University of Georgia and the U.S. Department of Energy.

References

Adams, F., 1974. *In* E. W. Carson, ed. *The plant root and its environment*, 441–482. University Press of Virginia, Charlotteville.

Adams, F., and P. J. Hathcock. 1984. Soil Sci Soc Am J 48:1305 1309.

Adams, F., and Z. F. Lund. 1966. Soil Sci 101:193–198.
Adams, F., and B. L. Moore. 1983. Soil Sci Soc Am J 47:99–102.
Adams, F., and R. W. Pearson. 1967. *In* R. W. Pearson and F. Adams, eds. *Soil acidity and liming,* 161–206. Amer Soc Agron, Madison, WI.
Akitt, J. W., N. N. Greenwood, and G. D. Lester. 1971. J Chem Soc (A) 1971:2450–2457.
Alva, A. K., D. G. Edwards, C. J. Asher, and F. P. C. Blamey. 1986a. Soil Sci Soc Am J 50:133–137.
Alva, A. K., D. G. Edwards, C. J. Asher, and F. P. C. Blamey. 1986b. Soil Sci Soc Am J 50:959–962.
Alva, A. K., M. E. Sumner, Y. C. Li, and W. P. Miller. 1989. Soil Sci Soc Am J 53:38–44.
Anderson, H. A., M. L. Barrow, V. L. Farmer, A. Hepburn, J. P. Russell, and J. D. Walker. 1982. J Soil Sci 33:125–136.
Anderson, M. A., and P. M. Bertsch. 1988. Soil Sci Soc Am J 52:1597–1602.
Ares, J. 1986. Soil Sci 141:399–407.
Arp, D., and R. Ouimet. 1986. Water, Air Soil Pollut 311:367–376.
Asp, H., B. Bengtsson, and P. Jensen. 1988. Plant and Soil 111:127–133.
Aveston, J. 1965. J. Chem. Soc. 1965:4438–4444.
Backes, C. A., and E. Tipping. 1987. Int J Environ Anal Chem 30:135–143.
Baker, J. P., and C. L. Schofield. 1980. Water Air Soil Pollut 18:289–309.
Baker, J. P., and C. L. Schofield. 1982. Water Air Soil Pollut 18:289–309.
Barner, R. B. 1975. Chem Geol 15:177–191.
Bartlett, R. 1988. Agron Abst Amer Soc Agron, Madison, WI.
Bartlett, R. J., and D. C. Riego. 1972a. Plant Soil 37:419–423.
Bartlett, R. J., and D. C. Riego. 1972b. Soil Sci 114:194–200.
Bartlett, R. J., D. S. Ross, and F. R. Magdoff. 1987. Soil Sci Soc Am J 51:1479–1482.
Bengtsson, B., H. Asp, P. Jensen, and D. Berggren. 1988. Physiol Plant 74:299–305.
Berggren, D., and G. Fiskesjo. 1987. Environ Toxicol Chem 6:771–779.
Bergkvist, B. 1987. Water Air Soil Pollut 33:131–154.
Bertsch, P. M. 1987a. Agron Abst, 166. Madison, WI.
Bertsch, P. M. 1987b. Soil Sci Soc Am J 51:825–828.
Bertsch, P. M. 1988. Agron Abst 1988:194.
Bertsch, P. M. 1989. *In* G. Sposito, ed. *Environmental chemistry of aluminum,* 87–116. CRC Press, Boca Raton, FL.
Bertsch, P. M., and M. A. Anderson. 1989. Anal Chem 61:535–539.
Bertsch, P. M., M. A. Anderson, and W. J. Layton. 1987. Book of Abstracts 194th National Meeting of the American Chemical Soc., New Orleans, LA. ACS, Washington, DC.
Bertsch, P. M., M. A. Anderson, and W. J. Layton. 1989. Mag Res Chem 27:283–287.
Bertsch, P. M., R. I. Barnhisel, G. W. Thomas, W. J. Layton, and S. L. Smith. 1986c. Anal Chem 58:2583–2585.
Bertsch, P. M., W. J. Layton, and R. I. Barnhisel. 1986b. Soil Sci Soc Am J 50:1449–1454.
Bertsch, P. M., G. W. Thomas, and R. I. Barnhisel. 1986a. Soil Sci Soc Am J 50:825–830.
Blamey, F. P., D. G. Edwards, and C. J. Asher. 1983. Soil Sci 136:197–207.
Blaser, P., and G. Sposito. 1987. Soil Sci Soc Am J 51:612–619.
Bloom, P. R., and M. S. Erich. 1989. In: G. Sposito, ed. *Environmental chemistry of aluminum*, 1–27. CRC Press, Boca Raton, FL.
Bloom, P. R., M. B. McBride, and R. M. Weaver. 1979. Soil Sci Soc Am J 43:488–493.
Bloom, P. R., R. M. Weaver, and M. B. McBride. 1978. Soil Sci Soc Am J 42:713–716.
Brassard, P., J. R. Kramer, P. Nosko, and A. Kershaw. 1988. Plant Cell Environ 11:863–873.

Brenes, E., and R. W. Pearson. 1973. Soil Sci 116:295–302.
Brown, D. J. A. 1983. Bull Environ Contam Toxicol 30:582–587.
Brown, L., S. J. Haswell, M. M. Rhead, P. O'Neill, and K. C. Bancroft. 1983. Analyst (London) 108:1511–1520.
Bull, K. R., and J. R. Hall. 1986. Environ Poll (B) 12:165–193.
Cameron, R. S., G. S. P. Ritchie, and A. D. Robson. 1986. Soil Sci Soc Am J 50:1231–1236.
Campbell, P. G. C., M. Bisson, R. Boagie, A. Tessier, P. J. Villeneuve. 1983. Anal Chem 55:2246–2251.
Campbell, P. G. C., D. Thomassin, and A. Tessier. 1986. Water Air Soil Pollut 30:1023–1032.
Childs, C. W., R. L. Parffitt, and R. Lee. 1983. Geoderma 29:139–155.
Cozzarelli, I. M., J. S. Herman, and R. A. Parnell. 1987. Water Resour Res 23:859–874.
Cronan, C. S., and C. L. Schofield. 1979. Science 204:304–306.
Cronan, C. S., W. J. Walker, and P. R. Bloom. 1986. Nature 324:140–143.
Cumming, J. R., R. T. Eckert, and L. S. Evans. 1985. Can J Bot 63:1099–1103.
Cumming, J. R., R. T. Eckert, and L. S. Evans. 1986. Can J For Res 16:864–867.
Dahlgren, R. A., C. T. Driscoll, and D. C. McAvoy. 1989. Soil Sci Soc Am J (in press).
David, M. B. 1986. Water Air Soil Pollut 29:415–424.
David, M. B., and C. T. Driscoll. 1984. Geoderma 33:297–318.
Dougan, W. K., and A. C. Wilson. 1974. Analyst (London) 99:413–420.
Driscoll, C. T. 1980. Ph.D. Thesis, Cornell University, Ithaca, NY.
Driscoll, C. T. 1984. Int J Environ Anal Chem 16:267–284.
Driscoll, C. T., W. A. Ayling, G. F. Fordham, and L. M. Oliver. 1989. Can J Fish Aquat Sci 46:258–267.
Driscoll, C. T., J. B. Baker, J. J. Bisogni, and C. L. Schofield. 1980. Nature 284:161–164.
Driscoll, C. T., J. P. Baker, J. J. Bisogni, and C. L. Schofield. 1984. *In* O. R. Bricker, ed. *Acid precipitation: Geological aspects,* 55–75. Butterworth, Stoneham, MA.
Driscoll, C. T., N. M. Johnson, G. E. Likens, and M. C. Feller. 1988. Water Resour Res 24:195–200.
Driscoll, C. T., and R. M. Newton. 1985. Environ Sci Tech 19:1018–1024.
Driscoll, C. T., N. van Breeman, and J. Mulder. 1985. Soil Sci Soc Am J 49:437–444.
Driscoll, C. T., N. van Breeman, J. Mulder, and M. van der Pol. 1983. VDI. Berichte Nr 500:349–364.
Driscoll, C. T., B. J. Wyskowski, C. C. Consentini, and M. E. Smith. 1987. Biogeochemistry 3:225–242.
Driscoll, C. T., B. J. Wyskowski, P. DeStaffan, and R. M. Newton. 1989. *In* T. Lewis, ed. *The environmental chemistry and toxicology of aluminum*, 1–26. Lewis Publishing.
Dufresne, A., and W. H. Hendershot. 1986. Can J Soil Sci 66:367–371.
Entry, J. A., K. Cromack, S. G. Stafford, and M. A. Castellano. 1987. Can J For Res 17:865–871.
Farmer, V. C., and A. R. Frasier. 1982. J Soil Sci 33:773–742.
Farmer, V. C., A. R. Frasier, and J. M. Tait. 1979. Geochim Cosmochim Acta 43:1417–1420.
Farmer, V. C., J. D. Russell, and M. C. Berrow. 1980. Geoderma 20:15–26.
Felmy, A. R., D. C. Girvin, and E. A. Jenne. 1983. *MINTEQ*—A computer program for calculating aqueous geochemical equilibria. U.S. EPA (68-03-3089)', Athens, GA.
Figura, P. and B. McDuffie. 1979. Anal Chem 51:120–125.
Figura, P., and B. McDuffie. 1980. Anal Chem 52:1433–1439.

Foy, C. D. 1984. *In* F. Adams, ed. *Soil acidity and liming*. Agron 12:57–97. Amer Soc Agron, Madison, WI.
Foy, C. D., W. A. Berg, and C. L. Dewald. 1987. Plant and Soil 99:39–46.
Foy, C. D., R. L. Chaney, and M. C. White. 1978. Ann Rev Plant Physiol 29:511–566.
Frink, C. R., and M. Peech. 1962. Soil Sci Soc Am Proc 26:346–347.
Frink, C. R., and M. Peech. 1963. Inorg Chem 2:473–478.
Gardiner, P. E., R. Schierl, and K. Kreutzer. 1987. Plant Soil 103:151–154.
Garrels, R. M., and M. E. Thompson. 1962. A J Sci 260:57–66.
Hall, R. J., C. H. Driscoll, and G. E. Likens. 1987. Freshwat Biol 18:17–43.
Hall, R. J., C. T. Driscoll, G. E. Likens, and J. M. Pratt. 1985. Limnol Oceanogr 30:212–220.
Hayden, P. L., and A. J. Rubin. 1976. *In* A. J. Rubin, ed. *Aqueous environmental chemistry of metals,* 317–381. Ann Arbor Science, Ann Arbor, MI.
Helgeson, H. C., T. H. Brown, A. Nigrini, and T. A. Jones. 1970. Geochim Cosmochim Acta 34:569–591.
Helgeson, H. C., R. M. Garrels, and F. T. MacKenzie. 1969. Geochim Cosmochim Acta 33:455–481.
Hendershot, W. H., A. Dufresne, H. Lolande, and F. Courchecne. 1986. Water Air Soil Pollut 31:231–237.
Henriksen, A., O. K. Skogheim, and B. O. Rosseland. 1984. Vatten 40:225–260.
Henriksen, A., B. M. Wathne, E. J. S. Rogeberg, S. A. Norton, and D. F. Brakke. 1988. Wat Res 22:1069–1073.
Hodges, S. C. 1987. Soil Sci Soc Am J 51:57–64.
Hooper, R. P., and C. Shoemaker. 1985. Science 229:463–465.
Hsu, P. H. 1963. Soil Sci 96:230–238.
Hue, N. V., G. R. Craddock, and F. Adams. 1986. Soil Sci Soc Am J 50:28–34.
Humphreys, F. R., and R. A. Truman. 1964. Plant and Soil 20:131–134.
James, B. R., C. J. Clark, and S. J. Riha. 1983. Soil Sci Soc Am J 47:893–897.
James, B. R., and S. J. Riha. 1984. Can J Soil Sci 64:637–646.
Jardine, P. M., and L. W. Zelazny. 1986. Soil Sci Soc Am J 50:895–900.
Jardine, P. M., and L. W. Zelazny. 1987a. Soil Sci Soc Am J 51:885–889.
Jardine, P. M., and L. W. Zelazny. 1987b. Soil Sci Soc Am J 51:889–892.
Jardine, P. M., L. W. Zelazny, and A. Evans. 1986. Soil Sci Soc Am J 50:891–894.
Johnson, N. M., C. T. Driscoll, J. S. Eaton, G. E. Likens, and W. H. McDowell. 1981. Geochim Cosmochim Acta 45:1421–1437.
Jones, B. F., V. C. Kennedy, and G. W. Zellweger. 1974. Water Resour Res 10:791–793.
Karlsson-Norrgren, L., I. Bjorklund, O. Ljungberg, and P. Runn. 1986. J Fish Disease 9:11–25.
Kennedy, V. C., G. W. Zellweger, and B. F. Jones. 1974. Water Resour Res 10:785–790.
Kerridge, P. C. 1969. Ph.D. Thesis, Oregon State University, Corvallis, OR.
Kerven, G. L., and D. G. Edwards. 1987. Proc. 9th Aust. Anal Chem. Conf. 831–834.
Kinraide, T. B., and D. R. Parker. 1987a. Physiol Planta 71:207–212.
Kinraide, T. B., and D. R. Parker. 1987b. Plant Physiol 83:546–551.
Kinraide, T. B., and D. R. Parker. 1989. Plant Cell Environ (in press).
Lalande, H., and W. H. Hendershot. 1986. Can J Fish Aquat Sci 43:231–234.
Lawrence, G. B., and C. T. Driscoll. 1988. Environ Sci Technol 22:1293–1299.
Lawrence, G. B., C. T. Driscoll, and R. D. Fuller. 1988. Water Resour Res 24:659–669.
Lawrence, G. B., R. D. Fuller, and C. T. Driscoll. 1986. Biogeochem 2:115–135.
Lawrence, G. B., R. D. Fuller, and C. T. Driscoll. 1987. J Environ Qual 16:383–390.

LaZerte, B. D. 1984. Can J Fish Aquat Sci 41:766–776.
LaZerte, B. D., C. Chun, and D. Evans. 1988. Environ Sci Technol 22:1106–1108.
Lindsay, W. L. 1979. *Chemical equilibria in soils*. John Wiley and Sons, New York.
Litaor, M. I. 1987. Geochim Cosmochim Acta 51:1285–1295.
Matsumoto, H., F. Hirasawa, S. Morimura, and E. Takahashi. 1976. Plant Cell Physiol 17:627–631.
May, H. M., P. A. Helnke, and M. L. Jackson. 1979. Chem Geol 24:259.
McAvoy, D. C. 1988. J Environ Qual 17:528–534.
McAvoy, D. C. 1989. Water Resour Res 25:233–240.
Mills, G. L., and J. G. Quinn. 1981. Marine Chem 10:93–102.
Moore, D. P. 1974. *In* E. W. Carson, ed. *The plant root and its environment,* 135–151. Univ Press of Virginia, Charlottesville, VA.
Muniz, I. P., and H. Leivestad. 1980. *In* D. Drablos and A. Tollen, eds. *Ecological impact of acid precipitation,* SNSF, Oslo, 320–321.
Neitzke, M., and M. Runge. 1985. Flora 177:237–249.
Noble, A. D., M. V. Fey, and M. E. Sumner. 1988c. Soil Sci Soc Am J 52;1651–1656.
Noble, A. D., M. E. Sumner, and A. K. Alva. 1988a. Soil Sci Soc Am J 52:1059–1063.
Noble, A. D., M. E. Sumner, and A. K. Alva. 1988b. Soil Sci Am J 52:1398–1402.
Nordstrom, D. K., and J. W. Ball. 1984. *In* C. J. M. Kramer and J. C. Duinker, eds. *Complexation of trace metals in natural waters,* 149–164. Martinus Nijhoff, Netherlands.
Nordstrom, D. K., and J. W. Ball. 1985. Science 232:54–56.
Nordstrom, D. K. and H. M. May, 1989. *In* G. Sposito ed., *Environmental Chemistry of Aluminum*, 29–53. CRC Press, Boca Raton, FL.
Nordstrom, D. K., L. N. Plummer, T. M. L. Wigley, T. J. Wolery, J. W. Ball, E. A. Jenne, R. L. Basset, D. A. Crerar, T. M. Florence, B. Fritz, M. Hoffman, G. R. Holdren, G. M. Lafon, S. V. Mattigod, R. E. McDuff, F. Morel, F. Reddy, G. Sposito, and J. Thrailkill. 1979. *In* E. A. Jenne, ed. *Chemical modeling in aqueous systems* Amer. Chem. Soc. Symp. Series 93:857–892.
Ohno, T., E. I. Sucoff, M. S. Erich, P. R. Bloom, C. A. Buschena, and R. K. Dixon. 1988. Environ Qual 17:666–672.
Okura, T., K. Gotto, and T. Votuyanagi. 1962. Anal Chem 34:581–582.
Omerod, S., N. Weatherly, P. French, S. Blake, and W. Jones. 1987. Annals Soc Zool Belg 117:435–447.
Parker, D. R., T. B. Kinraide, and L. W. Zelazny. 1989. Soil Sci Soc Am J (in press).
Parker, D. R., L. W. Zelazny, and T. B. Kinraide. 1988. Soil Sci Soc Am J 52:67–75.
Parks, G. A. 1972. Am Mineral 57:1163–1189.
Parthasarathy, N., and J. Buffle. 1985. Water Res 19:25–36.
Pavan, M. A., and F. T. Bingham. 1982. Soil Sci Soc Am J 46:993–997.
Pavan, M. A., F. T. Bingham, and P. F. Pratt. 1982. Soil Sci Soc Am J 46:1201 1207.
Plankey, B. J., and H. H. Patterson. 1987. Environ Sci Technol 21:595–601.
Plankey, B. J., and H. H. Patterson. 1988. Environ Sci Technol 22:1454–1459.
Plankey, B. J., H. H. Patterson, and C. S. Cronan. 1986. Environ Sci Tech 20:160–165.
Raddum, G. G., and A. Fjellheim. 1987. Annals Soc Zool Belg 11:77–87.
Rengasamy, P., and J. M. Oades. 1979. Aust J Soil Res 17:141–153.
Rogeberg, E. J. S., and A. Henriksen. 1985. Vatten 41:48–53.
Rost-Siebert, K. 1983. Allg Forst Zeitschr 27:686–689.
Royset, O., A. O. Staunes, G. Ogner, and G. Sjotveit. 1987. Int J Environ Anal Chem 29:141–149.

Ryan, D. K., and J. H. Weber. 1982a. Anal Chem 54:986–990.
Ryan, D. K., and J. H. Weber. 1982b. Environ Sci Technol 16:866–872.
Ryan, P. J., S. P. Gessel, and R. J. Zasoski. 1986a. Plant and Soil 96:239–257.
Ryan, P. J., S. P. Gessel, and R. J. Zasoski. 1986b. Plant Soil 96:259–272.
Saar, R. A., and J. H. Weber. 1980. Anal Chem 52:2095–2100.
Schecher, W. D., and C. T. Driscoll. 1987. Water Resour Res 23:525–534.
Schecher, W. D., and C. T. Driscoll. 1988. Water Resour Res 24:533–540.
Schofield, C. L., and R. J. Trojnar. 1980. *In* T. Y. Toribara, M. W. Miller, and P. E. Morrow, eds. *Polluted rain,* 341–366. Plenum Press, New York.
Seip, H. M., L. Muller, and A. Naas. 1984. Water Air Soil Pollut 23:81–95.
Shann, J. R., and P. M. Bertsch. 1988. Agron Abst 1988:249.
Shortle, W., and K. T. Smith. 1988. Science 240:1017–1018.
Shotyk, W., and G. Sposito. 1988. Soil Sci Soc Am J 52:1293–1297.
Smith, R. W. 1971. Adv Chem Ser 106:250–279.
Smith, R. W., and J. D. Hem. 1972. Geol Surv Water Supp Paper 1287-D:1–51.
Sposito, G. 1984. Soil Sci Soc Am J 48:531–536.
Suhayda, C. G., and A. Haug. 1986. Physiol Plant 68:189–195.
Sullivan, T. J., N. Christophersen, I. P. Muniz, H. M. Seip, and P. D. Sullivan. 1986. Nature 323:324–327.
Tanaka, A., T. Tadano, K. Yamamoto, and N. Kanamura. 1987. Soil Sci Plant Nutr 33:43–55.
Thornton, F. C., M. Shaedle, and D. J. Raynal. 1986. Can J For Res 16:892–896.
Truman, R. A., F. R. Humphreys, and P. J. Ryan. 1986. Plant Soil 96:109–123.
Tukey, H. B. 1966. Bull Torrey Bot Club 93:385–401.
Turner, R. C. 1969. Can J Chem 47:2521–2527.
Turner, R. C. 1971. Can J Chem 49:1688–1690.
Turner, R. C. 1976. Can J Chem 54:1910–1915.
Turner, R. C., and W. Sulaiman. 1971. Can J Chem 49:1683–1687.
Turner, R. S., A. H. Johnson, and D. Wang. 1985. J Environ Qual 14:314–323.
Ulrich, B., R. Mayer, and P. K. Khanna. 1980. Soil Sci 130:193–199.
Van Praag, H. J., and F. Weissen. 1985. Plant and Soil 83:331–338.
Van Praag, H. J., F. Weissen, M. Sougnez, S. Remy, and G. Carletti. 1985. Plant and Soil 83:339–356.
Vogelman, H. W. 1982. Nat Hist 91:8–14.
Wada, S. I., and K. Wada. 1980. J Soil Sci 31:457–467.
Wagatsuma, T., and Y. Ezoe. 1985. Soil Sci Plant Nutr 31:547–561.
Wagatsuma, T., and M. Kaneko. 1987. Soil Sci Plant Nutr 33:57–67.
White, R. E., L. O. Tiffin, and A. W. Taylor. 1976. Plant Soil 45:521–529.
Wright, R. J., V. C. Baligar, and S. F. Wright. 1987. Soil Sci 144:224–232.
Zhao, X. J., E. Sucoff, and E. J. Stadelmann. 1987. Plant Physiol 83:159–162.

Ryan, D. K., and J. H. Weber. 1982a. Anal Chem 54:986–990.
Ryan, D. K., and J. H. Weber. 1982b. Environ Sci Technol 16:866–872.
Ryan, P. J., S. P. Gessel, and R. J. Zasoski. 1986a. Plant and Soil 96:239–257.
Ryan, P. J., S. P. Gessel, and R. J. Zasoski. 1986b. Plant Soil 96:259–272.
Saar, R. A., and J. H. Weber. 1980. Anal Chem 52:2095–2100.
Schecher, W. D., and C. T. Driscoll. 1987. Water Resour Res 23:525–534.
Schecher, W. D., and C. T. Driscoll. 1988. Water Resour Res 24:533–540.
Schofield, C. L., and R. J. Trojnar. 1980. *In* T. Y. Toribara, M. W. Miller, and P. E. Morrow, eds. *Polluted rain*, 341–366. Plenum Press, New York.
Seip, H. M., L. Muller, and A. Naas. 1984. Water Air Soil Pollut 23:81–95.
Shann, J. R., and P. M. Bertsch. 1988. Agron Abst 1988:249.
Shortle, W., and K. T. Smith. 1988. Science 240:1017–1018.
Shotyk, W., and G. Sposito. 1988. Soil Sci Soc Am J 52:1293–1297.
Smith, R. W. 1971. Adv Chem Ser 106:250–279.
Smith, R. W., and J. D. Hem. 1972. Geol Surv Water Supp Paper 1827-D:1–51.
Sposito, G. 1984. Soil Sci Soc Am J 48:531–536.
Suhayda, C. G., and A. Haug. 1986. Physiol Plant 68:189–195.
Sullivan, T. J., N. Christophersen, I. P. Muniz, H. M. Seip, and P. D. Sullivan. 1986. Nature 323:324–327.
Tanaka, A., T. Tadano, K. Yamamoto, and N. Kanamura. 1987. Soil Sci Plant Nutr 33:43–55.
Thornton, F. C., M. Schaedle, and D. J. Raynal. 1986. Can J For Res 16:892–896.
Truman, R. A., F. R. Humphreys, and P. J. Ryan. 1986. Plant Soil 96:109–123.
Tukey, H. B. 1966. Bull Torrey Bot Club 93:385–401.
Turner, R. C. 1969. Can J Chem 47:2521–2527.
Turner, R. C. 1971. Can J Chem 49:1688–1690.
Turner, R. C. 1976. Can J Chem 54:1910–1915.
Turner, R. C., and W. Sulaiman. 1971. Can J Chem 49:1683–1687.
Turner, R. S., A. H. Johnson, and D. Wang. 1985. J Environ Qual 14:314–323.
Ulrich, B., R. Mayer, and P. K. Khanna. 1980. Soil Sci 130:193–199.
Van Praag, H. J., and F. Weissen. 1985. Plant and Soil 83:331–338.
Van Praag, H. J., F. Weissen, M. [illegible], J. Remy, and G. Carletti. 1985. Plant and Soil 83:339–356.
Vogelmann, H. W. 1982. Nat Hist 91:8–14.
Wada, S.-I., and K. Wada. 1980. J Soil Sci 31:457–467.
Wagatsuma, T., and Y. Ezoe. 1985. Soil Sci Plant Nutr 31:547–561.
Wagatsuma, T., and M. Kaneko. 1987. Soil Sci Plant Nutr 33:57–67.
White, R. E., L. O. Tiffin, and A. W. Taylor. 1976. Plant Soil 45:521–529.
Wright, R. J., V. C. Baligar, and S. F. Wright. 1987. Soil Sci 144:224–232.
Zhao, X. J., E. Sucoff, and E. J. Stadelmann. 1987. Plant Physiol 83:159–162.

Snowpack Storage of Pollutants, Release during Melting, and Impact on Receiving Waters

Dean S. Jeffries*

Abstract

Information on the snowpack content of major ions, trace metals, and organic contaminants (pesticides and total PCBs) has been reviewed and discussed. Although several limitations exist, regional snowpack surveys have been successfully used to delineate spatial trends in acidic deposition. In contrast to the annual anionic predominance of SO_4^{2-} in atmospheric deposition, NO_3^- is often of a similar magnitude or even greater than SO_4^{2-} in the snowpack in locations affected by acidic deposition. Trace metal concentrations are generally greater than tenfold higher at European and North American regional locations than in Arctic or Antarctic "background" sites. The dry deposited component of the total snowpack pollutant load is generally less significant (1% to 45%) than the wet-deposited component, although there is much variability among chemical parameters and locations. There is conflicting evidence on the premelt stability of snowpacks; stability is clearly governed by many factors, and the occurrence of unfrozen underlying soils may be very important. The net radiative energy flux is primarily responsible for melting. Delivery of meltwater is often greatly complicated by the snowpack mesostructure; ice layers and the development of within-pack pipe flow can make modeling of this process very difficult. Rain-on-snow events can be very important both hydrologically and chemically. Fractionation of the pollutants during normal snowpack metamorphosis gives early meltwater ion and metal concentrations that are five- to ten-fold greater than those in the parent snow. Major ions are lost from the snowpack at differing rates during melting, a process known as *preferential elution*. Springtime reductions in pH, acid-neutralizing capacity, and base cations observed in surface waters occur due to both simple dilution and the differential release of snowpack pollutants. In contrast, lake and stream concentrations of Al and NO_3^- typically increase, although the behavior of the latter is variable from location to location. Concentrations of SO_4^{2-} remain

---

*Rivers Research Branch, National Water Research Institute, Department of Environment, P.O. Box 5050, Burlington, Ontario, Canada L7R 4A6.

comparatively constant, an observation attributed to rapid exchange of this ion in the soil environment. In lakes, the effect of snowmelt is generally limited to a near surface layer, 1 to 3 m thick. The spring melt event may hold grave consequences for several species of aquatic biota; most reported fishkill events have occurred in Scandinavia.

I. Introduction

In areas where accumulation of a winter-long snowpack is a normal phenomenon, rapid melting during spring almost always yields the major hydrological event of the year. In northern latitudes, these areas often spatially coincide with large regional or point-source emitters of acidic precursors (SO_2 and NO_x) leading to the accumulation of a pollutant load within the snowpack. Sudden release of this load during melting may provide a chemical event that is also of great significance (Jeffries et al., 1979), particularly for aquatic biota that are either inherently intolerant to rapid changes in pH or are passing through sensitive life stages during the melt period (Gunn and Keller, 1984; Mierle et al., 1986). In some areas, drinking water quality may also be impaired (Liebfried et al., 1984). The short-term magnitude of the chemical event may be accentuated by differential elution of ionic pollutants (relative to water) from the snowpack (Johannessen and Henriksen, 1978) although several other climatic, hydrologic, and terrain-related factors can also be important in determining its ultimate ecological effect.

Late-winter sampling of the snowpack is a potential means of defining the regional pattern and magnitude of acidic deposition (e.g., Wright and Dovland, 1978). Sampling techniques are well established, having been developed earlier for the assessment of water equivalent and spring runoff forecasting (Colbeck, 1987a); furthermore, snowpack surveys are inexpensive compared to direct deposition monitoring. However, the utility of snow surveys is dependent on the assumption that the pollutant load is conservative; that is, it exhibits little loss or internal alteration prior to sampling. Because many of the regions of concern commonly experience short-term periods of melting in midwinter, this assumption is often not valid; therefore, in practice, snowpack surveys have found only limited application.

A need to understand the factors governing the storage and release of pollutants from the snowpack and the resulting ecological effect is clearly reflected in the burgeoning scientific literature on the subject; for example, see the review by Marmorek and others (1984) and portions of Marmorek and others (1986). Entire volumes are now being dedicated to defining the physics, chemistry, and hydrology of snowpacks (Jones and Orville-Thomas, 1987). My objective here is to present a concise summary of the most important findings associated with three aspects of snow and snowpack research: (1) definition of regional snowpack chemistry, (2) physical and chemical snowpack processes, and (3) snowmelt-induced effects on surface water chemistry.

II. Snowpack Chemistry

The chemical content of the snowpack is influenced by many factors, the most important of which are the magnitude of both wet and dry deposition and the major impact of melting and/or rain-on-snow episodes. Other factors that may play locally important roles include small- and large-scale orographic effects, type of forest cover, wind-induced snow redistribution, and existence of significant local sources of pollutants.

Snowpack surveys, particularly those conducted on a regional scale, are designed to minimize the potential variability introduced by these factors. Hence sampling protocols usually demand that integrated snow core sampling be conducted well before significant melting occurs and at sites (typically forest clearings) that both minimize the potential for pack alteration by natural organic debris and are remote from urban or industrial influences (Barrie and Vet, 1984). Multiple coring at each site is often employed to overcome snowpack heterogeneity. In all cases, the very low concentrations of some chemical parameters (particularly the trace metals) demand that great care be taken during sample collection, processing, and analysis to ensure that no contamination occurs. This may involve approaching the sampling point by foot, using specially constructed and cleaned coring devices (generally made from Plexiglas or other like substance), and wearing polyethylene gloves. Even these precautions are inadequate for investigations at background locations (Legrand and Delmas, 1987). Use of "clean laboratory" conditions for sample processing and analysis may also be appropriate for the analysis of certain constituents.

Objectives for snowpack chemical sampling range from definition of winter deposition in regional and subregional settings to investigations of hydrochemical processes at specific locations. Regional surveys occur in locations having a widespread snowpack that is permanent for several months each year, usually in Scandinavia, Canada, and portions of the northern United States. Snowpack studies in other locations are usually conducted in comparatively high-elevation settings.

For the purpose of this review, I have compiled chemical information from the literature, and it is important to know the criteria used for selecting the data presented. First, the data exclusively describe "snowpack" samples. The large body of information that exists within the literature devoted to atmospheric deposition is not included; consideration of it is beyond the scope of this chapter. Second, only snowpack data representative of regional conditions were included; snowpack chemistry in areas affected by point sources such as smelters (e.g., Jeffries and Snyder, 1981a; Phillips et al., 1986; Zajac and Grodzinska, 1982) or major diffuse sources such as urban centers (e.g., Landsberger et al., 1982; Lewis et al., 1983) was excluded. When extensive data tabulations were present in the references, parameter means were calculated for use in the following discussion. Third, only premelt snowpack chemistry was considered; however, several references present "new" and "old" data (usually to demonstrate the influence of dry deposition), and in these cases the "old" data were selected, as they better represent the pollutant load released during subsequent snowmelt. Finally, I did not intend to produce an exhaustive summary of the available snowpack chemistry. Therefore, sometimes I have arbitrarily used data from one reference over that contained in another when presentation of both would be redundant. Dewalle (1987) has also summarized snowpack data for ions and metals but has placed special emphasis on eastern North America.

A. Major Ions

Major ion concentrations in regional snowpacks have been summarized in Table 4–1. The most striking feature of these data is the large degree of variability that exists among regions and/or sites and the very large range of reported values—qualities in common with atmospheric deposition data. This large variability is present despite the fact that a significant portion of the overall raw data variability has been excluded by the data selection criteria noted above. For example, Barrie and Vet (1984), in a large regional survey that encompassed most of the southern Ontario and Quebec in Canada, found that the mean within-sampling-site coefficient of variation (*CV*) for ionic parameters ranged from 12% (NO_3^-) to 41% (Cl^-).

The deposition of marine aerosols to the snowpack must be considered in attempts to relate observed ionic composition to the long-range atmospheric transport of pollutants. Elevated concentrations of both Cl^- and Na^+ suggest a marine influence at one location in Antarctica (reference *d* in Table 4–1), the Northwest Territories location in northern Canada (*g*), the Scottish (*t, u*), Norwegian (*y, z*), and Japanese (*cc*) sites and probably the Welsh location (*s*). Estimation of ion concentrations in excess of the marine source is a routine calculation; for instance, Wright and Dovland (1978) corrected Norwegian survey data for this influence when assessing the SO_4^{2-} contribution from distant sources.

Once the data in Table 4–1 have been "corrected" for marine influences, the dominant cations are usually H^+, Ca^{2+}, and NH_4^+ and dominant anions are SO_4^{2-} and NO_3^-. As discussed by Tranter and others (1986), H^+, SO_4^{2-}, and NO_3^- concentrations are often closely related (often significantly correlated statistically) and indicative of acidic pollutant inputs. Hence remote locations (Arctic and Antarctic sites) and areas of North America and Europe receiving little long-range transport (e.g., far western United States) have exceedingly low values for these three parameters, whereas the highest regional values occur in the northeastern United States, southern Ontario and Quebec, Britain, and central Europe. In particular, Wright and Dovland (1978) specifically evaluated north-south variability in the Norwegian snowpack composition and showed that the generally higher H^+, NO_3^-, and sea-salt-corrected SO_4^{2-} concentrations observed in the south were directly related to long-range input of acidic pollutants from major European emission areas. Calcium present in snow samples probably arises from dissolution of the mineral particulate matter incorporated into the snowpack, and the variability for this parameter evident in Table 4–1 reflects, at least in part, variability in the amount of incorporated "dust" from location to location.

Specific exceptions to above generalizations can usually be traced to the influence of a local factor influencing the snowpack chemistry. For example, Jones (1984) sampled snowpacks located in a boreal forest setting in southern Quebec, and leaching of basic cations (particularly K^+) from the organic debris incorporated into the snow at this location was considered to influence substantially the pack chemistry. Factor analysis of these data clearly separated ions primarily originating from long-range transport from those present due to local factors (Jones, 1985). Similarly, Cadle and others (1986) reported that snow collected under a red pine canopy had significantly lower H^+ and NO_3^- than snow in an open field; deciduous canopies exhibited negligible influences, however. The influence of the forest on snowpack composition cannot be taken lightly; much of the terrain receiving the greatest acidic deposition in North America and Europe is forested.

Consideration of the relative magnitudes of SO_4^{2-} and NO_3^- in Table 4–1 suggests that NO_x emissions are a significant source of acidity in the snowpack in many locations, and, in fact, NO_3^- predominates over SO_4^{2-} in most North American sites (Johannes et al., 1980; Barrie and Vet, 1984; Semkin and Jeffries, 1986a). This observation contrasts with the fact that emission strengths for SO_4^{2-} precursors are known to be greater, and weighted annual deposition concentrations invariably show SO_4^{2-} as dominant. The relatively greater importance of NO_3^- in the winter snowpack, compared to summer and annual deposition, arises from the strong seasonal cycle in SO_4^{2-} deposition (high in summer, low in winter) and the comparative seasonal constancy of NO_3^- deposition (Semkin and Jeffries, 1986b; Summers and Barrie, 1986). Hypothesized reasons to explain the seasonal cycle in SO_4^{2-} range from decreased atmospheric SO_2 oxidation rates in winter (Summers and Barrie, 1986) to a reduced natural (biogenic) source of S during the winter (Nriagu et al., 1987). Whatever the reason, it is clear that during snowmelt the snowpack may deliver higher concentrations of NO_3^- to the receiving environment than it experiences at any other time of year. This observation is important for subsequent discussion of snowmelt-related, episodic acidification of surface waters.

B. Trace Metals

Snowpack concentrations for metals normally found in trace amounts in natural waters are presented in Table 4–2. The quantity of published data is far more restricted for trace metals than it is for major ions, although there is a fairly substantial literature on metals in deposition (Jeffries and Snyder, 1981b; Galloway et al., 1982). Table 4–2 specifies "total" metal concentrations because all the references cited used methods (usually graphite furnace AAS of acidified samples) that operationally define this designation. Measurement of metal levels typical of those reported for snowpacks in remote areas (Table 4–2) require extraordinary and exacting methodologies (Legrand and Delmas, 1987).

Table 4–1. Major ion concentrations (μeq L^{-1}) in premelt snowpacks from remote, North American, European, and Japanese locations.

Location	H^+	Ca^{2+}	Mg^{2+}	Na^+	K^+	NH_4^+	SO_4^{2-}	Cl^-	NO_3^-
Remote Areas									
Ellesmere Island[a]	5.6			1.1					
Greenland[b]							1.9		1.4
Greenland[c]		<0.4	0.1	0.3	0.0		3.6		
Antarctica[d]		1.5	7.7			0.9	3.5	26.2	0.6
Antarctica[e]	3.4		0.2	0.6	0.0	0.2	1.5	1.3	1.4
North America									
British Columbia[f]	7.2	1.5				1.9			4.1
NW Territories[g]	22.4	6.8	3.1	11.7	<0.5		<13.8	12.4	
NW Ontario[h]	13.1	18	5.7	10.0	2.4	2.8			14.7
Central Ontario[i]	24.6	4.4	1.6	4.7	0.6	8.1	12.3	4.5	21.8
S-Central Ontario[j]	45.0	14				13.5	19.0	6.8	36
South Quebec[k]	33.2	3.9	1.5	2.7	5.0	5.8	24.2	11.7	26.7
Eastern Canada[l]	40.1	3.6	<0.9	<2.1	0.6	8.8	18.1	<3.5	33.1
Washington, Oregon, California[m]	3	2.1	0.7	3.0	0.5	5.7	2.9	6.2	1.8
Utah[n]	0.7						22.8		7.1
Minnesota[o]	17.7	8.0	1.5	6.1	3.1	7.9	14.4	6.8	16.4
N Michigan[p]	27.7	7.6	4.3	13.1	1.0	6.1	17.0	10.3	21.1
Pennsylvania[q]	55.9	62.4	9.9	17.4		19.1	36.8	22.3	64.6
New York[r]	54						50.1	<0.3	45.8

Europe									
Wales[s]	129	14	11	30	5		78	69	64
Scotland[t]	31.6	3.5	13	52.3	1.3		50	28.8	5.5
Scotland[u]	137		18	76.4			55.3	111	40.5
France[v]		5.8	4.0	2.4	3.1				
FRG[w]	52							33.9	58.1
Czechoslovakia[x]	26	54	13	7.2	7.5	24	115	36	40.8
S Norway[y]	48.0	27	8.8	50.6	13	36	65	62	39
Norway (S of 65°)[z]	15.1	1.0	6.6	28.3	1.8	4.7	14	27	9.4
Norway (N of 65°)[z]	11.3	5.6	23	108	3.4	2.4	25	110	4.7
N Sweden[aa]	4	13		50	<10	4.5	17	27	4
N Sweden[bb]	30.0	2	3	12		0.8	23	12	17
Japan[cc]		30	39	137	5.8				

Data are mean or midrange of values selected from each reference. "Less than" data in the references were taken at the nominal values when determining means; such means are indicated by "<." Consult text for other data selection criteria.

[a] Koerner and Fisher (1982). [b] Neftel et al. (1985). [c] Davidson et al. (1981). [d] Gjessing (1984). [e] Legrand and Delmas (1984). [f] McBean and Nikleva (1986). [g] Welch and Legault (1986). [h] Barica and Armstrong (1971). [i] Semkin and Jeffries (1986). [j] Muskoka-Haliburton only; Jeffries and Snyder (1981a). [k] Jones (1984). [l] Barrie and Vet (1984). [m] Laird et al. (1986). [n] Messer (1983). [o] Munger (1982). [p] Cadle et al. (1984b). [q] DeWalle et al. (1983). [r] Galvin and Cline (1978). [s] Reynolds (1983). [t] Thomas and Morris (1985). [u] Tranter et al. (1986). [v] Batifol and Boutron (1984). [w] Schrimpff (1980). [x] Babiakova and Bodis (1986). [y] Johannessen and Henriksen (1978). [z] Gjessing et al. (1976); see also Wright and Dovland (1978). [aa] Bjarnborg (1983). [bb] Ross and Granat (1986). [cc] Suzuki (1982).

Table 4-2. Total trace metal concentrations ($\mu g\ L^{-1}$) in premelt snowpacks from various remote, North American, and European locations.

Location	Cu	Ni	Zn	Pb	Cd	Fe	Mn	Al
Remote Areas								
Arctic[a]	0.034	0.060		0.037	0.002			
Greenland[b]	0.045	<0.13	<0.27	0.106	<0.013	<2.9	<0.035	
Antarctica[c]	0.025		0.072	0.028	0.010		0.006	
North America								
Eastern Canada[d]	<0.6	2.1		3.2		21	1.5	21
Ontario[e]				<2.7	<0.1			
Washington, Oregon, California[f]	0.44			0.33	0.12	0.92	1.0	3.2
Europe								
S. Norway[g]	13		63	30	3.4		17.0	
Norway (S of 65°)[h]	2.1		9.6	3.4	0.45			15
Norway (N of 65°)[h]	4.1		7.3	2.7	0.43			15
N Sweden[i]	0.38		3.1	1.6	0.03	12	1.0	
USSR[j]	11		30	5.0	0.9			

Data are mean or midrange of values selected from each reference. "Less than" data in the references were taken at the nominal values when determining means; such means are indicated by "<." Consult text for other data selection criteria.

[a] Mart (1983). [b] Davidson et al. (1981). [c] Boutron (1982). [d] Barrie and Vet (1984). [e] Murphy and Robertson (1979). [f] Samples filtered; Laird et al. (1986). [g] Johannessen and Henriksen (1978). [h] Gjessing et al. (1976); see also Wright and Dovland (1978). [i] Ross and Granat (1986). [j] Zolotareva (1984).

Snowpack metal concentrations at regional sites in northeastern North America and Europe are generally 10- to 1,000-fold greater than those reported for remote areas. This is true of all the metals in Table 4–2, both those having a major lithologic as well as anthropogenic source (i.e., Fe, Mn, Al in mineral dust or fly ash) and those principally released by anthropogenic activities (see enrichment factor discussion in Jeffries and Snyder, 1981b, and Barrie and Vet, 1984). Snowpack concentrations of trace metals are generally similar to those found in deposition samples (Barrie and Vet, 1984), so that the larger database available from deposition-monitoring networks may provide useful surrogate information in situations where there are no snow data.

C. Synthetic Organics

Information on snowpack concentrations of synthetic organic contaminants is extremely limited but includes some data for various pesticides and total PCBs (Table 4–3) and also for chlorophenols in Finnish snow (Paasivirta et al., 1985a, 1985b). Many of the data are for very remote locations, presumably having been collected to demonstrate the long-range transport of these compounds. In general, pesticide and PCB concentrations are very low. The paucity of data for synthetic organic contaminants and trace metals in snow points to a continuing research need.

III. Snowpack Processes

A. Chemical Stability of the Snowpack

As noted above, most regional surveys of snowpack composition conduct sample collection prior to significant melting in order to minimize the potential for chemical alteration. In many locations near the geographic limits of uninterrupted winter snowpack development (areas that also often receive acidic deposition), the occurrence of a midwinter thaw is the rule rather than the exception, and this must be considered when assessing the accumulation of a pollutant load. Apart from the obvious degradation that accompanies above-freezing air temperatures, the experimental evidence on snowpack chemical stability suggests that many site-specific factors play a role. Furthermore, dry deposition to an existing snowpack can be important in increasing the chemical load present. The following discussion on snowpack "stability" refers only to preservation of its chemical character; the normal process of physical metamorphism (within-pack flow of water vapor, snow crystal growth, and compaction) are not included here, even though Colbeck (1981) has shown that exsolving of chemical impurities to crystal surfaces occurs during recrystallization. Snow metamorphism has been reviewed by Colbeck, (1987b).

Dry deposition to a snowpack has been investigated through several field experimental studies. Cadle and others (1984a) quantified dry inputs of ions by water extraction of the "dry-side" bucket of a wet-only Wong precipitation

Table 4-3. Concentration of selected pesticides and total PCBs (ng L^{-1}) in snowpack samples from various locations. Sampling and analytical methodologies and detection limits vary greatly among the studies referenced.

Compound[a]	Ellesmere Island[b]	Canadian Arctic[c]	Ontario[d]	Ontario[e]	FRG[f]	Antarctica[g]
α-BHC	<1–18	0.43–8.72	0.5–1.5		16–30	
δ-BHC	<1–8	0.22–4.08	nd–0.4	<1	700–1,300	1.5–4.9
α-Chlordane	<1–2	<0.13–0.40				
δ-Chlordane	<2–2	<0.13–0.48				
op'-DDT	<1–4			<1–2		
pp'-DDT	<1–2			<1–4		
Σ-DDT			0.3–1.9			0.009–0.016
α-Endosulphan	<1–2	<0.13–1.34	nd			
HEOD	<2–4	<0.13–1.39	nd	<1–1		
Methoxyclor			0.1–5.8			
HCB			nd–0.1	<1		
Σ-PCB		<0.05–1.67	18–43	<10		0.160–1.000

[a] δ-BHC = lindane; HEOD = dieldrin; HCB = hexchlorobenzene; nd = below-detection data for those references not specifying detection limits.
[b] McNeely and Gummer (1984), [c] Gregor et al. (in press), [d] Strachan and Huneault (1979). [e] Murphy and Robertson (1979); data presented only when values reported at 2 or more sampling sites. [f] Schrimpff (1980). [g] Tanabe et al. (1983).

collector. This procedure, although providing apparently reasonable results, had a large uncertainty due to the problems known to exist from overcollection of particles in a bucketlike container (Ibrahim et al., 1983). Later studies (Cadle et al., 1985, 1986) used specially prepared snow surfaces (snow contained in a 2-cm-high bucket) to measure dry deposition directly from the difference in concentration between the pre- and postexposure snow (exposure period was 3 to 4 days). An earlier study (Forland and Gjessing, 1975) also used short-term changes (3 days) in the concentration of surface snow to infer dry deposition. Semkin and Jeffries (1988) estimated winter season dry inputs by comparing cumulative wet deposition with the measured snowpack load and later with cumulative snowmelt from a 1 m^2 snow lysimeter. The results of studies in rural locations by Cadle and others (1986) and Semkin and Jeffries (1988) are shown in Table 4–4. The dry component of total chemical deposition to the snowpack is, in general, smaller than the wet component; however, there is substantial variability among the ions. These studies, plus the fact that cumulative wet deposition of ions is less than the observed snowpack ion content (Figure 4–1) whereas cumulative bulk deposition is greater than the snowpack content, indicate that dry inputs significantly modify the overall pollutant load stored in the snowpack. The dry component is much more important near pollutant source areas (Dasch and Cadle, 1986).

Other studies have attempted to estimate dry deposition by determining deposition velocities (V_d) for the various chemical species to a snow surface (e.g., Ibrahim et al., 1983; Granat and Johansson, 1983; Bales et al., 1987). Multiplication of measured air concentrations by V_d gives the dry input value. The strong temperature dependence of V_d for some chemical species (particularly the gaseous S and N species) may help to explain the wide range in values for H^+, SO_4^{2-}, and NO_3^- in Table 4–4.

Conflicting field evidence exists concerning the premelt chemical stability of snowpacks. Elgmork and others (1973) carefully examined snow cores collected in southern Norway that contained visible gray layers that corresponded to high-concentration snowfall events. They concluded that the snowpack was stable throughout the winter because a similar pattern of gray layers could be identified from core to core on a regional basis, and the ionic composition of corresponding layers was reasonably constant. Cadle and others (1984a, 1984b) similarly

Table 4–4. Dry deposition as a percentage of total atmospheric deposition for selected ions in rural areas in north Michigan[a] and central Ontario.[b]

Ion	North Michigan	Central Ontario
H^+	1-13	17-22
NH_4^+	14-25	4-17
Ca^{2+}	19-45	5
SO_4^{2-}	11-30	11-13
Cl^-	15-44	28

[a] Cadle et al. (1986).

[b] Semkin and Jeffries (1988).

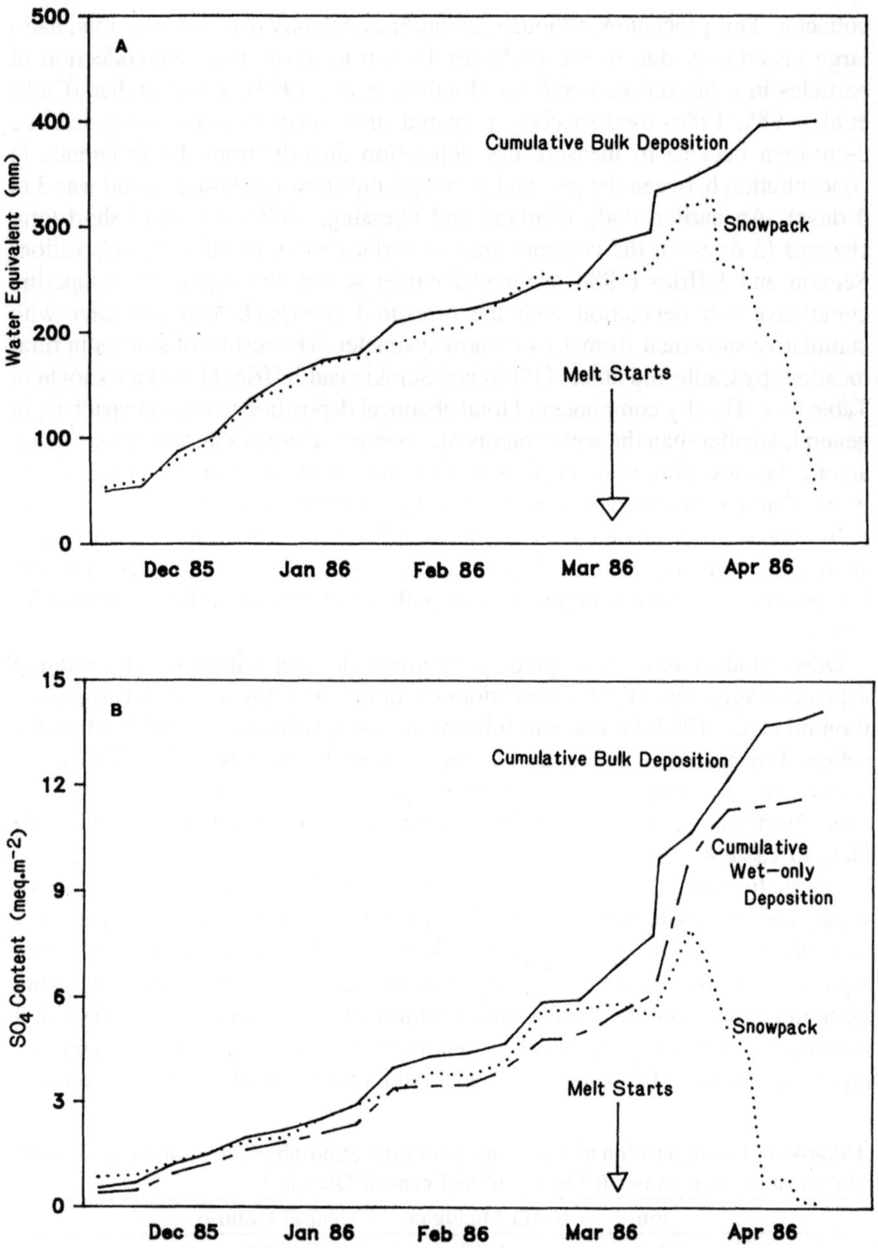

Figure 4–1. Comparison of mass loadings of water (mm) and SO_4^{2-} (meq m^{-2}) in the snowpack with cumulative atmospheric deposition for the 1985/1986 winter season in the Turkey Lakes Watershed (data from Semkin and Jeffries, 1988; similar data presentation and/or interpretation is given by Cadle et al., 1984a, 1984b). Snowpack SO_4^{-2} content prior to melting is generally greater than cumulative wet-only deposition and less than cumulative bulk deposition.

examined the compositional constancy through time in 10-cm subsections of snow cores in northern Michigan. They also concluded that the snowpack is chemically stable prior to the first melt. Cadle and others (1984a, 1984b) also inferred stability for both the premelt water and ion content through comparison of measured cumulative deposition with the standing load in the snowpack, an approach also used by Semkin and Jeffries (1988). Figure 4–1 demonstrates typical results for water equivalent and SO_4^{2-}; adding dry deposition to the cumulative wet deposition or correcting the cumulative bulk deposition for overcollection of the dry component (Ibrahim et al., 1983) yields nearly coincidental premelt curves for SO_4^{2-}, for example, similar to water. Finally, analysis of the isotopic composition of the snowpack profile through time in Switzerland (Stichler et al., 1981) demonstrated chemical stability throughout the winter in an alpine location.

There is also literature demonstrating premelt instability of snowpacks. Schemenauer and others (1985) present snowpack temperature profiles for a central Ontario location. In late January, temperature increased uniformly from −7°C at the surface to near 0°C at the ground-snow interface. This situation is typical of the area; heavy snow in early winter usually insulates the ground from freezing. In such a situation, a convective heat flux from the underlying soils can cause substantial water equivalent loss from the snowpack (Motoyama et al., 1986; English et al., 1987). Loss of ions at the same time may explain the observation of premelt snowpack alteration reported by Jeffries and Snyder (1981a) and Babiakova and Bodis (1986). Jones and Bisson (1984) also reported loss of ions from a southern Quebec snowpack prior to the onset of melting. By comparing the quality of meltwaters draining from lysimeters that contained varying amounts of forest litter, Jones (1987) hypothesized that NO_3^- and NH_4^+ loss from the snow in a boreal forest site in southern Quebec was associated with microbiological activity on the organic debris. Further discussion can also be found in Jones (1985) and Jones and Sochanska (1985). Page (1987) also reported cases of ion migration in alpine snowpacks in France, although it is not clear how much of this can be attributed to melt episodes; perhaps of greater interest is his speculation of the upward movement of ions from the soil into the snow. The conflicting observations on snowpack stability noted above point to the poor understanding that exists concerning the multiple factors and interactions that control snowpack storage and release of pollutants.

B. Snowpack Hydrology

The impact of snowmelt on associated terrestrial and aquatic ecosystems is a function of meltwater-generating mechanisms and subsequent within-pack and below-pack flowpaths. Several evaluations of snow surface energy exchange (e.g., Male and Granger, 1981), often in the context of model development (Goodison et al., 1986; Price, 1987; Stein et al., 1987), show that the net radiative energy flux is most responsible for snowmelt production. The ideal case of simple gravity flow of meltwater through the snowpack has been described (Colbeck and Anderson, 1982; Jordan, 1983a), and several predictive models have been developed (e.g., Jordan, 1983b). Under such conditions, meltwater movement is rather slow (Jordan, 1983a, reported average meltwater wave front velocity of 0.22 m h^{-1}), and chemical exchange and homogenization among the snowpack strata will occur (e.g., the isotopic evidence presented by Stichler et al., 1981).

Snowpack delivery of meltwater is complicated by several other factors, however. Diurnal freeze–thaw cycles can delay the appearance of meltwater at the base of the pack, and models have been modified to account for this (Bengtsson, 1982a, 1982b). Ice or high-density snow layers contained within the pack can strongly influence the spatial delivery of meltwater. For example, English and others (1986, 1987) have shown that substantial lateral diversion of meltwater occurred in a hillside snowpack in central Ontario, the net result being that the lower portions of the hillslope received a greater water and chemical loading than the upper portion. Marsh and Woo (1985) similarly observed the strong influence of ice layers on ground-level meltwater delivery in the Arctic and have developed a descriptive multiflowpath model. Jones (1985) has shown that various mesostructural characteristics of the snowpack (including ice layers) can lead to very heterogeneous meltwater flow patterns. In particular, pipeflow can develop around organic debris (twigs, shrub and tree stems, etc.) that is capable of rapidly delivering a large proportion of the meltwater to spatially limited locations at the ground-snow interface; also, vertical pipeflow develops within the drip zone of the forest canopy. Furthermore, as the pack becomes saturated, the snow mesostructure can change very rapidly. These observations point out the difficulty that exists in interpreting data on snowpack decay in the nonideal but normal field situation.

Rainfall can also strongly influence water delivery from a melting snowpack by dramatically increasing its free water content and thereby altering the normal flow patterns (not to mention any rain-induced mesostructural changes, as above). Semkin and Jeffries (1986a) reported that rainfall caused an almost immediate response in water delivery from a snow lysimeter, and Brown and others (1985) used isotopic analyses to show that almost all the water leaving a snowpack after rain events was, in fact, rainwater.

C. Chemical Loss from the Snowpack

Johannessen and Henriksen (1978) first described the fractionation of pollutants in the snow and occurrence of high concentrations of both ionic and metal species within the early melt fractions. The physical processes yielding these observations were described by Colbeck (1981) and are directly analogous to the well-established chemical procedures for solute purification through recrystallization. During normal snowpack metamorphism involving grain coalescence and recrystallization, impurities within the snow tend to segregate at the grain surfaces because they are not easily incorporated into the crystalline lattice. Later, when liquid water begins to pass through the snowpack, these segregated impurities are readily dissolved and appear in high concentrations in the first melt fractions. The entire process can be repeated with subsequent freeze–thaw events, each displaying high pollutant levels in the early melt; however, the absolute magnitude of the concentration maxima must necessarily decline with each event. Colbeck also noted that the rate at which the impurities are removed depends on the atmospheric conditions under which the snow was deposited, the degree and type of snow metamorphism, and the sequence and/or intensity of the liquid water supply, be it from melting or rainfall. There have been many publications verifying the occurrence of this differential fractionation and release of pollutants (e.g., Johannes et al., 1980; Cadle et al., 1984a, 1987; Semkin and Jeffries, 1986a, 1988; Babiakova and Bodis, 1986; Tranter et al., 1986; Brimblecombe et al., 1987; Davies et al., 1987). Models have also been successfully developed to describe this phenomenon (Leung and Carmichael, 1984; Hibberd, 1984).

An example of ion fractionation and snowpack release of SO_4^{2-} and the confounding influence of rainfall is illustrated in Figure 4–2. The SO_4^{2-} concentration of early meltwater exiting a field snow lysimeter is up to tenfold greater than that in the parent snowpack. In many temperate areas, rainfall regularly occurs during the snowmelt period, and this can dramatically influence the concentration of ions leaving the base of the snowpack. As shown in Figure 4–2A (1984/1985 snow season), the SO_4^{2-} concentration of snow lysimeter output waters can quickly respond to rainfall and achieve a value that approximates the rain concentration. This almost instantaneous response of the lysimeter output to a rain event corroborates the observations of Brown and others (1985) that rain often passes directly through the snowpack without substantial interaction. Semkin and Jeffries (1986a, 1988) have calculated snowpack mass balances that show that rainfall inputs during snowmelt can account for a large proportion (up to approximately 50%) of the solute flux departing the snowpack. When there is little rain (e.g., Figure 4–2B, 1985/1986 snow season), the lysimeter output versus time plot is similar to the snowmelt concentration plots of Johannessen and Henriksen (1978).

The differential process of impurity concentration and release has been termed *preferential elution*. There is a sequence of ion elution as illustrated in Figure 4–3: $SO_4^{2-} > NO_3^- > H^+ > Cl^-$. The percent of total ion output as a function of water output can also be read directly from Figure 4–3. This elution sequence was reported by Tranter and others (1986) for the Scottish snowpack; in fact, the comparatively high sea salt component in this location tends to accentuate the development of the late-eluting NaCl fraction. Data presented by Babiakova and Bodis (1986) in Czechoslovakia are similar as well. Those ions that are more soluble in ice (i.e., incorporate more easily into the ice crystal lattice) are the least mobile upon melting (Davies et al., 1987). In contrast, the earliest-eluting ions are those that are most strongly partitioned to the ice crystal surface during the fractionation process accompanying snow metamorphosis.

So-called acidic shock potential models have been developed that use a simple empirical expression of the differential release of snowpack acidity (Wilson and Barrie, 1981; Agnew et al., 1982; Goodison et al., 1986). These models generally have the intent of predicting acidic shock potential by using a climatic probability

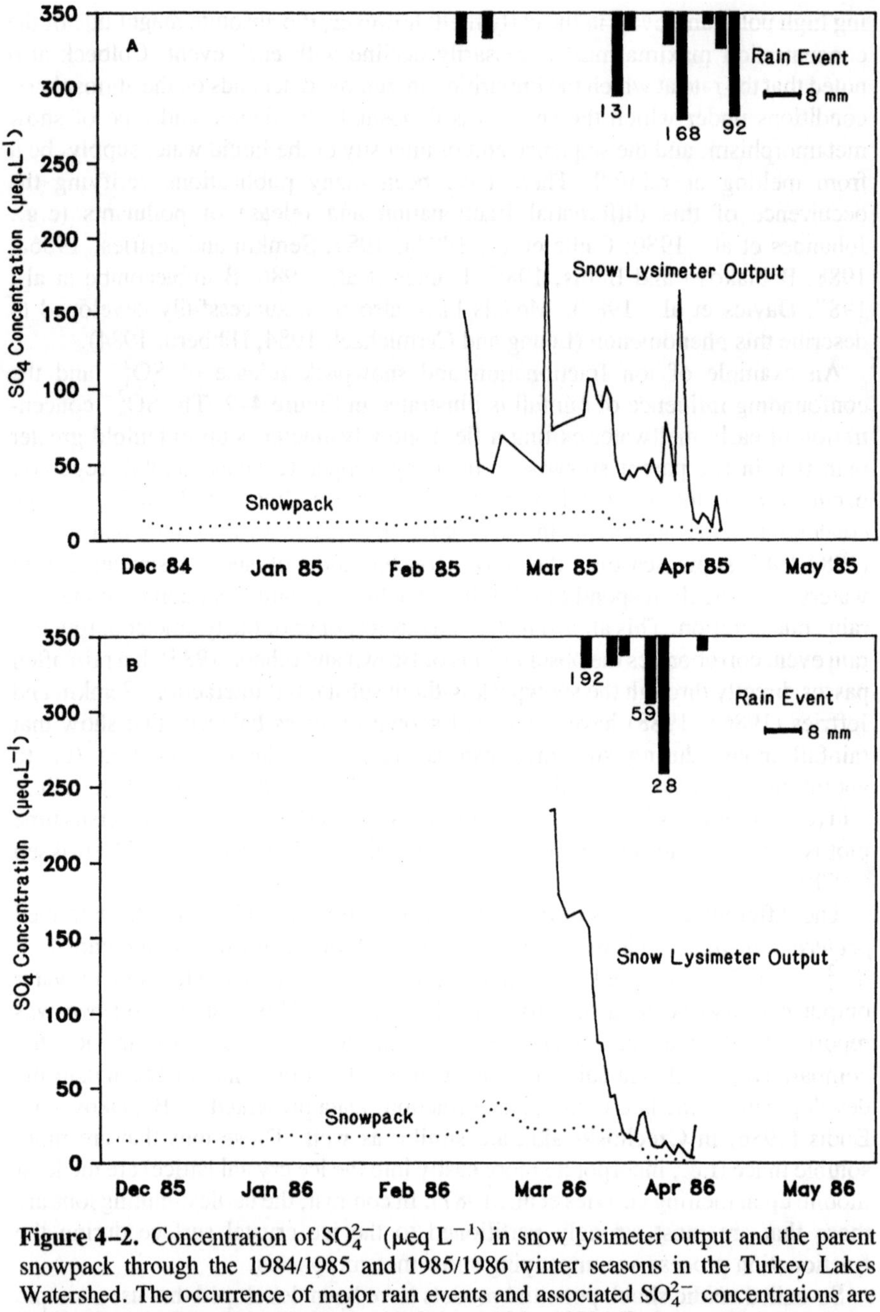

Figure 4–2. Concentration of SO_4^{2-} (μeq L^{-1}) in snow lysimeter output and the parent snowpack through the 1984/1985 and 1985/1986 winter seasons in the Turkey Lakes Watershed. The occurrence of major rain events and associated SO_4^{2-} concentrations are also indicated. (Data from Semkin and Jeffries, 1986a, 1988.) Early snowmelt concentrations are five-to ten-fold greater than the levels in the parent snowpack. (See also similar presentation by Johannessen and Henriksen, 1978.) Rainfall can dramatically affect the concentration of snow lysimeter output.

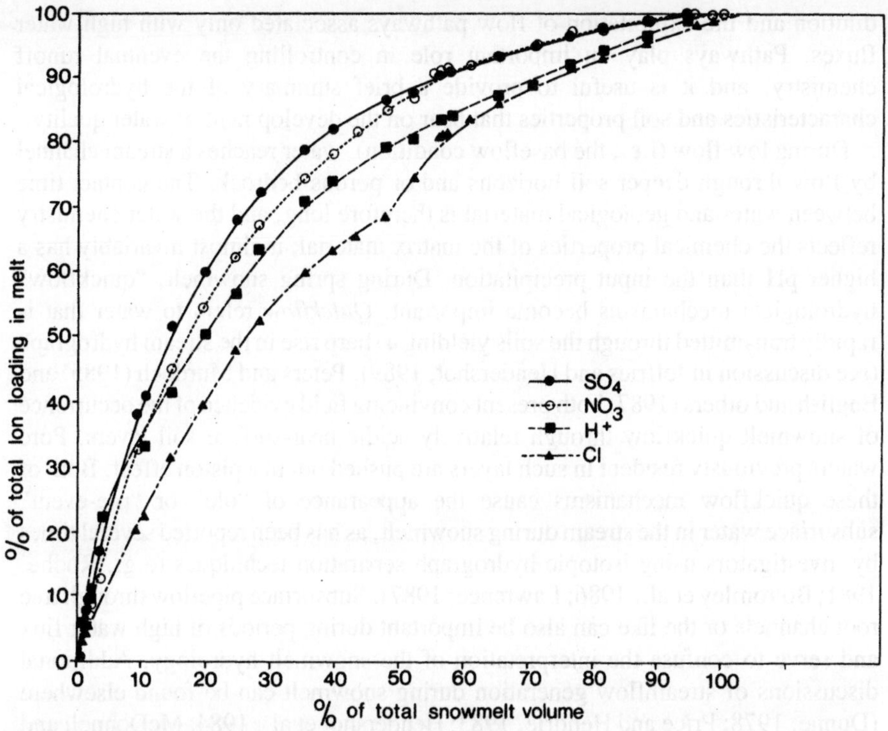

Figure 4–3. Percent loss of H^+, SO_4^{2-}, NO_3^-, and Cl^- as a function of percent water loss from the snowpack in the Turkey Lakes Watershed during the 1986 snowmelt. Preferential elution of the ions from the snowpack ($SO_4^{2-} > NO_3^- > H^+ > Cl^-$) accounts for the lack of coincidence of the lines. Approximately 50% of the major ion loading was lost from the snowpack with the first 20% of melt. (Data from Semkin and Jeffries, 1988. Similar data presentation and/or interpretation also given by Tranter et al., 1986; and Babiakova and Bodis, 1986.)

scheme and employ an exponential factor to simulate the differential elution of acidity. The numeric value of the factor is usually based on the results presented by Johannessen and Henriksen (1978).

IV. Snowmelt Effects

A. Surface Water Chemistry

The differential release of acids during snowmelt occurs at a time when the runoff waters have a reduced ability to neutralize them. Even in the absence of an elevated pollutant load in the snowpack, spring melt is a major hydrochemical event that results in rapid changes in the concentration of many chemical species due to both dilution and the exploitation of flow pathways associated only with high water fluxes. Pathways play an important role in controlling the eventual runoff chemistry, and it is useful to provide a brief summary of the hydrological characteristics and soil properties that bear on the development of water quality.

During low flow (i.e., the baseflow condition), water reaches a stream channel by flow through deeper soil horizons and/or porous bedrock. The contact time between water and geological material is therefore long, and the water chemistry reflects the chemical properties of the matrix material; it almost invariably has a higher pH than the input precipitation. During spring snowmelt, "quickflow" hydrological mechanisms become important. *Quickflow* refers to water that is rapidly transmitted through the soils yielding a sharp rise in the stream hydrograph (see discussion in Jeffries and Hendershot, 1989). Peters and Murdoch (1985) and English and others (1987) both present convincing field evidence of the occurrence of snowmelt quickflow through relatively acidic near-surface soil layers. Pore waters previously resident in such layers are pushed out in a piston effect. Both of these quickflow mechanisms cause the appearance of "old" or "pre-event" subsurface water in the stream during snowmelt, as has been reported several times by investigators using isotopic hydrograph separation techniques (e.g., Rodhe, 1981; Bottomley et al., 1986; Lawrence, 1987). Subsurface pipeflow through tree root channels or the like can also be important during periods of high water flux and serve to confuse the interpretation of the snowmelt hydrology. Additional discussions of streamflow generation during snowmelt can be found elsewhere (Dunne, 1978; Price and Hendrie, 1983; Hendershot et al., 1984; McDonnell and Taylor, 1987).

Chemical alteration of the water leaving the snowpack within the terrestrial basin to yield what is eventually observed in the streams is a function of (among others) the permeability of the underlying materials, the amount of underlying soil material (i.e., controlling the overall potential for soil–water interaction), the geochemistry of the soils, the water input rate, and the quickflow flow path (e.g., water flowing laterally through organic or A soil horizons will achieve a different final composition from those flowing through B horizons). In the event that the substrate immediately under the snowpack is impermeable (or at least incapable of passing the water at the rate supplied), then overland flow will occur and the water reaching the stream channel will exhibit a composition that reflects the ion fractionation mechanisms discussed above. In fact, such cases have been observed (e.g., Skartveit and Gjessing, 1979; Johannessen et al., 1980) in which catchment outflows in Norway exhibit H^+ and SO_4^{2-} concentration peaks that precede hydrograph peaks. This effect is due to the extremely thin to absent soils that overlay silicate bedrock at the observation sites. Similar results have also been reported for upland terrain in Scotland (Morris and Thomas, 1985).

Soil permeability at the snow interface can also be influenced by freezing, a condition that takes one to two forms—honeycomb or concrete (Post and Dreibelbis, 1942). The former is the normal condition in most temperate, forested terrain (Price and Hendrie, 1983) and leads to no significant decrease in the soil infiltration capacity; the latter normally occurs in subarctic and arctic climates

(English, 1984) and often produces dramatic overland flow during spring melt. These situations are of lesser concern because the arctic snowpack generally contains only a small load of acidic pollutants. There are a few reported observations of the occasional development of concrete frost in temperate areas leading to the direct discharge of snowmelt waters into streams via overland flow (Price and Hendrie, 1983; Pierson and Taylor, 1985); they normally occur in boggy terrain or may occasionally develop in response to heavy winter rain followed by severe cold. Rainfall reduces the insulating properties of a snowpack (Schemenauer et al., 1985), thereby allowing development of concrete ice layers at the snowpack base (Price and Hendrie, 1983).

Overall, most of the temperate forested terrain in Europe and North America that is threatened by potential snowmelt effects experiences snowpack and soil conditions that permit easy infiltration of meltwaters, thereby allowing significant chemical interactions of meltwaters with the soil matrix within the water pathway. Ion exchange reactions proceed very rapidly relative to the rate of water movements for even the most extreme cases of subsurface flow. Hence, under such conditions, the degree and timing of the pH depression are related more to the water flux and path than to the onset of melting; that is, H^+ peak concentrations often occur well after the initiation of melting and are often coincident with hydrographic peaks (Jeffries et al., 1979; one of three cases presented by Johannessen et al., 1980; Siegel, 1981; Jeffries and Semkin, 1983; Bjarnborg, 1983; Cadle et al., 1984b; Jacks et al., 1986).

Meltwater undergoes immediate and large changes in composition in those areas where soil infiltration predominates during snowmelt (Barry and Price, 1987). Alkalinity (or acid-neutralizing capacity = ANC) typically decreases as flow increases. Although very early increases in base cations (Ca^{2+}, Mg^{2+}, Na^+, K^+) and ANC concentrations may occur in streams at the initiation of melt as pre-event soil pore waters are pushed out (Johannessen et al., 1980), a substantial portion of the subsequent ANC reduction is simply due to dilution as reflected by coincident reductions in base cations (Galloway et al., 1980, 1987; Semkin and Jeffries, 1988). However, some of the ANC loss is usually due to strong acids, as reflected in variations in NO_3^- and/or SO_4^{2-} concentrations and an excessive depression in pH. Decreases in pH are typically on the order of 1 pH unit or less (Jeffries et al., 1979; Galloway et al., 1980; Siegel, 1981; Keller, 1983; Bjarnborg, 1983; Henriksen et al., 1984), although there are examples of much more dramatic changes of >2 units (e.g., Jacks et al., 1986). The stream pH values rarely approach those of the early snowmelt fraction, however.

Once the primary influence of dilution in producing pH and ANC depressions is taken into account, the relative responsibility of NO_3^- or SO_4^{2-} for further declines is often evaluated by consideration of the temporal variability of these ions. For example, Galloway and others (1980, 1987) noted that reduced pH corresponded with increased NO_3^- in the Adirondacks, whereas SO_4^{2-} remained relatively constant, albeit at much higher overall concentrations than NO_3^-. The implication is that NO_3^- is strongly influencing the occurrence of the pH depression at this location but is doing so under the influence of a greater and relatively constant acidification by SO_4^{2-}. The comparatively constant SO_4^{2-} concentrations that are observed in many locations during spring runoff reflect rapid exchange processes occurring in the soil environment (Dahl et al., 1979); SO_4^{2-} adsorbed to soil particles during earlier portions of the year must be easily released as the dilute meltwaters pass through. This is likely the only explanation that allows the observed spring melt SO_4^{2-} concentration "buffering" in streams while still permitting balanced annual input–output budgets for many catchments in glacial terrains (Rochelle and Church, 1987).

The behavior of NO_3^- concentrations during spring melt is much more variable from location to location. In contrast to Galloway and others (1980, 1987), Cadle and others (1987) observed only a minor increase in NO_3^- at the beginning of the melt, which rapidly decreased to near zero as the melt progressed; the differences are most likely related to the relative nitrification potential in upper soil horizons and, less likely, differing nutritional requirements for N exerted by the forests in question. Galloway and others (1987) calculated a NO_3^- mass balance for spring melt runoff and hypothesized that soil nitrification account for the apparent excess output at one of their two study locations. If NO_3^- input is retained within the catchment, then it acts as a net source of ANC rather than a sink, and in fact most basins do retain a significant portion of their total NO_3^- on an annual basis (Hemond and Eshleman, 1984).

Streams commonly exhibit elevated concentrations of Al during snowmelt (see review by Jeffries and Hendershot, 1989). This phenomenon reflects both routing of subsurface stormflow through near-surface soil horizons with attendant ion exchange (Hendershot et al., 1985) and desorption of Al from the stream bottom substrate (Norton et al., 1987). In chronically acidified but not yet acidic catchments, total Al concentrations commonly increase from tens to hundreds of $\mu g\ L^{-1}$, although there are several reports of concentrations exceeding 1 $mg\ L^{-1}$. Such levels may be dangerous to aquatic biota although the concentration of the inorganic monomeric fraction is a better indicator in this regard. Henriksen and others (1984) reported fish mortality during spring runoff in the Vikedal River of Norway in which the inorganic monomeric Al species increased to only 50 $\mu g\ L^{-1}$.

Several models of varying complexity have been developed to predict short-term changes in catchment hydrology and runoff chemistry (e.g., Christophersen et al., 1983; Chen et al., 1983; Christophersen et al., 1984; Bergstrom et al., 1985; Lam et al., 1986). Compartmentalizing the subsurface flow hydrology (number of compartments varies from model to model) permits simulation of stream flow generation. Many of them employ the "mobile anion" concept as the main mechanism for transporting cations through soils into streams; adsorption-desorption, ion exchange, and weathering reactions are typically included. The models provide realistic simulations of snowmelt conditions in catchment outflows and have been a most useful tool for integrating the many divergent and often competing processes operating at this time of year.

Analogous to streams, the chemical effect of snowmelt on lakes is primarily controlled by the physical processes of meltwater dispersion under ice. Studies investigating these processes show that drainage basin characteristics (size and location of stream inflows), lake morphology, and residence time as well as climate (melt rates, rainfall, etc.) are important in determining meltwater dispersion (Bergmann and Welch, 1985; Bengtsson, 1986). The lower density of cold runoff waters (relative to the under-ice waters) means that the portion of the lake affected by snowmelt is restricted to a relatively narrow layer immediately under the ice. The incoming waters spread out across the lake, and much of the snowpack output may in fact exit the lake via the outflow without significant interaction. Thickness of the under-ice layer is very often on the order of 1 m (Gunn and Keller, 1985; Bergman and Welch, 1985), although other studies report some interaction to a depth of approximately 3 m (Jeffries and Semkin, 1983; Hasselrot et al., 1987).

Chemical composition of the spring melt runoff layer in the lake is very similar to that observed in the streams. Compared to premelt lake water, reduced pH, ANC, and base cations are observed (Jeffries and Semkin, 1983; Charette et al., 1984), and elevated NO_3^- and Al may also occur (Hasselrot et al., 1987; Jeffries and Hendershot, 1989). Spatial heterogeneity is often observed in the layer due to the location of inflows and the influence of groundwater seepage. Another feature of spring inputs to lakes is the material trapped in the ice and snow directly on the lakes' surfaces. Ice decay and eventual incorporation of these chemicals occurs at such a time (i.e., late spring) that they are not immediately flushed from the lake. This input has been shown as important for the nutrient budgets (particularly P) of some lakes (Adams et al., 1979; Premo et al., 1985) and explains the swift rise in primary production that often occurs in lakes soon after ice breakup (Rask et al., 1985).

B. Aquatic Biology

A substantial literature exists describing the impact of snowmelt-induced changes in surface water chemistry on aquatic biota; however, it is outside the scope of this chapter to review this material. Therefore, the following short description with citation of important references is included for information and to identify the other major and active area of snowmelt research.

One of the principal driving forces behind early acidic deposition investigations was to determine the cause of fish population loss and observed fishkills in lakes and rivers. Much of the early work was in Scandinavia. In particular, dramatic fishkills were often associated with spring runoff (Leivestad and Muniz, 1976; Henriksen et al., 1984), and pH and Al levels in the water appear to be the most critical factors (Baker and Schofield, 1982). Damage to fish populations by acidic deposition has been reviewed by Rosseland and others (1986) for Scandinavia and by Haines and Baker (1986) for the northeastern United States. In Canada, Harvey and Whelpdale (1986) demonstrated that snowmelt runoff in south-central Ontario caused fish mortality, and Gunn and Keller (1984) have shown that lake trout sac fry may be affected by spring melt because spawning beds are characteristically located in shallows within the acidified lake layer discussed above. Finally, a whole spectrum of nonfish aquatic biota are also susceptible to the influence of acidic deposition (reviewed by Mierle et al., 1986). Marmorek and others (1984, 1986) provide comprehensive reviews of the effects of snowmelt and other episodic chemical events on aquatic biota.

Acknowledgment

R. G. Semkin has contributed to the production of this review both through his research and editorial efforts. I am grateful for this support.

References

Adams, W. P., M. C. English, and D. C. Lasenby. 1979. Wat Res 13:213–215.

Agnew, T. A., E. Wilson, L. A. Barrie, J. D. Reid, and D. Faulkner. 1982. Proc Can Hydrol Symp, Fredericton, New Brunswick, 597–614. National Research Council Canada.

Babiakova, G., and D. Bodis. 1986. Proc Symp Modelling Snowmelt-Induced Processes, Budapest, Hungary, IAHS Publ No 155:271–281.

Baker, J. D., and C. L. Schofield. 1982. Water Air Soil Pollut 18:289–309.

Bales, R. C., M. P. Valdez, G. A. Dawson, and D. A. Stanley. 1987. *In* H. G. Jones and W. J. Orville-Thomas, eds. *Seasonal snowcovers: Physics, chemistry, hydrology,* 289–297. D. Reidel Publishing Co., Dordrecht, Netherlands.

Barica, J., and F. A. J. Armstrong. 1971. Limnol Oceanogr 16:891–899.

Barrie, L. A., and R. J. Vet. 1984. Atmos Environ 18:1459–1469.

Barry, P. J., and A. G. Price. 1987. *In* H. G. Jones and W. J. Orville-Thomas, eds. *Seasonal snowcovers: Physics, chemistry, hydrology,* 501–503. D. Reidel Publishing Co., Dordrecht, Netherlands.

Batifol, F. M., and C. F. Boutron. 1984. Atmos Environ 18:2507–2515.

Bengtsson, L. 1982a. Nordic Hydrol 13:1–12.

Bengtsson, L. 1982b. Cold Regions Sci Technol 6:73–81.

Bengtsson, L. 1986. Nordic Hydrol 17:151–170.

Bergmann, M. A., and H. E. Welch. 1985. Can J Fish Aquat Sci 42:1789–1798.

Bergstrom, S., B. Carlsson, and G. Sandberg. 1985. Nordic Hydrol 16:89–104.

Bjarnborg, B. 1983. Hydrobiologia 101:19–26.

Bottomley, D. J., D. Craig, and L. M. Johnston. 1986. J Hydrol 88:213–234.

Boutron, C. 1982. Atmos Environ 16:2451–2459.

Brimblecombe, P., S. L. Clegg, T. D. Davies, D. Shooter, and M. Tranter. 1987. Water Res 21:1279–1286.

Brown, R. M., A. G. Price, and W. Workman. 1985. Annal Glaciol 7:148.

Cadle, S. H., and J. M. Dasch. 1987. *In* H. G. Jones and W. J Orville-Thomas, eds. *Seasonal snowcovers: Physics, chemistry, hydrology,* 299–320. D. Reidel Publishing Co., Dordrecht, Netherlands.

Cadle, S. H., J. M. Dasch, and N. E. Grossnickle. 1984a. Atmos Environ 18:807–816.

Cadle, S. H., J. M. Dasch, and N. E. Grossnickle. 1984b. Water Air Soil Pollut 22:303–319.

Cadle, S. H., J. M. Dasch, and R. V. Kopple. 1986. Atmos Environ 20:1171–1178.

Cadle, S. H., J. M. Dasch, and R. V. Kopple. 1987. Environ Sci Technol 21:295–299.

Cadle, S. H., J. M. Dasch, and P. A. Mulawa. 1985. Atmos Environ 19:1819–1827.
Charette, J. Y., H. G. Jones, W. Sochanska, and J. M. Gauthier. 1984. Proc Can Hydrol Symp, Quebec, Vol 1, 201–220. National Research Council Canada No 24633.
Chen, C. W., S. A. Gherini, R. J. M. Hudson, and J. D. Dean. 1983. *The integrated lake-watershed acidification study*, Volume 1: *Model principles and application procedures*. Electric Power Research Institute Rep No EA-3221, Palo Alto, CA.
Christophersen, N., L. H. Dymbe, M. Johannssen, and H. M. Seip. 1983. Ecological Modelling 21:35–61.
Christophersen, N., S. Rustad, and H. M. Seip. 1984. Phil Trans Roy Soc Lond B305:427–439.
Colbeck, S. C. 1981. Water Resourc Res 17:1383–1388.
Colbeck, S. C. 1987a. J Glaciol Special Issue:60–65.
Colbeck, S. C. 1987b. *In* H. G. Jones and W. J. Orville-Thomas, eds. *Seasonal snowcovers: Physics, chemistry, hydrology,* 1–35. D. Reidel Publishing Co., Dordrecht, Netherlands.
Colbeck, S. C., and E. A. Anderson. 1982. Water Resourc Res 18:904–908.
Dahl, J. B., C. Quenild, H. M. Seip, and O. Tollan. 1979. *Investigation of changes in calcium and sulphate in melt- and rain-water on minicatchments using radioactive tracers,* SNSF project Rep No IR 49/79. Norwegian Institute for Water Research, Oslo. 65 p.
Dasch, J. M., and S. H. Cadle. 1986. Water Air Soil Pollut 29:297–308.
Davidson, C. I., L. Chu, T. C. Grimm, M. A. Nasta, and M. P. Qamoos. 1981. Atmos Environ 15:1429–1437.
Davies, T. D., P. Brimblecombe, M. Tranter, S. Tsiouris, C. E. Vincent, P. Abrahams, and I. L. Blackwood. 1987. *In* H. G. Jones and W. J. Orville-Thomas, eds. *Seasonal snowcovers: Physics, chemistry, hydrology,* 337–392. D. Reidel Publishing Co., Dordrecht, Netherlands.
Dewalle, D. R. 1987. *In* H. G. Jones and W. J. Orville-Thomas, eds. *Seasonal snowcovers: Physics, chemistry, hydrology,* 255–268. D. Reidel Publishing Co., Dordrecht, Netherlands.
DeWalle, D. R., W. E. Sharpe, J. A. Izbicki, and D. L. Wirries. 1983. Water Resourc Bull 19:993–1001.
Dunne, T. 1978. *In* M. J. Kirby ed. *Hillslope hydrology,* 227–293. Wiley, Chichester, Great Britain.
Elgmork, K., A. Hagen, and A. Langeland. 1973. Environ Pollut 4:41–52.
English, M. C. 1984. Proc Can Hydrol Symp No 15, Quebec, Vol 1, National Research Council Canada No 24633, 329–341.
English, M. C., D. S. Jeffries, N. W. Foster, R. G. Semkin, and R. W. Hazlett. 1986. Water Air Soil Pollut 31:27–34.
English, M. C., R. G. Semkin, D. S. Jeffries, P. W. Hazlett, and N. W. Foster. 1987. *In* H. G. Jones and W. J. Orville-Thomas, eds. *Seasonal snowcovers: Physics, chemistry, hydrology,* 467–499. D. Reidel Publishing Co., Dordrecht, Netherlands.
Forland, E. J., and Y. T. Gjessing. 1975. Atmos Environ 9:339–352.
Galloway, J. N., G. R. Hendrey, C. L. Schofield, N. E. Peters, and A. H. Johannes. 1987. Can J Fish Aquat Sci 44:1595–1602.
Galloway, J. N., C. L. Schofield, G. R. Hendrey, N. E. Peters, and A. H. Johannes. 1980. *In* D. Drablos and A. Tollan, eds. *Ecological impact of acid precipitation,* 264–265. Proc Int Conf Ecol Impact Acid Precip, Sandefjord, Norway.
Galloway, J. N., J. D. Thornton, S. A. Norton, H. L. Volchok, and R. A. N. McLean. 1982. Atmos Environ 16:1677–1700.

Galvin, P. J., and J. A. Cline. 1978. Atmos Environ 12:1163–1167.

Gjessing, E., T. Dale, M. Johannessen, C. Lysholm, and R. F. Wright. 1976. Regionale snoudersokelser vinteren 1974–75. SNSF project Rep No TN 22/76. Norwegian Institute for Water Research, Oslo. 65 p.

Gjessing, Y. 1984. Atmos Environ 18:825–830.

Goodison, B. E., P. Y. T. Louie, and J. R. Metcalfe. 1986. Water Air Soil Pollut 31:131–138.

Granat, L., and C. Johansson. 1983. Atmos Environ 17:191–192.

Gregor, D. J., R. L. Thomas, and W. D. Gummer. (in press.) Environ Sci Technol.

Gunn, J. M., and W. Keller. 1984. Can J Fish Aquat Sci 41:319–329.

Gunn, J. M., and W. Keller. 1985. Anal Glaciol 7:208–212.

Haines, T. A., and J. P. Baker. 1986. Water Air Soil Pollut 31:605–629.

Harvey, H. H., and D. M. Whelpdale. 1986. Water Air Soil Pollut 30:579–586.

Hasselrot, B., I. B. Andersson, I. Alenas, and H. Hultberg. 1987. Water Air Soil Pollut 32:341–362.

Hemond, H. F., and K. N. Eshleman. 1984. Water Resourc Res 20:1718–1724.

Hendershot, W. H., A. Dufresne, H. Lalande, and R. K. Wright. 1985. Proc 42d Eastern Snow Conference, Montreal, Quebec, 58–68.

Hendershot, W. H., H. Lalande, and A. Dufresne. 1984. Water Pollut Res J Can 19:11–25.

Henriksen, A., O.K. Skogheim, and B. D. Rosseland. 1984. Vatten 40:255–260.

Hibberd, S. 1984. J Glaciol 30:58–65.

Ibrahim, M., L. A. Barrie, and F. Fanaki. 1983. Atmos Environ 17:781–788.

Jacks, G., E. Olofsson, and G. Werme. 1986. Ambio 15:282–285.

Jeffries, D. S., C. M. Cox, and P. J. Dillon. 1979. J Fish Res Board Can 36:640–646.

Jeffries, D. S., and W. H. Hendershot. 1989. *In* G. Sposito, ed. *Environmental chemistry of aluminum,* 279–301. CRC Press, Boca Raton, FL.

Jeffries, D. S., and R. G. Semkin. 1983. Proc Int Symp Acid Precip, Lindau, FRG, VDI-Bericte, VDI-Verlag GmbH, Dusseldorf, 500:377–386.

Jeffries, D. S., and W. R. Snyder. 1981a. Proc 38th Eastern Snow Conference, Syracuse NY, 11–22.

Jeffries, D. S., and W. R. Snyder. 1981b. Water Air Soil Pollut 15:127–152.

Johannes, A. H., J. N. Galloway, and D. E. Troutman. 1980. *In* D. Drablos and A. Tollan, eds. *Ecological impact of acid precipitation,* 260–261. Proc Int Conf Ecol Impact Acid Precip, Sandefjord, Norway.

Johannessen, M., and A. Henriksen. 1978. Water Resourc Res 14:615–619.

Johannessen, M., A. Skartveit, and R. F. Wright. 1980. *In* D. Drablos and A. Tollan, eds. *Ecological impact of acid precipitation,* 224–225. Proc Int Conf Ecol Impact Acid Precip, Sandefjord, Norway.

Jones, H. G. 1984. Proc 41st Eastern Snow Conference, Washington, DC, 126–138.

Jones, H. G. 1985. Annal Glaciol 7:161–166.

Jones, H. G. 1987. *In* H. G. Jones and W. J. Orville-Thomas, eds. *Seasonal snowcovers: Physics, chemistry, hydrology,* 531–574, D. Reidel Publishing Co., Dordrecht, Netherlands.

Jones, H. G., and M. Bisson. 1984. Verh Internat Verien Limnol 22:1786–1792.

Jones, H. G., and W. J. Orville-Thomas, 1987. *Seasonal snowcovers: Physics, chemistry, hydrology,* NATO ASI Series C: Mathematical and Physical Sciences Vol. 211. D. Reidel Publishing Co., Dordrecht, Netherlands. 746 p.

Jones, H. G., and W. Sochanska. 1985. Annal Glaciol 7:167–174.

Jordan, P. 1983a. Water Resourc Res 19:971–978.

Jordan, P. 1983b. Water Resourc Res 19:979–985.

Keller, W. 1983. J Great Lakes Res 9:425–429.

Koerner, R. M., and D. Fisher. 1982. Nature 295:137–140.
Laird, L. B., H. E. Taylor, and V. C. Kennedy. 1986. Environ Sci Technol 20:275–290.
Lam, D. C. L., S. Boregowda, A. G. Bobba, D. S. Jeffries, and G. G. Patry. 1986. Water Air Soil Pollut 31:149–154.
Landsberger, S., R. E. Jervis, S. Aufreiter, and J. C. Van Loon. 1982. Chemosphere 11:237–247.
Lawrence, J. R. 1987. Water Resourc Res 23:519–521.
Legrand, M. R., and R. J. Delmas. 1984. Atmos Environ 18:1867–1874.
Legrand, M. R., and R. J. Delmas. 1987. *In* H. G. Jones and W. J. Orville-Thomas. eds. *Seasonal snowcovers: Physics, chemistry, hydrology,* 225–254. D. Reidel Publishing Co., Dordrecht, Netherlands.
Leibfried, R. T., W. E. Sharpe, and D. R. DeWalle. 1984. J Am Water Works Assoc (March):50–53.
Leivestad, H., and I. P. Muniz. 1976. Nature 259:391–392.
Leung, W. K. S., and G. R. Carmichael. 1984. Water Air Soil Pollut 21:141–150.
Lewis, J. E., T. R. Moore, and N. J. Enright. 1983. Water Air Soil Pollut 20:2–22.
Male, D. H., and R. J. Granger. 1981. Water Resourc Res 17:609–627.
Marmorek, D. R., G. Cunningham, M. L. Jones, and P. Bunnel. 1984. *Snowmelt effects related to acidic deposition: A structured review of existing knowledge and current research activities*. ESSA Environmental and Social Systems Analysts Ltd., Vancouver, BC 115 p.
Marmorek, D. R., K. W. Thorton, J. P. Baker, D. P. Bernard, and B. Reuber. 1986. *Acidic episodes in surface waters: The state of the science*. Final report for the U.S. Environmental Protection Agency, Environmental Research Laboratory, Corvallis, OR. 232 p.
Marsh, P., M.-K. Woo. 1985. Water Resourc Res 21:1710–1716.
Mart, L. 1983. Tellus 35B:131–141.
McBean, G. A., and S. Nikleva. 1986. Atmos Environ 20:1161–1164.
McDonnell, J. J., and C. H. Taylor. 1987. Atmosphere-Ocean 25:251–266.
McNeely, R., and W. D. Gummer. 1984. Arctic 37:210–223.
Messer, J. J. 1983. Atmos Environ 17:1051–1054.
Mierle, G., K. Clark, and R. France. 1986. Water Air Soil Pollut 31:593–604.
Morris, E. M., and A. G. Thomas. 1985. J Glaciol 31:190–193.
Motoyama, H., D. Kobayashi, and K. Kojima. 1986. Jap J Limnol 47:165–176.
Munger, J. W. 1982. Atmos Environ 16:1633–1645.
Murphy, K. L., and J. L. Robertson. 1979. *International Environmental Consultants report for Environment Canada*. Burlington, Ontario. 64 p.
Neftel, A., J. Beer, H. Oeschger, F. Zurcher, and R. C. Finkel. 1985. Nature 314:611–613.
Norton, S. A., A. Henriksen, B. M. Wathne, and A. Veidel. 1987. *Proc Internat Symp on Acidification and Water Pathways,* vol. 1, 249–258. Bolkesjo, Norway. Norwegian National Committee for Hydrology, Oslo. Bolkesjo, Norway.
Nriagu, J. O., D. A. Holdway, and R. D. Coker. 1987. Science 237:1189–1192.
Paasivirta, J., K. Heinola, T. Humppi, A. Karjalainen, J. Knuutinen, K. Mantykoski, R. Paukku, T. Piilola, K. Surma-Aho, J. Tarhanen, L. Welling, H. Vihonen, and J. Sarkka. 1985a. Chemosphere 14:469–491.
Paasivirta, J., M. Knuutila, R. Paukku, and S. Herve. 1985b. Chemosphere 14: 1741–1748.
Page, Y. 1987. *In* H. G. Jones and W. J. Orville-Thomas, eds. *Seasonal snowcovers: Physics, chemistry, hydrology,* 281–288. D. Reidel Publishing Co., Dordrecht, Netherlands.

Peters, N. E., and P. S. Murdoch. 1985. Water Air Soil Pollut 26:387–402.
Phillips, S. F., D. L. Wotton, and D. B. McEachern. 1986. Water Air Soil Pollut 30:253–261.
Pierson, D. C., and C. H. Taylor. 1985. Can J Fish Aquat Sci 42:1979–1985.
Post, F. A., and F. R. Dreibelbis. 1942. Soil Sci Soc Am Proc 7:95–104.
Premo, B. J., C. D. McNabb, F. C. Payne, T. R. Batterson, J. R. Craig, and M. Siami. 1985. Hydrobiologia 122:231–241.
Price, A. G. 1987. *In* H. G. Jones and W. J. Orville-Thomas, eds. *Seasonal snowcovers: Physics, chemistry, hydrology,* 151–165. D. Reidel Publishing Co., Dordrecht, Netherlands.
Price, A. G., and L. K. Hendrie. 1983. J Hydrol 64:339–356.
Rask, M., L. Arvola, and K. Salonen. 1985. Aqua Fennica 15:41–46.
Reynolds, B. 1983. Atmos Environ 17:1849–1851.
Rochelle, B. P., and M. R. Church. 1987. Water Air Soil Pollut 36:61–74.
Rodhe, A. 1981. Nordic Hydrol 12:21–30.
Ross, H. B., and L. Granat. 1986. Tellus 38B:27–43.
Rosseland, B. O., O. K. Skogheim, and I. H. Sevaldrud. 1986. Water Air Soil Pollut 30:65–74.
Schemenauer, R. S., P. W. Summers, H. A. Wiebe, and K. G. Anlauf. 1985. Annal Glaciol 7:185–190.
Schrimpff, E. 1980. *In* D. Drablos and A. Tollan, eds. *Ecological impact of acid precipitation,* 130–131. Proc Int Conf Ecol Impact Acid Precip, Sandefjord, Norway.
Semkin, R. G., and D. S. Jeffries. 1986a. Water Air Soil Pollut 31:215–221.
Semkin, R. G., and D. S. Jeffries. 1986b. Water Pollut Res J Can 21:474–485.
Semkin, R. G., and D. S. Jeffries. 1988. Can J Fish Aquat Sci, 45(Suppl 1):38–46.
Siegel, D. I. 1981. Water Resourc Res 17:238–242.
Sigg, A., A. Neftel, and F. Zurcher. 1987. *In* H. G. Jones and W. J. Orville-Thomas, eds. *Seasonal snowcovers: Physics, chemistry, hydrology,* 269–279. D. Reidel Publishing Co., Dordrecht, Netherlands.
Skartveit, A., and Y. T. Gjessing. 1979. Nordic Hydrol 10:141–154.
Stein, J., D. L. Kane, M. Prevost, R. Barry, and A. P. Plamondon. 1987. *In* H. G. Jones and W. J. Orville-Thomas, eds. *Seasonal snowcovers: Physics, chemistry, hydrology,* 167–178. D. Reidel Publishing Co., Dordrecht, Netherlands.
Stichler, W., W. Rauert, and J. Martinec. 1981. Nordic Hydrol 12:297–308.
Strachan, W. M. J., and H. Huneault. 1979. J Great Lakes Res 5:61–68.
Summers, P. W., and L. A. Barrie. 1986. Water Air Soil Pollut 30:275–383.
Suzuki, K. 1982. Jap J Limnol 43:102–112.
Tanabe, S., H. Hidaka, and R. Tatsukawa. 1983. Chemosphere 12:277–288.
Thomas, A. G., and E. M. Morris. 1985. *In* I. Johansson, ed. *Hydrological and hydrogeochemical mechanisms and model approaches to the acidification of ecological systems,* 121–129. Nordic Hydrological Programme Rep 10, NFRs Committee for Hydrology, Uppsala, Sweden.
Tranter, M., P. Brimblecombe, P., T. D. Davies, C. E. Vincent, P. W. Abrahams, and I. L. Blackwood. 1986. Atmos Environ 20:517–525.
Welch, H. E., and J. A. Legault. 1986. Can J Fish Aquat Sci 43:1104–1134.
Wilson, E. E., and L. A. Barrie. 1981. Proc 38th Eastern Snow Conference, Syracuse NY, 23–32.
Wright, R. F., and H. Dovland. 1978. Atmos Environ 12:1755–1768.
Zajac, P. K., and K. Grodzinska. 1982. Water Air Soil Pollut 17:269–280.
Zolotareva, B. N. 1984. Water Air Soil Pollut 21:71–76.

Buffering of pH Depressions by Sediments in Streams and Lakes

Stephen A. Norton,* Jeffrey S. Kahl,† Arne Henriksen,‡ and Richard F. Wright‡

Abstract

We have evaluated buffering of pH depressions in streams by sediments by examination of water chemistry data from experimental acidifications of small streams that have low acid-neutralizing capacity, low pH, and a variable history of impact from acidic precipitation. Neutralization of acidity is dominated by the release of Ca and Al from stream sediments and vegetation and by dissolution of aluminum hydroxide solids. Aluminum is released from vegetation, primarily below pH 5, by cation exchange and possibly dissolution of a poorly crystallized or amorphous $Al(OH)_3$ phase. This cation exchange reservoir is easily depleted. Inorganic substrates release Al by dissolution of an $Al(OH)_3$ phase. The reservoir for $Al(OH)_3$ is large in streams with a pH between 5 and 6.

Lake sediments neutralize acidity in overlying lake water primarily by (1) release of Ca by desorption (cation exchange), and (2) SO_4 reduction and net storage of S in the sediment. Nitrate reduction may also be significant. Paleolimnological assessment of the effectiveness of these two processes for recently acidified lakes indicates that the total *increase* in sediment alkalinity generation is in the range of 0 to 5 $\mu eq\ L^{-1}$ lake water for typical drainage lakes. The importance of sediment-generated alkalinity increases as the ratio of lake area:watershed area increases. Batch studies and microcosm studies of alkalinity generation greatly overestimate the role of these two processes, especially base cation release from sediments.

I. Introduction

Acidic deposition on terrestrial and aquatic ecosystems elicits a variety of chemical responses that tend to neutralize the acidic components. Terrestrial

---

*Department of Geological Sciences, University of Maine, Orono, ME 04469, USA.

†Department of Environmental Protection, Augusta, ME 04333, USA.

‡Norwegian Institute for Water Research, Box 333, 0314 Oslo, Norway.

processes include throughfall–canopy interaction, increased chemical weathering, increased release of base cations from the soil cation exchange complex, adsorption of anions (largely SO_4), titration of dissociated organic acids, and dissolution of secondary soil mineral phases. Each of these processes may be thought of as a line of defense against chronic acidification. To the degree that these processes keep up with the inputs of acids, downstream ecosystems are protected. Some of these processes, such as SO_4 adsorption by soils in glaciated terrain, are easily exhaustible. Others, such as chemical weathering, are capable of producing acid-neutralizing capacity (ANC), alkalinity, on a steady-state basis.

Terrestrial ecosystems produce acidity, including carbonic acid and organic acid acidity, on a steady-state basis. Much of this acidity operates only within the soil and thus serves to translocate certain elements within the soil but not to export from the ecosystem. Such is generally the case for Al and Fe in acidic soils. Typically these two elements are moved downward in the soil profile, where they are precipitated when the pH of soil waters rises due to additional chemical weathering, and metabolism and precipitation of dissolved organic carbon.

Under certain circumstances, carbonic acid and organic acidity with its complexing capacity may be sufficient to transport Fe and Al into surface waters along with base cations and bicarbonate alkalinity. Downstream, the pH typically rises due to loss of dissolved organic acid through metabolism or photooxidation, loss of CO_2, and dilution by higher pH groundwaters. If stream water enters a lake, additional processes may occur to raise the pH still further, precipitating Fe and Al either as metal-humic complexes or as hydroxides. Thus the ecosystem acts as a chromatograph with respect to these two elements. In soft water systems, sensitive to acidic precipitation, base cations remain largely in solution as the pH rises.

Acidic precipitation shifts chemical processes toward mobilization of base cations and other metals and reduces alkalinity and pH. It also shifts the locus of precipitation or stabilization of certain metals further downstream in the hydrologic sense. For example, Cronan and Schofield (1978) demonstrated that the presence of the extra acidity due to acidic precipitation and in the presence of a mobile anion (SO_4) results in the export of Al from soils that historically accumulated Al. Similarly, various workers have demonstrated that the export of base cations is accelerated from soils when they are leached by more acidic solutions (Skeffington, 1986). Although some neutralization of the excess acidity is achieved by increased export of base cations and other metals, groundwaters emerge with lowered pH and alkalinity. The next line of defense consists of interactions with stream substrates, followed downstream by in-lake processes.

This chapter evaluates a series of experiments that were designed to explore the release of metals from streambeds in response to episodic acidification, the precursor for chronic acidification of an ecosystem. We evaluate the ability of stream substrates to buffer stream water against pH variations. We then evaluate the extent to which lakes can buffer pH changes through processes such as the release of cations and reductions of SO_4 and NO_3.

II. Stream Sediments

Upon emerging as surface water, groundwater typically degasses excess CO_2 acquired during passage through the forest floor or B horizon (Norton and Henriksen, 1983). Dissolved organic carbon, also derived from the forest floor, is partly dissociated as a weak organic acid; it typically decreases in concentration during the passage of groundwater from the forest floor through the E and B horizons as a result of metabolism of the DOC (Cronan and Aiken, 1985). Furthermore, chemical weathering increases the pH, possibly also resulting in the precipitation of some organic matter. The emerging groundwater, in equilibrium with elevated CO_2 pressures in soil air, reequilibrates at a higher pH in streams. In addition to degassing effects, mixing downstream with more alkaline waters causes a shift in cation exchange equilibria, favoring higher base saturation of stream sediments; that is, a greater proportion of exchange sites become occupied by base cations.

The ability of suspended sediments, bedload, and stationary stream sediments to buffer stream chemistry has been documented by both empirical (Kennedy, 1965) and experimental studies (e.g., Bencala, 1983, 1984; Bencala et al., 1983, 1984, 1987; Bencala and Walters, 1983). Cation exchange dominates the short-term buffering of base cation concentrations. However, the effectiveness with which this process neutralizes acidic episodes and buffers pH changes that are related to variations in discharge has only recently been explored.

A. Norris Brook, New Hampshire

Hall and others (1980, 1987) conducted both short-term and long-term stream acidification in a first-, second-, and third-order brook in the Norris Brook watershed in Hubbard Brook Experimental Forest, New Hampshire. The short-term experiments (approximately 8 hours) were unconfounded with changes in hydrology with time. These experiments consisted of adding HCl to depress the pH to about 4 (measured 15 m downstream) and of addition of $AlCl_3$ to increase Al to concentrations as high as 4 mg L^{-1}. The streambed consisted of fine sand to boulders, with small amounts of attached algae.

In the three experiments with HCl addition, neutralization was dominated by the release of Al and Ca, with the amount of neutralization decreasing with increasing stream order. For example, in the first-order stream, approximately 450 μeq HCl L^{-1} were added to the stream. Fifteen meters downstream increases in Al and Ca concentrations reached 175 and 125 μeq L^{-1}, respectively, with a pH of 4 (100 μeq L^{-1}). Only 150 μeq L^{-1} HCl were required to depress the pH to 4 at a downstream distance of 15 m for the third-order stream. There the increases in Al and Ca were about 15 and 50 μeq L^{-1}, respectively. The release of Al appears to decline slightly with time, even at constant acidic loading but remained between 10 to 15 μeq L^{-1}, similar to values achieved in the long-term (6-month) acidification experiment conducted further downstream (Hall et al., 1980). The data are not sufficient to determine whether the increase of cation concentrations is from cation exchange, dissolution of precipitation phases [e.g., $Al(OH)_3$], or increased chemical weathering. Some increase in chemical weathering is suggested by the observed increase in SiO_2 concentrations downstream, although not in a stoichiometric relationship with other mineral-forming elements. Desorption of SiO_2 from other phases is probable.

The addition of $AlCl_3$ to the stream and the resulting water chemistry provide significant insight into the mechanisms operating. In all cases the amount of Al added caused saturation. The resulting hydrolysis produced H^+, thereby lowering the pH and thus promoting the desorption of Ca from substrates. In addition, some Al would exchange for Ca and Mg on the various stream substrates. Downstream changes thus included decreases in pH and dissolved Al and an increase in the major base cations previously occupying ion exchange positions in the stream bed (Hall et al., 1987).

B. Vikedalselva Tributary, Norway

Henriksen and others (1984, 1988) acidified a small, unnamed tributary to the River Vikedalselva in southwestern Norway. The pH of the stream was lowered incrementally from about 6 to about 4. Four separate experiments of about 15 hours total duration were conducted over 3 days. Sulfuric acid was added in amounts up to 20 times greater than was necessary to titrate the water alone to pH 4. The neutralization of the acid was effected by the release of labile Al in concentrations up to 2,500 $\mu g\ L^{-1}$; the sum of H^+, decreased alkalinity, and increased Ca^{+2}, Mg^{+2}, Na^{+2}, and Al^{+++} was equivalent to the sulfate added. Neutralization above pH 5 was dominated by base cation release and reaction with bicarbonate in the water, whereas Al release dominated below 5.

C. Nant Mynydd Trawsnant, Wales

Nant Mynydd Trawsnant is a soft water stream ($Ca^{+2} = 4.7\ mg\ L^{-1}$) in mid-Wales that was experimentally acidified for 24 hours to pH 4.2 to 4.5 (Ormerod et al., 1987). The streambed consisted of sand and gravel. A stretch of 200 m of stream was acidified and monitored. Below, Al was also added to the water. In the acidified-only reach, no significant increase in dissolved Al was noted. The Al above the acidification was 46 $\mu g\ L^{-1} \pm 15$; in the acidified section, the stream Al was 52 $\mu g\ L^{-1} \pm 8$. Also, Ca^{+2} remained unchanged. The lack of vegetation in the streambed probably precluded significant amounts of exchangeable Ca^{+2} being present. This is also suggested by the observation that the added Al had no effect in displacing Ca from stream substrates. The high ambient pH of the stream (7.10 ± 0.30) suggests that the groundwater pH was not acidic (greater than pH 5). Thus little Al would be carried from the soil environment in solution to be precipitated or adsorbed in the streambed after CO_2 degassing.

D. Bonnabekken, Norway

Norton and others (1987) acidified a small episodically acidic stream just north of Oslo, Norway. The pH of the stream water was lowered with sulfuric acid from about 6 to the mid-4s in four separate experiments. The first two were separated by 2 days and involved a 70-m reach of the stream. The second set of two experiments separated by 2 days occurred 2 weeks later and involved 135 m of stream, the lower 70 m of which were used in the first two experiments. Neutralization of the acidic additions was dominated by the release of Al and Ca (Table 5–1). The total acidic neutralization can be estimated from Table 5–1. For example, at 7^{10} into the experiment at 135 m below the point of acidic addition, the excess strong acid (as sulfate) was 73 μeq L^{-1}. Twenty-eight μeq L^{-1} were present as H^+, 9 as extra Ca, 1 as extra Mg, 12 as extra Al, and the rest titrated the HCO_3 alkalinity. In the first and third experiments, Al was released for a few hours from an easily mobilized but small pool of Al in the sediment. This release was followed by slower release of Al over the entire period of the experiments. The relationship between pH and dissolved labile Al (largely Al^{+++} at the pH of the experiments) was not constant at the lower end of the experimental reach (Figure 5–1). As the experiment progressed, the apparent solubility of $Al(OH)_3$ decreased with time. Cation exchange of Al for H may be partly responsible for the resultant water chemistry, and the decline of dissolved Al with constant pH may reflect partial depletion of an exchanger. During 44 hours of acidification, Al release was 0.5 g m^{-2} of stream bottom. At the same time, Ca was released, probably by cation exchange; this cation exchange dominated neutralization for the first 6 hours, but the exchange reservoir became depleted. As would be expected, little Al was released at a pH over 5. At lower pH and after 6 hours of acidification, neutralization was dominated by the release of Al. Downstream, progressive release of Ca resulted in a rise of pH sufficient to cause a loss of Al from solution (Table 5–1), probably by precipitation of $Al(OH)_3$. The maximum Ksp for Al calculated as the dissolution of aluminum hydroxide

$$3H^+ + Al(OH)_3 = Al^{+++} + 3H_2O$$

is about $10^{8.6}$.

E. Sogndal, Norway

In the summer of 1987, Norton and others (in preparation) acidified a small, unnamed stream under Sogndal in western Norway. Precipitation in the region averages about 4.9 with more acidic episodes associated with elevated NO_3^{-1} and SO_4^{-2} (Wright, 1987). Additionally, streams in the area are subjected to acidification caused by the sea-salt effect (Skartveit, 1981; Norton et al., 1987; Wright et al., 1988). Preacidification stream pH was 6, typical for a highly sensitive stream with low alkalinity. The discharge at the time of the experiment was approximately 15 L min^{-1} throughout the experimental reach. The reach was a

Table 5-1. Chemistry of the stream Bonnabekk, Nordmarka, Norway, during artificial acidification.

														(R=AL=AL=IL)				
4^{45}	(10/10/86)	pH	Δ (H^+)	Ca (mg/L)	Δ (μeq)	Mg (mg/L)	Δ (μeq)	K (mg/L)	Δ (μeq)	SO_4 (mg/L)	Δ (μeq)	TOC (mg/L)	Δ (mg/L)	R=AL (μg/L)	Δ (μeq/L)	AL=IL (μg/L)	Fe (μg/L)	Zn (μg/L)
Above	acid addition	5.75	(0)	1.78	(0)	0.34	(0)	0.23	(0)	5.2	(0)	3.41	(0)	91	(0)	70	168	10
35 m	below acid addition	4.57	(25.1)	1.91	(6.5)	0.36	(1.7)	0.24	(0.3)	8.5	(68.8)	3.69	(0.28)	197	(14)	50	193	10
85 m	below acid addition	4.73	(16.8)	2.03	(12.5)	0.37	(2.5)	0.24	(0.3)	8.4	(66.7)	4.20	(0.79)	171	(11.3)	48	200	20
135 m	below acid addition	4.87	(11.7)	2.14	(18.0)	0.38	(3.3)	0.25	(0.5)	8.4	(66.7)	4.30	(0.89)	168	(11)	48	173	20
7^{10}	(10/10/86)																	
35 m	below acid addition	4.48	(31.3)	1.86	(4)	0.35	(0.8)	0.23	(0)	8.8	(75)	4.17	(0.76)	192	(13.8)	48	—	—
85 m	below acid addition	4.55	(26.4)	1.90	(6)	0.35	(0.8)	0.24	(0.3)	8.7	(72.9)	4.15	(0.74)	227	(16.8)	55	—	—
135 m	below acid addition	4.52	(28.4)	1.95	(8.5)	0.36	(1.1)	0.24	(0.3)	8.7	(72.9)	4.05	(0.64)	181	(12.4)	48	168	—

From Norton et al. (1987).

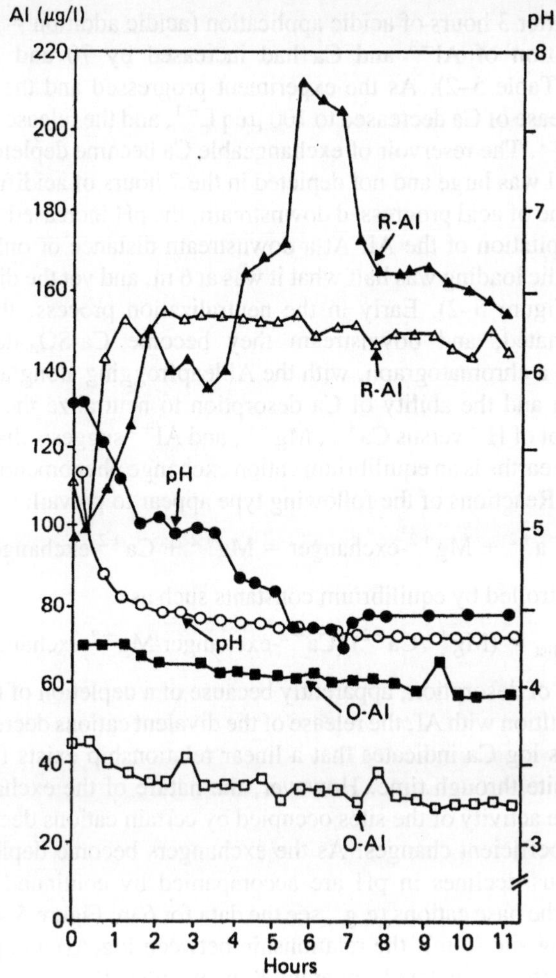

Figure 5–1. Chemistry of the stream Bonnabekk, Nordmarka, Norway, 135 m downstream from the point of H_2SO_4 additions. (From Norton et al., 1987.)

series of pools and riffles with a bed of mosses, sand, and gravel. Vegetation covered nearly half of the streambed, which was up to 1 m wide. Sulfuric acid was added at increasing dosages until concentrations were approximately 560 μeq L^{-1}. Downstream dilution of the acid was extensive (Table 5–2) because of low flow and pool capacity. It took nearly 6 hours for steady-state concentrations to be achieved just 6 m downstream. With no neutralization of the added acid, the expected pH should have been about 3.7 at 6 m downstream. Samples were taken at hourly frequency at distances of 6, 16, 26, 40, 56, 66, 79, 88, 104, and 120 m.

Neutralization was dominated by the release of Al and Ca. For example, at 6 m downstream after 3 hours of acidic application (acidic addition = 400 μeq L^{-1}), the concentration of Al^{+3} and Ca had increased by 70 and 150 μeq L^{-1}, respectively (Table 5–2). As the experiment progressed and the acidic loading increased, release of Ca decreased to 100 μeq L^{-1}, and the release of Al increased to 300 μeq L^{-1}. The reservoir of exchangeable Ca became depleted rapidly. The reservoir of Al was large and not depleted in the 7 hours of acidification.

As the plume of acid progressed downstream, the pH increased, which resulted in the reprecipitation of the Al. At a downstream distance of only 40 m after 7 hours, the acidic loading was half what it was at 6 m, and yet the dissolved Al was virtually 0 (Figure 5–2). Early in the neutralization process, the waters were Al-SO_4-dominated, and downstream they become Ca-SO_4-dominated. The stream acts as a chromatograph, with the Al leapfrogging along according to the acidic loading and the ability of Ca desorption to neutralize the acidity to pHs above 5. A plot of H^+ versus Ca^{++}, Mg^{++}, and Al^{+3} suggests that initial release of the alkaline earths is an equilibrium cation exchange phenomenon above pH 5.7 (Figure 5–3). Reactions of the following type appear to prevail:

$$Ca^{+2} + Mg^{+2}\text{-exchanger} = Mg^{+2} + Ca^{+2}\text{-exchanger}$$

which are controlled by equilibrium constants such as

$$K_{dist} = (Mg^{+2}/Ca^{+2})\ (Ca^{+2}\text{-exchanger}/Mg^{+2}\text{-exchanger})$$

After a period of desorption, apparently because of a depletion of the reservoir as well as competition with Al, the release of the divalent cations decreases. A plot of log Mg versus log Ca indicates that a linear relationship exists for the samples taken at one site through time. However, the nature of the exchanger changes. Apparently the activity of the sites occupied by certain cations decreases, and the distribution coefficient changes. As the exchangers become depleted in Ca and Mg, continuous declines in pH are accompanied by continued but decreased desorption of the base cations (e.g., see the data for 6 m; Figure 5–4). At the most acidic sampling site (6 m), the relationship between log(Al) and pH has a slope near three, with a calculated apparent Ksp of about 9.5, assuming that total dissolved Al is Al^{+3}. Most importantly, the relationship remains unchanged through the experiment, demonstrating the much larger capacity of the stream substrates to neutralize acidity through the release of Al. The calculated solubility of a hypothetical $Al(OH)_3$ phase ranges from about $10^{9.1}$ to $10^{9.7}$, in excess of well-crystallized gibbsite, but less than that of amorphous $Al(OH)_3$. Further downstream the apparent Ksp is closer to 10, indicating that reprecipitation of $Al(OH)_3$ is occurring.

Evidence against chemical weathering playing a major role in these chemical changes in the stream water includes very fast kinetics, noncongruent but constant stoichiometry for the base cations with a very large excess of Al, a lack of increased concentrations of H_4SiO_4 (less than 1 ppm variation in the experiment), reversibility of reactions, and the depletion of Ca and Mg from substrates within hours of the beginning of the experiments but sustained release of Al.

Table 5-2. Chemistry of an artificially acidified stream at Sogndal, Norway, at selected distances downstream from the point of H_2SO_4 addition at 1-hour intervals.

	Distance	Time (h) from start							
	(M)	0	1	2	3	4	5	6	7
pH	0	6.21	—	—	—	—	—	—	6.28
	6	—	—	—	4.64	4.47	4.44	4.37	4.35
	16	—	6.08	5.93	5.78	5.81	5.01	5.03	4.82
	26	—	—	—	5.96	5.90	6.00	5.79	5.68
	40	—	—	6.14	6.10	6.11	5.98	6.06	6.03
	56	—	—	—	—	6.38	6.14	6.19	6.15
	66	—	—	—	—	6.36	6.30	6.26	6.26
	79	—	—	—	—	—	6.26	6.34	6.30
	88	—	—	—	—	—	—	—	6.17
	104	—	—	—	—	—	—	—	6.20
	120	—	—	—	—	—	—	—	6.10
Ca (mg L^{-1})	0	0.52	—	—	—	—	—	—	0.54
	6	—	—	—	3.47	3.43	3.01	2.77	2.55
	16	—	1.35	2.65	3.75	5.45	4.79	4.63	4.20
	26	—	—	—	2.76	3.35	4.12	4.55	4.83
	40	—	—	0.94	1.73	2.29	2.90	3.41	3.81
	56	—	—	—	—	2.20	2.75	3.05	3.46
	66	—	—	—	—	0.86	1.28	1.61	2.07
	79	—	—	—	—	—	0.88	1.13	1.51
	88	—	—	—	—	—	—	—	1.08
	104	—	—	—	—	—	—	—	0.53
	120	—	—	—	—	—	—	—	0.42
SO_4 (mg L^{-1})	0	0.9	—	—	—	—	—	—	1.0
	6	—	—	—	19.0	23.0	26.0	28.0	28.0
	16	—	5.3	9.7	13.5	16.5	19.0	20.5	22.0
	26	—	—	—	10.6	13.0	16.0	16.5	18.5
	40	—	—	2.9	7.7	9.5	11.5	13.0	15.0
	56	—	—	—	—	5.7	7.8	9.2	10.7
	66	—	—	—	—	4.0	5.7	7.2	8.8
	79	—	—	—	—	—	4.1	5.3	6.7
	88	—	—	—	—	—	—	—	4.9
	104	—	—	—	—	—	—	—	1.9
	120	—	—	—	—	—	—	—	1.2
R-Al (μg L^{-1})	0	<10	—	—	—	—	—	—	<10
	6	—	—	—	620	1440	2120	2550	2820
	16	—	<10	<10	<10	50	—	430	990
	26	—	—	—	<10	<10	<10	<10	18
	40	—	—	<10	<10	<10	<10	<10	<10
	56	—	—	—	—	<10	<10	<10	<10
	66	—	—	—	—	<10	<10	<10	<10
	79	—	—	—	—	—	<10	<10	<10
	88	—	—	—	—	—	—	—	<10
	104	—	—	—	—	—	—	—	<10
	120	—	—	—	—	—	—	—	<10

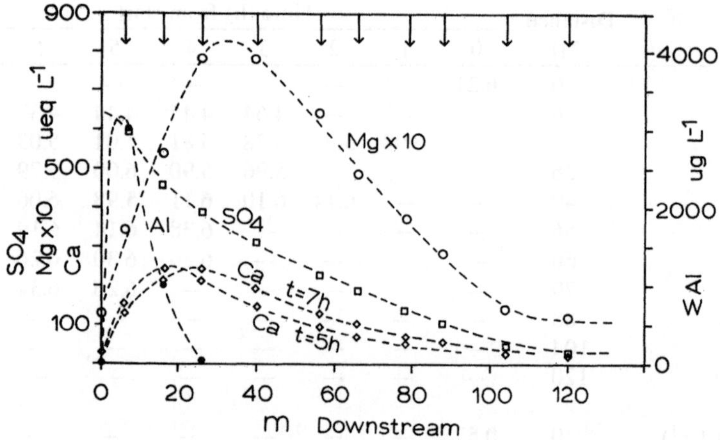

Figure 5–2. Chemistry of the artificially acidified stream at Sogndal, Norway, at selected distances downstream from the point of H_2SO_4 addition, 7 hours after the start of acidification. The 5-hour chemistry is shown also for Ca.

F. Discussion

In geologic terrains where chemical weathering is slow and alkalinity production is low, groundwater typically has a pH less than 5. The low pH is due to carbonic and organic weak acids. The effect of carbonic acid is shown in Figure 5–5. Solutions of "distilled water" change chemistry along the path 1–2–3 as they are subjected to

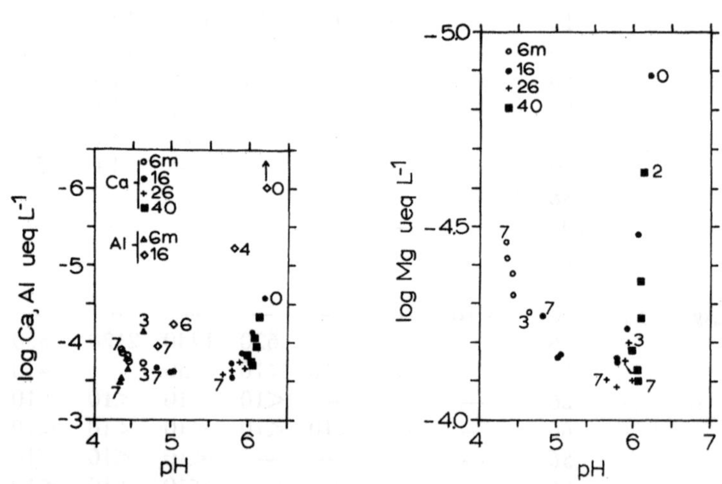

Figure 5–3. Ca, Al, and Mg concentration of the acidified stream at Sogndal, Norway, as a function of distance downstream from the point of acidic addition (shown in different symbols) and at different times. Small numbers indicate the start of sampling (in hours) and the end for each sampling site.

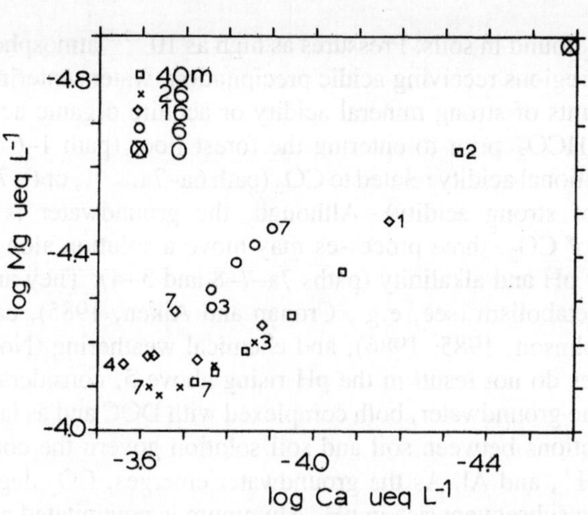

Figure 5–4. The relationship between Ca and Mg in stream water during acidification of the stream at Sogndal, Norway, as a function of distance downstream from the point of acidic addition (shown in different symbols) and at different times. Small numbers indicate the start (in hours) and the end, for each sampling site.

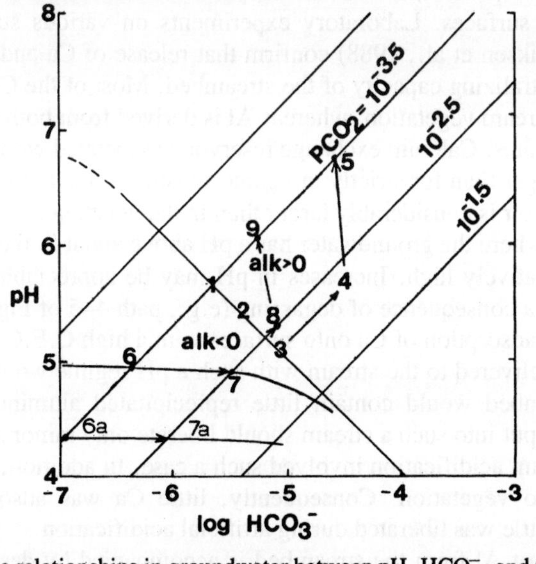

Figure 5–5. The relationships in groundwater between pH, HCO_3^-, and CO_2 modified by strong acidic content, chemical weathering, and cation exchange. (After Norton and Henriksen, 1983.)

the high PCO_2 found in soils. Pressures as high as $10^{-1.5}$ atmospheres can yield a pH of 4.6. In regions receiving acidic precipitation, waters entering the soil have variable amounts of strong mineral acidity or acquire organic acidity, reducing their pH and HCO_3^- prior to entering the forest floor (path 1–6–6a), thereafter acquiring additional acidity related to CO_2 (path 6a–7a, . . ., or 6–7, depending on the amount of strong acidity). Although, the groundwater is subject to an overpressure of CO_2, three processes may move a solution along a CO_2 isobar toward higher pH and alkalinity (paths 7a–7–8 and 3–4). They are loss of DOC because of metabolism (see, e.g., Cronan and Aiken, 1985), cation exchange (Reuss and Johnson, 1985, 1986), and chemical weathering (Norton, 1980). If these processes do not result in the pH rising above 5, considerable Al may be dissolved in the groundwater, both complexed with DOC and as labile Al. Cation exchange reactions between soil and soil solution govern the concentrations of base cation, H^+, and Al. As the groundwater emerges, CO_2 degasses from the solution with a subsequent rise in pH. Aluminum is precipitated and/or adsorbed on the streambed, and base cations are subject to cation exchange. The cation exchange capacity of inorganic stream sediments in glaciated terrains is quite low because of the lack of appreciable clay mineral material. Consequently, adsorption of Ca and Al by inorganic substrates is small. However, precipitation of $Al(OH)_3$ is apparently common. Dissolution of this material during stream acidification appears to involve material of increasingly greater insolubility with time, as shown by experiments in streams that do not have a history of recent acidic episodes. We interpret this to be caused by increasing crystallinity of the aluminum hydroxide being dissolved.

If vegetation is present, Al and Ca may be occluded within or adsorbed or precipitated on surfaces. Laboratory experiments on various stream substrate materials (Henriksen et al., 1988) confirm that release of Ca and Al constitutes most of the neutralizing capacity of the streambed. Most of the Ca is apparently released from stream vegetation, whereas Al is derived from both vegetation and inorganic substrates. Calcium exchange reservoirs associated with the vegetation appear to be larger than for strictly inorganic substrates; the amount of Al in the inorganic reservoir is considerably larger than in the vegetation.

For systems where the groundwater has a pH above about 5, the Ca concentration may be relatively high. Increases in pH may be appreciable for emerging groundwater as a consequence of degassing (e.g., path 4–5 of Figure 5–5). This would promote adsorption of Ca onto surfaces with a high C.E.C. However, the amount of Al delivered to the stream with such a pH regime would be minimal. Thus the streambed would contain little reprecipitated aluminum hydroxide. Strong acidic input into such a stream should liberate only minor amounts of Al. The Welsh stream acidification involved such a case. In addition, the streambed had virtually no vegetation. Consequently, little Ca was adsorbed onto the substrate, and little was liberated during artificial acidification.

Mobilization of Al from the streambed is accompanied by desorption of Ca. Above a pH of 5, neutralization of acidity is dominated by Ca. In all experiments discussed above, the capacity of the streams to neutralize acidity through release of

Ca was quite limited. The neutralization by Ca release diminished within a time frame of a few hours to days. Below pH 5, neutralization was dominated by the release of Al, and the reservoirs appear to be substantial compared to that for Ca. The amount of Al released per unit area of stream bottom exceeds what might be expected from cation desorption alone. Although the stoichiometry of desorption and dissolution are essentially the same with respect to pH, the neutralization by Al appears to be linked to the progressive dissolution of progressively less soluble (better crystallized?) $Al(OH)_3$.

For streams chronically below pH 5, less Al becomes precipitated on substrates because the degassing of groundwaters does not produce such a significant rise in pH (Reuss and Johnson, 1986). As water progresses downstream to higher-order streams, the pH typically increases because of thicker soils and longer residence time of groundwaters in soil prior to their emergence as surface flow. Consequently, there is a trend toward precipitation of Al and adsorption of Ca onto substrates. Thus an episode of acidification in a first-order stream may translocate Al and Ca to a second- or third-order stream. In this manner streams act as cation chromatographs, with the acid acting as the eluent. If the acidic loading is greater than the rate of alkalinity production in the mineral soils, chronic acidification in a stream reach may ultimately follow episodic acidification. The reach of chronically acidic stream then extends downstream with time under elevated acidic loading.

The time frame for the loss of neutralizing capability of stream substrates through dissolution of $Al(OH)_3$ and the desorption of Ca has not been established for any of the studied sites and would presumably be site specific. The longest experimental acidification was that conducted by Hall and others (1980). During their 6-month acidic addition to a third-order stream, the release of Al and Ca during low flow was equivalent to about 10 and 15 to 20 $\mu eq\ L^{-1}$, respectively. A very small fraction of the increase noted over the 120-m-long acidified reach may be attributable to Al and Ca in groundwater entering the stream between the point of acidic addition and the sampling point 120 m downstream. However, it seems clear that even with a pH depression to 4 the stream provides significant neutralization for a year or more. The pH of the acidified stream must have been below 3.9 (based on sulfate concentrations) just below the acidic addition, where no neutralization had yet occurred. Whereas natural pH depressions are initially less likely to be so severe for such prolonged periods, neutralization by such a stream should last for a few to tens of years. To some unknown extent, sediment with acquired $Al(OH)_3$ may be eroded from the stream and deposited within a lake, or fresh sediment may be added to a stream, reducing the effects described here. Stretches of stream now receiving groundwater bearing Al may have historically had influxes of groundwater with sufficiently high pH so that they received no Al. Incipient soil acidification in effect moves Al from the B horizon of podsolic soils into the streams where it can act as an ameliorator of episodic acidification. This change in capacity to release Al during episodes of acidification is shown schematically in Figure 5–6.

Given that stream substrates may damp pH depressions, they may also retard

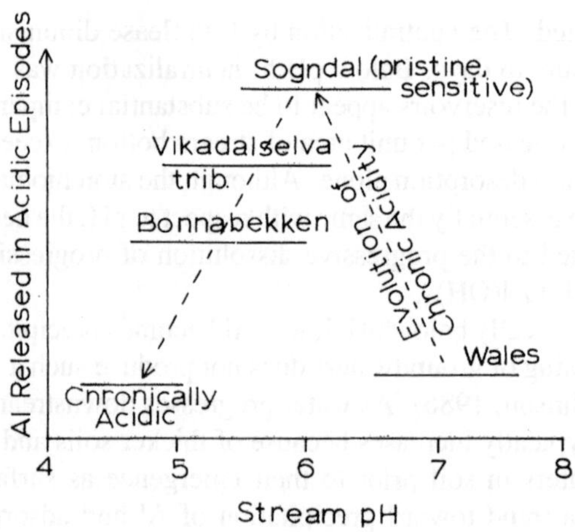

Figure 5–6. Schematic relationship between pH of stream systems and Al release during acidic episodes.

recovery from acidic pulses due to the regeneration of Ca and (to a lesser extent) Al to the cation exchangers. In the absense of vegetation in the streambed, acidic episodes should be more severe but also shorter lived.

III. Lake Sediments

It has long been well known that lake sediments are a significant source of alkalinity. This can be readily established by evaluation of interstitial water chemistry profiles (Figure 5–7), lakewater column chemistry profiles over time (Figure 5–8), or performing alkalinity budgets for lakes (e.g., Schindler, 1986). Alkalinity generation is dominated by release of Ca and Mg from the sediments, net sulfate and nitrate reduction in the sediments, and other reduction processes. On an areal basis, alkalinity generation at the sediment–water interface ranges in importance from much less than to greater than terrestrial alkalinity generation. The importance of alkalinity generation in the lake diminishes relative to that of the terrestrial watershed as the ratio of watershed area:lake area increases. Alkalinity generation m^{-2} by lake sediment is commonly comparable to that in the watershed.

Alkalinity generation by in-lake processes is a natural phenomenon. Of considerable interest is to what extent are strong acidic inputs directly to the lake from the atmosphere or indirectly from the watershed neutralized by *increased* alkalinity production in lakes. This situation is analagous to the *increased* terrestrial alkalinity production from increased chemical weathering or more likely increased base cation desorption from soils and stream sediments (see above). Sulfate adsorption by soils, biological storage of reduced S in any part of the

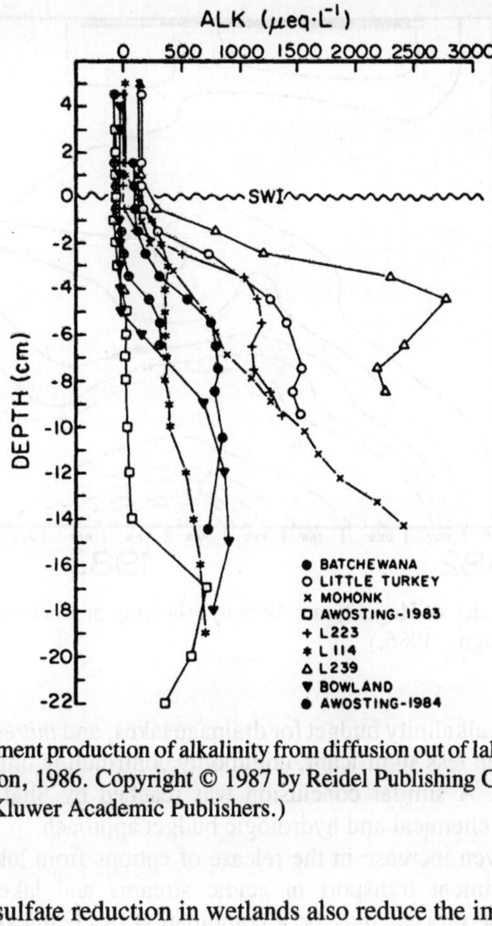

Figure 5–7. Sediment production of alkalinity from diffusion out of lake sediments. (After Schiff and Anderson, 1986. Copyright © 1987 by Reidel Publishing Company. Reprinted by permission of Kluwer Academic Publishers.)

ecosystem, and sulfate reduction in wetlands also reduce the impact of increased inputs of strong acids. These three processes have in-lake equivalents.

Increased alkalinity production occurs in lakes by increased (over prepollution values) sulfate reduction and storage of S in the sediment (or S can escape downstream as particulate material). The sediment storage component has been evaluated by S budgets for lakes (Shafran and Driscoll, 1987). Such estimates are somewhat uncertain because the flux of dry-deposited S to the system and groundwater fluxes are commonly not known or are estimated by difference. Alternatively, total S storage in sediments may be estimated by peeper studies such as those by Carignan and Tessier (1985). Such studies are site specific and typically short term. They indicate that S storage in sediments may be on the order of 0 to 15 μeq cm^{-2} yr^{-1}. For seepage lakes with long residence times, the upper part of this range would contribute most of the alkalinity observed in lakes receiving acidic precipitation. Paleolimnological studies are also site specific but offer a long-term evaluation of net S accumulation in sediments. Norton and others (1988) have evaluated sediment cores from 16 lakes with greatly differing hydrology and sulfate inputs. They conclude that sulfate reduction is relatively

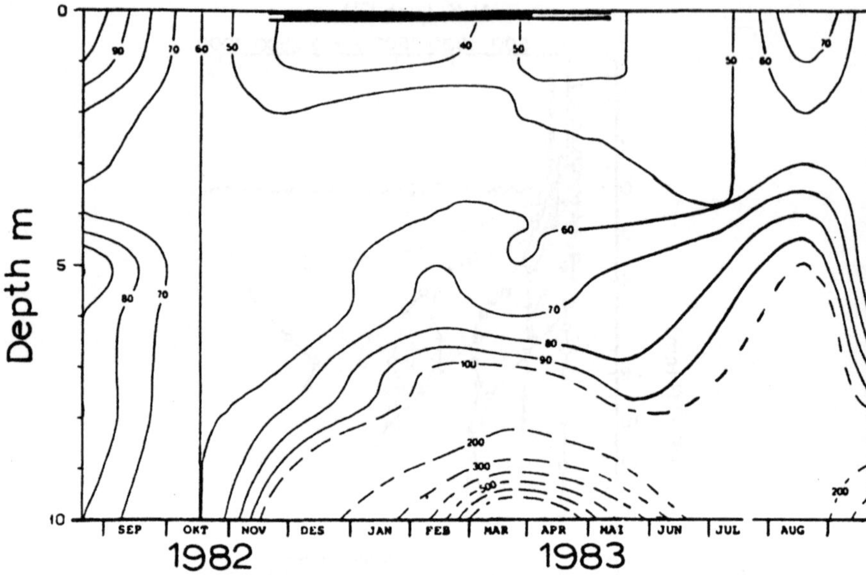

Figure 5–8. Chemistry of Hogelitjonn, Norway, showing production of alkalinity from sediments. (From Hegna, 1986.)

unimportant in the alkalinity budget for drainage lakes, and *increases* in the rate of production are even less significant, commonly contributing only a few μeq L^{-1} yr^{-1} (Table 5–3). A similar conclusion was reached by Shafran and Driscoll (1987) based on a chemical and hydrologic budget approach.

The acidity-driven increase in the release of cations from lake sediment may occur during sediment transport in acidic streams and lake water or after sedimentation. The process may be a combination of chemical weathering and cation exchange.

Experimental studies of these processes include batch studies, in situ microcosm studies, laboratory microcosm studies, and whole-lake acidification studies. Batch studies (Oliver and Kelso, 1983; Baker et al., 1985) have the advantage of enabling the execution of short-term experiments. They give qualitative information about processes but exaggerate both the rates and overall effectiveness of the processes. This is because of the atypically low pH used in experiments and the low ratio of sediment surface area to water volume. For example, Oliver and Kelso (1983) estimated that the ANC of the top 1 cm of sediment from a core had 1.3 times the ANC of the water column. However, this conclusion was based on the invalid assumption that sediment had a density of 2.55 (0% H_2O). Their studies indicate that neutralization by base cation release is dominated by Ca and secondarily by Mg. These and other batch studies are reviewed in Molot (1986).

In situ (Baker et al., 1985; Schindler et al., 1980) and laboratory microcosm studies, the latter commonly using intact sediment–water cores (Hongve, 1978; Molot, 1986; Kahl and Norton, 1983; Norton, 1983), have also evaluated the ANC of sediments. The experiments realistically depict water column:sediment column

Table 5-3. Production of alkalinity from SO_4 reduction and storage of S in sediments.

Lake (core state)	Background acc. rate of S (μmol/cm^2/y) [equivalent alkalinity (μeq/L) (drainage)]	Maximum acc. rate of S (μmol/cm^2/y) [equivalent alkalinity (μeq/L) (drainage)]	Increase in acc. rate of S adjusted for sedimentation (μmol/cm^2/y) [equivalent alkalinity (μeq/L) (drainage)]
Mud, C2 (ME)	0.32	0.63	0.20
	[0.2]	[0.4]	[0.1]
Mud (ME)	0.58	1.17	0.63
	[0.4]	[0.7]	[0.4]
Little Long, C2 (ME)	1.5	2.45	1.34
	[3.3]	[5.3]	[2.9]
Little Long (ME)	1.4	4.4	2.2
	[3.0]	[9.6]	[4.8]
Salmon (ME)	0.6	0.95	0.5
	[0.8]	[1.2]	[0.6]
Tilden (ME)	1.7	2.1	0.2
	[6.8]	[8.4]	[0.8]
Haystack, C2 (VT)	0.86	2.0	0
	[1.2]	[9.3]	[0]
Haystack, C3 (VT)	0.85	2.3	0
	[1.2]	[10.7]	[0]
Haystack, C5 (VT)	0.9	2.3	0.3
	[1.2]	[10.7]	[1.4]
Big Moose, C2 (NY)	1.2	5.2	3.9
	[1.2]	[5.4]	[4.0]
Deep, C2 (NY)	0.9	12.5	11.
	[2.1]	[29.]	[26.]
Queer, C1 (NY)	2.3	12.5	8.
	[6.1]	[32.9]	[21.]
U. Wallface (NY)	0.7	1.5	0.6
	[1.3]	[2.8]	[1.1]
Andrus (MI)	4.0	5.5	2.8
	[6.7]	[9.2]	[4.7]
McNearney, 2A (MI)	1.2	1.2	0
	[6.1]	[6.1]	[0]
Dunnigan (MN)	1.3	2.3	1.1
	[9.3]	[6.1]	[7.9]
Hustler (MN)	3.0	7.7	3.7
	[6.5]	[16.]	[7.6]
Black Joe (WY)	0.6	1.4	−0.3
	[0.4]	[0.1]	[0.0]
Deep (WY)	0.9	1.2	0.7
	[1.8]	[2.4]	[1.3]
Hobbs, C1 (WY)	0.6	0.8	0.2
	[0.3]	[0.4]	[0.1]
Hobbs, C2 (WY)	0.7	0.9	0.4
	[0.3]	[0.5]	[0.2]

areal relationships but yield site-specific conclusions. It is possible to carefully monitor chemistry of the water and maintain a constant pH. Experiments approaching 1 year in length (Oliver and Kelso, 1983; Kahl and Norton, 1983) with pH depressions to below 4 indicate that release of Ca and Mg dominates neutralization processes, although Al may be mobilized in significant amounts. Many studies indicate elevated release of Fe, Mn, Zn, and other trace elements at or below the sediment–water interface (e.g., Schindler et al., 1980). However, redox-sensitive elements are typically oxidized and reprecipitated in the oxygenated water column, whereas some other important alkalinity-producing reactions such as the release of NH_4 are reversed by biological uptake (Schiff and Anderson, 1986). Some processes such as SO_4 reduction are partly reversed on a seasonal basis (Rudd et al., 1986).

Kahl and Norton (1983) demonstrated that loss of base cations from core microcosms during artificial acidification was sufficiently great to alter measurably the chemistry of the remaining sediment. This was true for all elements for which the rate of release was large with respect to the size of the sediment reservoir (Ca, Mg, and Zn in their experiments). These finding were consistent with the suggestion by Hansen and others (1982) that declines in the concentration of these elements in recently deposited lake sediment could be interpreted in term of acidification of the water column.

Whole-lake acidification experiments, such as that of Schindler and others (1980) at the Experimental Lakes Area in western Ontario, give considerable insight into the various possible within-lake mechanisms of response to elevated strong acidic loading. These are summarized by Cook and others (1986). In Lake 223, over an 8-year period, they found that alkalinity generation was dominated by SO_4 reduction (85%), followed by release of Ca from the sediment (19%). Net production of all other studied components ranged from +5% to −8% of the total alkalinity budget for within-lake processes. Whereas it appears that production of alkalinity by SO_4 reduction in Lake 223 is unusually high for oligotrophic lakes (Norton et al., 1988), Ca release appears to be a likely candidate as an important contributor to alkalinity.

A. Paleolimnological Studies

We have evaluated increases in the alkalinity production, based on the sedimentary record of seven lakes. On the basis of water chemistry, all the lakes are sensitive to the effects of acidic precipitation. The PIRLA project (Charles and Whitehead, 1988) studied cores from these seven lakes located in Northern New England, the Adirondack Mountains of New York, the upper midwest Great Lakes states, and Florida in order to evaluate the rate and extent of recent acidification. Sediment cores were retrieved from the deep part of the lakes and sectioned according to methods described in Charles and Whitehead (1986). The cores were dated using ^{210}Pb (Eakins and Morrison, 1978) and the CRS model of Appleby and Oldfield (1978). Metal chemistry was determined using the methods of Buckley and Cranston (1971) in conjunction with flame and flameless atomic absorption spectroscopy and inductively coupled emission spectroscopy. The data obtained enable us to determine the flux of material to the sediment in μg cm^{-2} yr^{-1}. Our calculations of *increased* alkalinity production are based on the following arguments.

Prior to recent acidification of the lakes, the deposition of Ca (and Mn and other elements whose abundance is controlled largely by inorganic processes) was in a constant proportion to TiO_2, which is derived principally from geologic sources within the drainage basin. As the net sedimentary flux of TiO_2 varied through time, the flux of CaO generally varied proportionally. Thus the TiO_2:CaO ratio remains approximately constant.

After the commencement of lake acidification, empirical and experimental evidence (cited above) suggest that Ca desorbs from sedimenting or already sedimented material. As a consequence, the CaO:TiO_2 concentration ratio (and accumulation ratio) should decrease in the uppermost sediment. A typical example is shown in Figure 5–9. This phenomenon has been observed in the sediment record of many lakes where diatom pH reconstructions suggest recent acidification (Charles et al., 1988) and is the case for the lakes presented here.

Acidification leads to a deficit of CaO, relative to TiO_2, in the sediment accumulation record. We have calculated the deficit as follows:

1. (Accumulation rate for TiO_2 for a recent sediment interval) (Accumulation rate for CaO for background conditions/Accumulation rate for TiO_2 for background conditions) = Expected accumulation rate for CaO for the recent sediment level, assuming no mobilization due to acidification.
2. Expected accumulation rate for CaO − Measured accumulation rate = CaO deficit (μg cm^{-2} yr^{-1}).

Figure 5–10 shows the calculated deficit for CaO from lakes that are presently acidic (based on water chemistry; see Charles and Whitehead, 1988) or recently acidified based on diatom pH reconstructions (Davis et al., 1988; Charles et al., 1988; Kingston et al., 1988). Table 5–4 gives data relevant to these lakes. Background sediment characteristics were picked subjectively, such that the sediment chemistry was not unusual over the depth interval selected and that levels selected as background are deep enough so as not to be affected by acidification commencing as early as 1900.

The CaO deficits are a function of two factors: the gross sedimentation rate and the degree of acidification. Deficits range from virtually zero (Mud and McNearney) to about 100 μg cm^{-2} yr^{-1} (Little Long). On an equivalent basis, this range is from 0 to 4 μeq cm^{-2} yr^{-1}. The effect on water quality of this *increased* alkalinity generation may be computed knowing the watershed area to lake area (Table 5–4) and making two assumptions: a water yield of 1 m in all watersheds and the CaO deficit measured in the one core is representative of the entire area of the lake. Both assumptions are probably overestimates. The errors compensate.

One example is discussed for clarity. The CaO deficit for Big Moose Lake sediment dated 1960 is about 55 μg cm^{-2} yr^{-1}. This is equivalent to 2 μeq cm^{-2} yr^{-1} (Figure 5–10). For 1 m of precipitation (yield) falling directly on the lake, this

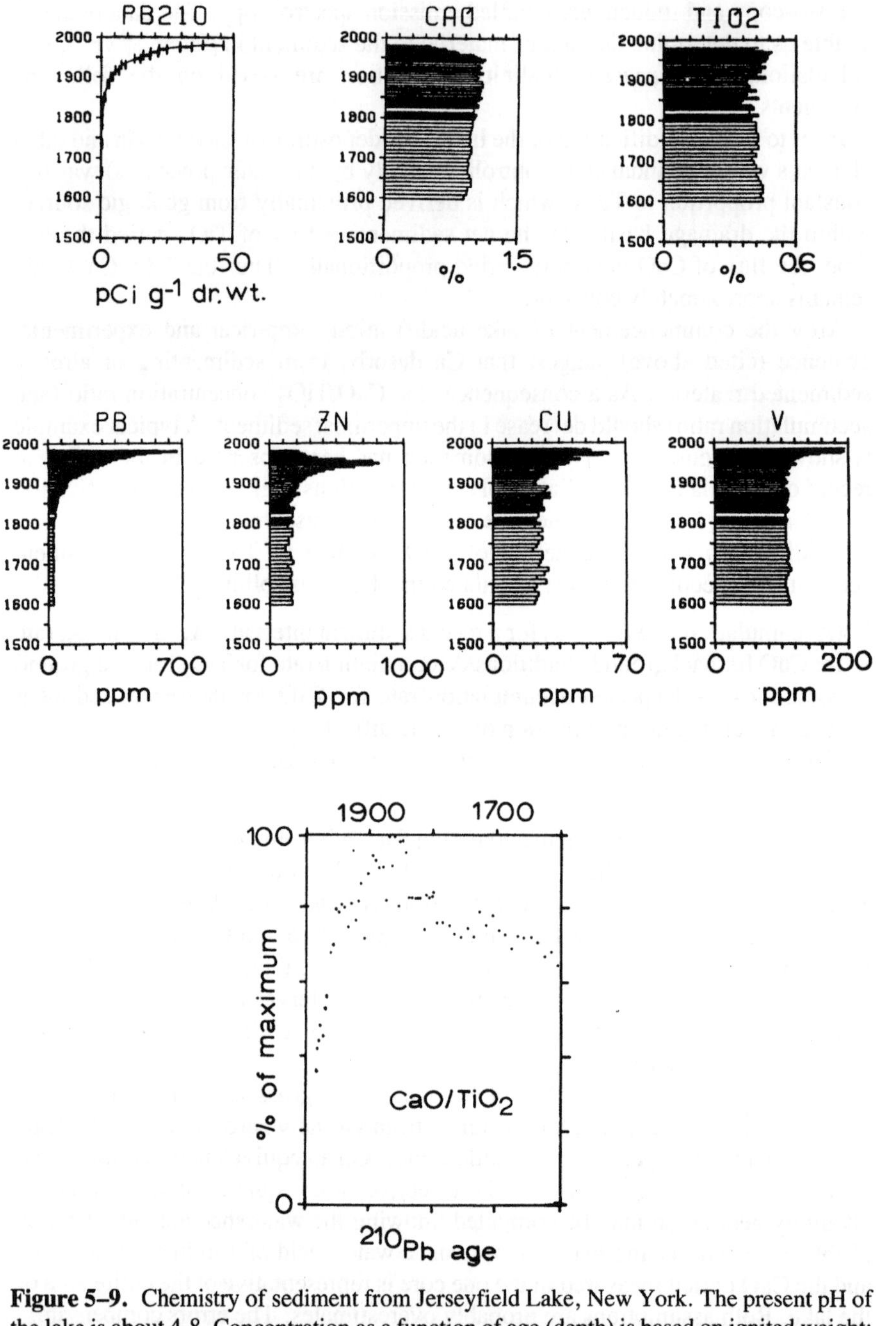

Figure 5–9. Chemistry of sediment from Jerseyfield Lake, New York. The present pH of the lake is about 4.8. Concentration as a function of age (depth) is based on ignited weight; ^{210}Pb profile is based on dried weight. Profiles for the pollution indicators Pb, Zn, Cu, and V are in ppm ignited weight. The $CaO:TiO_2$ ratio declines to 40% of the maximum value reached in the last 400 years. (From Norton, 1985.)

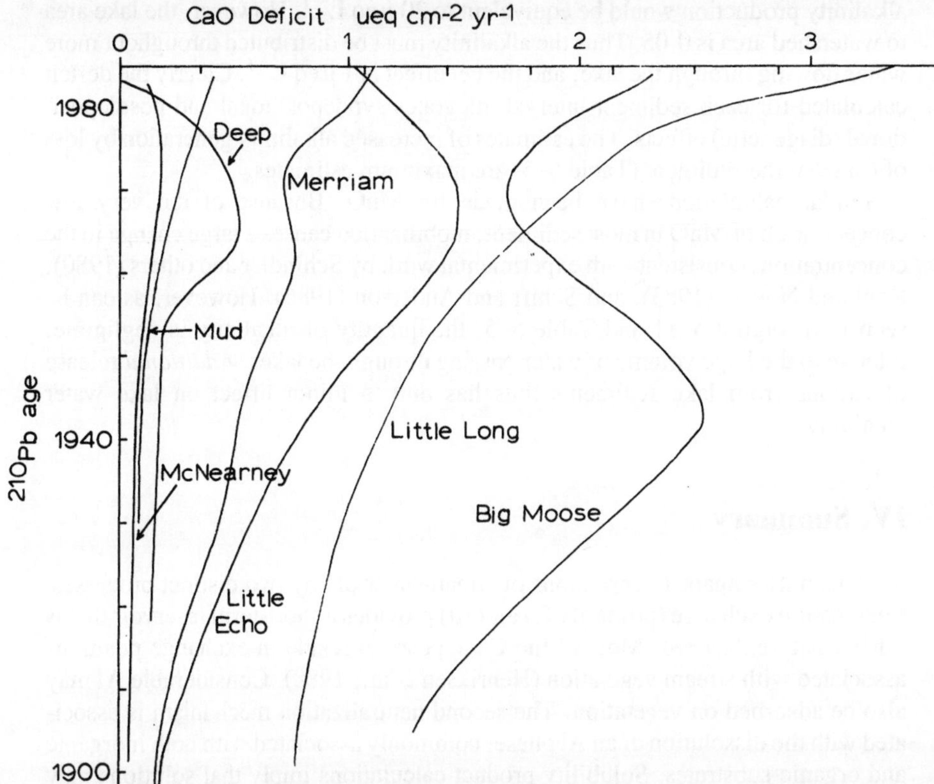

Figure 5–10. CaO deficit (μg cm^{-2} yr^{-1}; μeq cm^{-2} yr^{-1}) in ^{210}Pb-dated sediment from acidic or acidified lakes in Maine, New York, and Minnesota.

Table 5–4. Characteristics of acidic or acidified lakes for which sediment alkalinity generations calculations have been made.

Lake (core state)	Depth (m)	pH	SO_4^{-2} (μeq L^{-1})	Lake area Watershed area	Background data
Little Long (ME)	24.5	5.56	79	0.11	1823-1850
Mud (ME)	15.0	4.58	97	0.02	1700-1750
Big Moose (NY)	22.0	5.07	132	0.05	1825-1850
Deep (NY)	24.0	4.67	143	0.15	1825-1875
Little Echo (NY)	5.2	4.28	77	1.00	1750-1800
Merriam (NY)	5.2	4.88	89	0.13	1800-1850
McNearney (MN)	8.1	4.41	151	0.34	1750-1800

alkalinity production would be equivalent to 20 μeq L^{-1}. However, the lake area to watershed area is 0.05. Thus the alkalinity must be distributed throughout more water flowing through the lake, and the net effect is 1 μeq L^{-1}. Clearly the deficit calculated for each sediment interval integrates syndepositional and postdepositional (diagenetic) effects. The estimates of increased alkalinity generation by loss of Ca from the sediment (Table 5–5) are maximum estimates.

Similar calculations have been made for MnO. Because of the very low concentration of MnO in most sediment, mobilization causes a large change in the concentration, consistent with experimental work by Schindler and others (1980), Kahl and Norton (1983), and Schiff and Anderson (1986). However, as can be seen from Figure 5–11 and Table 5–5, the quantity of alkalinity is negligible, relative to the huge volume of water passing through the lakes. *Additional* release of cations from lake sediments thus has only a minor effect on lake water alkalinity.

IV. Summary

Streams buffer against depressions of streamwater pH by two distinct processes: First, cation exchange (primarily Ca and Al) provides a short-term reservoir that is fairly easily exhausted. Most of the Ca appears to reside in exchange positions associated with stream vegetation (Henriksen et al., 1988). Considerable Al may also be adsorbed on vegetation. The second neutralization mechanism is associated with the dissolution of an Al phase, commonly associated with both inorganic and organic substrates. Solubility product calculations imply that solutions may supersaturate with respect to well-crystallized Al phases such as gibbsite. The amount of Al in stream sediment appears to be large for streams that have a long history of acidic groundwater input but that have a pH greater than 5. We suggest that more alkaline and chronically acidic streams lack this neutralization mechanism. Increases in pH are also resisted by a recharging of the depleted exchange surfaces. Low-order streams are thus somewhat protected against episodes of

Table 5-5. Increased alkalinity generation from loss of Ca and Mn from sediments caused by acidification.

	From CaO (μeq L^{-1})		From MnO (μeq L^{-1})	
Lake (core state)	Surface sediment	Maximum (year)	Surface sediment	Maximum
Little Long (ME)	1.80	2.00 (1981)	0.10	0.10
Mud (ME)	0.04	0.04 (1983)	0.00	0.00
Big Moose (NY)	1.20	1.30 (1943)	0.02	0.05
Deep (NY)	0.03	0.50 (1962)	—	—
Little Echo (NY)	5.50	7.60 (1960)	0.20	0.30
Merriam (NY)	0.14	0.14 (1983)	—	—
McNearney (MN)	0.17	0.37 (1974)	—	—

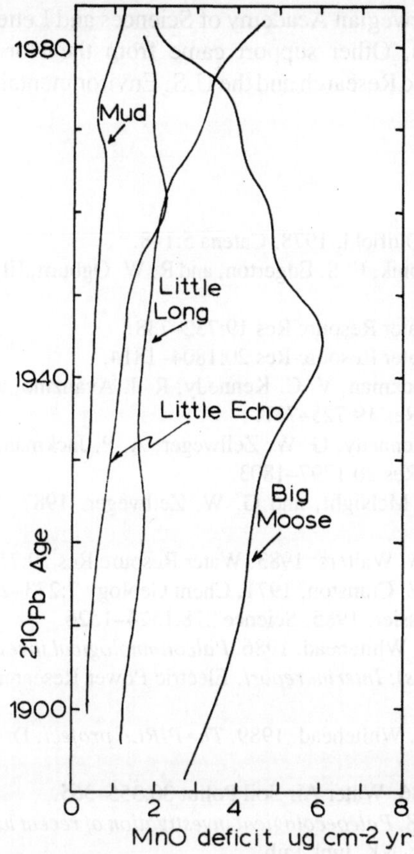

Figure 5-11. MnO deficit (μg cm^{-2} yr^{-1}; μeq cm^{-2} yr^{-1}) in ^{210}Pb-dated sediment from acidic or acidified lakes in Maine and New York.

acidification until the Al and Ca reservoirs are exhausted. At that time the stream may become chronically acidic.

Increases in the rate of production of alkalinity by lake sediments as a consequence of acidic precipitation are dominated by two processes: SO_4 reduction and storage within the sediment, and the release of cations, primarily Ca. The paleolimnological record shows that both of these processes increase in lakes that receive increased excess SO_4 and that undergo acidification. However, although alkalinity production from sediments may be significant in lakes, *increases* in production are important to water quality only in lakes with small drainage basins.

Acknowledgments

This work was supported in part by the Surface Water Acidification Project (SWAP) via the Royal Society of London, the Norwegian Ministry of the

Environment, The Norwegian Academy of Sciences and Letters, and the Swedish Academy of Sciences. Other support came from the Norwegian Council for Industrial and Scientific Research and the U.S. Environmental Protection Agency.

References

Appleby, P. G., and F. Oldfield. 1978. Catena 5:1–8.

Baker, L. A., P. L. Brezonik, E. S. Edgerton, and R. W. Ogburn, III. 1985. Water Air Soil Pollut 25:215–230.

Bencala, K. E. 1983. Water Resourc Res 19:732–738.

Bencala, K. E. 1984. Water Resourc Res 20:1804–1814.

Bencala, K. E., A. P. Jackman, V. C. Kennedy, R. J. Avanzino, and G. W. Zellweger. 1983. Water Resourc Res 19:725–731.

Bencala, K. E., V. C. Kennedy, G. W. Zellweger, A. P. Jackman, and R. J. Avanzino. 1984. Water Resourc Res 20:1797–1803.

Bencala, K. E., D. M. McNight, and G. W. Zellweger. 1987. Water Resources Res 23:827–836.

Bencala, K. E., and R. A. Walters. 1983. Water Resourc Res 19:718–724.

Buckley, D. E., and R. E. Cranston. 1971. Chem Geology 7:273–284.

Carignan, R., and A. Tessier. 1985. Science 228:1524–1526.

Charles, D. F., and D. R. Whitehead. 1986. *Paleolimnological investigation of recent lake acidification (protocols): Interim report*. Electric Power Research Institute, Palo Alto, CA.

Charles, D. F., and D. R. Whitehead. 1989. *The PIRLA project*. Dr. W. Junk, Dordrecht, The Netherlands.

Charles, D. F., et al. 1986. Water Air Soil Pollut 30:355–365.

Charles, D. F., et al. 1988. *Paleoecological investigation of recent lake acidification in the Adirondack Mountains, NY*. Junk Pub.

Cook, R. B., C. A. Kelly, D. W. Schindler, and M. A. Turner. 1986. Limnol Oceanogr 31:134–148.

Cronan, C. S., and G. R. Aiken. 1985. Geoch Cosmoch Acta 49:1697–1705.

Cronan, C. S., and C. L. Schofield. 1978. Science 204:304–306.

Davis, R. B., et al. 1988. *Paleoecological investigations of recent lake acidification in northern New England*. Junk Pub.

Eakins, J. D., and R. T. Morrison. 1978. Intern Appl Rad Isot 29:531–536.

Hall, R. J., C. T. Driscoll, and G. E. Likens. 1987. Freshwater Biology 18:17–43.

Hall, R. J., G. E. Likens, S. B. Fiance, and G. R. Hendrey. 1980. Ecology 61:976–989.

Hanson, D. W., S. A. Norton, and J. S. Williams. 1982. Water Air Soil Pollut 18:227–239.

Hegna, K. 1986. Unpublished cand. scient. thesis, Department of Biology, University Oslo, Norway. 151 p. and appendices.

Henriksen, A. 1982. *Acid Rain Research Rep*. Norway Institute of Water Research, Oslo. 50 p.

Henriksen, A., O. K. Skogheim, and B. O. Rosseland. 1984. Vatten 40:255–260.

Henriksen, A., B. M. Wathne, E. J. S. Rogeberg, S. A. Norton, and D. F. Brakke. 1988. Water Res 22:1069–1073.

Hongve, D. 1978. Verh Internat Verein Limnol 20:743–748.

Kahl, J. S., and S. A. Norton. 1983. *Impact and mobilization of metals in acid-stressed lake watersheds*. Land and Water Resources Center, University of Maine, Orono, ME. 70 p.

Kennedy, V. C. 1965. U.S. Geol Survey Prof Paper 433-D, 28 p.
Kingston, J. L., 1988. *Paleoecological investigation of recent lake acidification in the northern Great Lakes states*. Junk Pub, Dordrecht, The Netherlands.
Molot, L. A. 1986. Water Air Soil Pollut 27:297–304.
Norton, S. A. 1980. *In* D. S. Shriner, ed. *Potential environmental and health consequences of atmospheric sulfur deposition,* 521–531. Ann Arbor Science, Ann Arbor, MI.
Norton, S. A. 1983. *The chemical role of lake sediments during lake acidification and de-acidification*. Proc 11th Nordic Symp on Sed. Universitet i Oslo, 7–19.
Norton, S. A. 1985. *In* D. D. Adams and W. Page, eds. Plenum, New York. 95–107.
Norton, S. A., and A. Henriksen. 1983. Vatten 39:346–354.
Norton, S. A., A. Henriksen, B. M. Wathne, and A. Veidel. 1987. *In* Proc *Acidification and water pathways,* 1:249–258. UNESCO, Oslo.
Norton, S. A., A. Henriksen, R. F. Wright, and D. F. Brakke. 1987. *In* Ext Abs of Intern Workshop Geochem and Monit in Rep Basins. Geol Survey, Prague.
Norton, S. A., J. S. Kahl, R. F. Wright, and R. B. Davis. 1985. Ext. Abs. Am Chem Soc Am Ann Mtg, Chicago.
Norton, S. A., M. J. Mitchell, J. S. Kahl, and C. F. Brewer. 1988. Water Air Soil Pollut 39:33–45.
Norton, S. A., R. F. Wright, T. Frogner, and E. Rogeberg. (in preparation.) Neutralization of acid additions to a small stream in Norway by stream substrates.
Oliver, B. G., and J. R. M. Kelso. 1983. Water Air Soil Pollut. 20:379–389.
Ormerod, S. J., P. Boole, C. P. McCahon, N. S. Weatherley, D. Pascoe, and R. W. Edwards. 1987. Freshwater Biology 17:341–356.
Reuss, J. O., and D. W. Johnson. 1985. J Environ Qual 14:26–31.
Reuss, J. O., and D. W. Johnson. 1986. *Acid deposition and the acidification of soils and water*. Springer-Verlag, New York.
Rudd, J. W. M., et al. 1986. Limnol Oceanogr 31:1267–1280.
Schafran, G. C., and C. T. Driscoll. 1987. Environ Sci Technol 21:988–993.
Schiff, S. L., and R. F. Anderson. 1986. Water Air Soil Pollut 31:941–948.
Schindler, D. W. 1986. Water Air Soil Pollut 30:931–944.
Schindler, D. W., et al. 1980. Can J Fish Aquatic Sci 37:342–354.
Skartveit, A. 1981. Nordic Hydrology 12:65–80.
Skeffington, R. A. 1986. Water Air Soil Pollut 31:891–900.
Wright, R. F. 1987. Acid Rain Research Rept 13/1987. Norway Institute of Water Research, Oslo. 90 p.
Wright, R. F., S. A. Norton, D. F. Brakke, and T. Frogner. 1988. Nature 334:422–424.

Mitigation of Acidic Conditions in Lakes and Streams

D.B. Porcella,* C.L. Schofield,† J.V. DePinto,‡ C.T. Driscoll,§ P.A. Bukaveckas,|| S.P. Gloss,# and T.C. Young‡

Abstract

The use of calcitic limestone to decrease water acidity and increase divalent cations for maintenance of conditions conducive to desirable aquatic organisms is practiced on a wide scale. Experience gained from operational and research programs has shown no significant deleterious effects of treatment with fine powders of calcitic limestone. This material has many advantages; for example, appropriate particle size of limestone can be selected to design site-specific cost-effective treatment of streams and lakes. Two major uncertainties still exist: the effect of episodic acidification on fish and invertebrates and the levels of fish production attainable with long-term liming.

I. Overview of Mitigation Efforts

Low-pH streams and lakes occur in regions where mineral weathering and release of basic cations (Ca^{++}, Mg^{++}, Na^{+}, K^{+}) through exchange or the retention of strong acidic anions ($SO_4^{=}$, NO_3^{-}) by adsorption and biologic reduction processes are insufficient to neutralize acidic inputs. Sources of acidic inputs include anthropogenic activities, such as fossil fuel burning and acidic mine drainage, and natural processes that produce organic acids NO_x, and SO_x. Mitigation strategies for surface water quality management are directed at decreasing the acidic supply

*Ecological Studies Program, Electric Power Research Institute, Palo Alto, CA, USA.

†Department of Natural Resources, Cornell University, Ithaca, NY, USA.

‡Department of Civil and Environmental Engineering, Clarkson University, Potsdam, NY, USA.

§Department of Civil Engineering, Syracuse University, Syracuse, NY, USA.

||Department of Biology, Indiana University, Bloomington, IN, USA.

#Wyoming Water Resources Center, University of Wyoming, Laramie, WY, USA.

from these activities and/or changing water quality by neutralization of acid or increased supply of basic cations. The primary objective of these strategies is the maintenance of water quality suitable for the support of fish populations. Several types of treatment can consume acid or increase basic cation supply: nutrient additions (e.g., Dillon et al., 1979), changed land use practices (e.g., Ormerod and Edwards, 1985), and direct addition of base (e.g., Hultberg and Anderson, 1982; Driscoll et al., 1982).

The most common treatment strategy for acidic surface waters is *liming,* a term referring to base addition. Water quality managers are particularly interested in liming, and it is an appropriate time to summarize information on proven liming methods, models, and the experience gained from studies in lakes and streams. At present, water quality managers have two major questions about liming surface waters: What are the cost-effective methods of liming different habitats? What are the ecological effects of liming?

Base treatment neutralizes acidity in surface water and increases the supply of basic cations, thereby improving water quality and fostering the presence of a broader array of organisms than found in acidic waters. Surface waters have been treated with many different base materials including furnace slag (Hultberg and Anderssen, 1982), hydrated lime [$Ca(OH)_2$] (Blake, 1981), and soda ash (Na_2CO_3) (Lindmark, 1982, 1984). Limestone is the preferred material because (1) it neutralizes acidity without causing excessively high or rapidly changing pH; (2) it increases acid-neutralizing capacity (ANC) with high buffer capacity at typical ambient pH; (3) it increases the amount of divalent cations, which are physiologically important to fish and may competitively inhibit the uptake of toxic metal ions; (4) it provides flexibility in treatment design owing to its pH-solubility relationship; and (5) it is inexpensive, readily available, and usually contains few toxic contaminants.

The common limestone minerals are calcite ($CaCO_3$) and dolomite [$CaMg(CO_3)_2$]. Pure forms are not common, and most limestones vary significantly in content of $CaCO_3$ and $MgCO_3$, as well as other constituents. The presence of other elements, such as phosphorus, sulfur, aluminum, iron, and trace metals, is often associated with silicate content (Barber, 1984), owing to clays in the limestone. Elemental analyses of limestones should always be performed to avoid limestone with excessive contamination.

Calcite is the most commonly used material in treating surface waters, largely because its rate of dissolution is about twice that of dolomite, and its efficiency of neutralization is greater, despite dolomite's greater potential neutralizing capacity. On a weight basis, pure dolomite equals 108% of the capacity of calcite because magnesium is lighter than calcium, but the ultimate potential of dolomite is never reached under typical liming conditions (e.g., pH 4.5 to 6). Because the common, reactive liming mineral for aquatic ecosystems is calcite, most of the remaining discussion focuses on calcite.

Liming is an established practice, with its beginnings two millennia ago during the Roman era (Barber, 1984). In the United States, soil liming with agricultural grades of limestone is extensive; for example, soil liming increased tenfold between 1935 and 1950, reaching 30 million metric tons per year (Barber, 1984). As in the United Kingdom (Ormerod and Edwards, 1985), soil liming in the United States has varied quantitatively with government subsidies and other incentives, ranging between 20 and 35 million metric tons per year during the 30 years after 1950 (Barber, 1984).

Treatment of surface waters for control of acidity began in Norwegian fish hatcheries in the 1920s (Rosseland and Skogheim, 1984). Recently, Bengtsson and others (1980) and Lessmark and Thornelof (1986) described the extensive liming program of surface waters in Sweden where more than 3,000 lakes and 100 streams had been treated by 1985. Operational liming has been conducted by several state agencies (Blake, 1981; Kretser and Colquhuon, 1984; Britt et al., 1984) and, more recently, by a U.S. privately funded corporation (Brocksen et al., 1987), and the Norwegian government (Johannessen and Hindar, 1987). Research has been conducted in Canada (Scheider and Brydges, 1984; Booth et al., 1986), the United Kingdom (Brown et al., 1987; Underwood et al., 1987), Finland (Björkqvist and Weppling, 1987), the United States (Schreiber and Rago, 1984; Porcella, 1987), and Sweden (Hultberg and Grennfelt, 1986). These projects provide results important to water quality managers and contribute to understanding ecosystem structure and function by providing an opportunity for studying interactions among chemical constituents and organisms following an environmental perturbation.

II. Conceptual Basis for Liming Surface Waters

When powdered limestone is added to an aquatic system (see Figure 6–1), the fine particles rapidly dissolve in the water column, moderate-sized particles remain suspended where they may subsequently dissolve or wash out of the system, and the coarsest particles settle to the bottom. The dissolved calcite can consume acidity directly or it may remain as cations and ANC to be later washed out, or react with hydrogen ion on sediment exchange sites. A mass balance on the limestone added to lake surfaces, performed within a few days of treatment, typically shows that some water column particles and ions are lost in the outflow. Generally this is negligible, depending on hydrologic conditions during treatment, but this can lead to an effective way of treating downstream water bodies, including other lakes. Laboratory measurements have shown that a certain fraction of suspended and settled particles may become inactivated by coating with humic complexes and metal precipitates (Sverdrup et al., 1984). Inactivation reduces the dissolution rate to near zero.

Major factors affecting limestone dissolution are $CaCO_3$ equivalence (reactive base as determined by acid titration), particle size, and water chemistry (pH, pCO_2, Ca^{++}). The content of other constituents affects equivalence; moisture content can reach 10% by weight, and the presence of clays and other inclusions may reduce the reactivity per unit weight of material (Barber, 1984).

Particle size is the major factor affecting settling velocity, which controls the

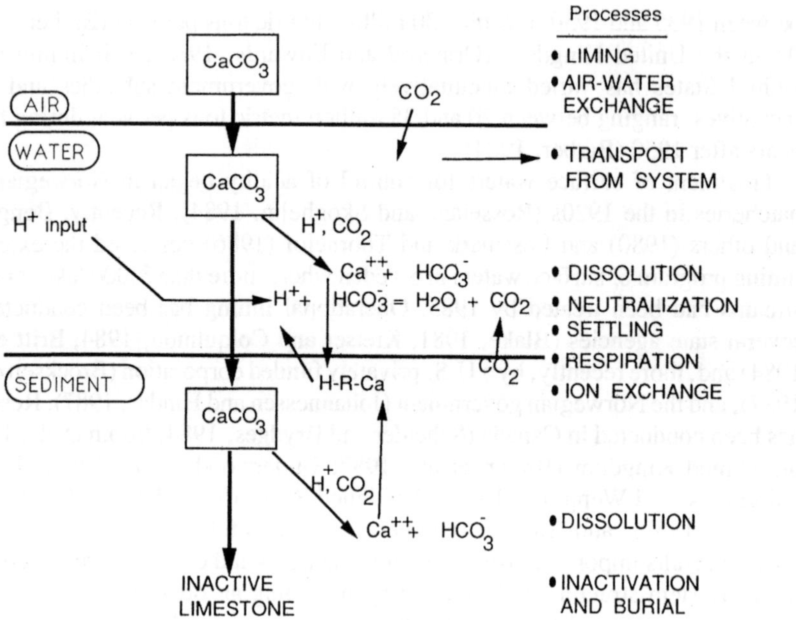

Figure 6–1. Simplified concept for calcite reactions in an acidic surface water system.

vertical distribution of particles in lakes and streams according to Stokes law (e.g., Sverdrup et al., 1984), but vertical dispersion and thermal stratification can cause significant deviation from Stokes law settling, especially for particles less than 5 μm diameter (DePinto et al., 1987). In lakes, use of limestone preparations dominated by large particles can result in the deposition of a large fraction of the application onto the bottom sediments, with proportionally less neutralization of the water column.

Surface area, a function of particle size, plays a role in controlling the rate of dissolution. For a given water chemistry, the rate of dissolution in field applications primarily depends on the limestone composition and its particle size distribution. Purer calcite will dissolve faster than that containing dolomite or significant other impurities. Generally, calcite is added to lakes or streams as finely ground particles, usually less than 0.2 mm in diameter. The small particle size permits rapid dissolution by providing additional surface area per unit weight and at the same time reduces the settling velocity so that particles remain suspended longer, giving more time for dissolution. The method of application can affect the dissolution efficiency; for example, uneven distribution of the chemical across the water body can lead to clumping, which inhibits dissolution and increases the velocity of settling. Mixing the dry powder with water to form a slurry can speed the dissolution of calcite as well as minimize clumping and maximize uniform distribution (Scheffe et al., 1986b; Booth et al., 1986; Warfvinge et al., 1984). Dispersant chemicals can further reduce aggregation of calcite particles in lakes (Scheffe et al., 1986b; Molot et al., 1986) and streams (Abrahamsen and Matzow, 1984).

Water chemistry, particularly the constituents pH, ANC, calcium, and aqueous CO_2 (Barton and Vatanatam, 1976; Plummer et al., 1978, 1979), affects the rate of dissolution. The rapid reaction of calcite in an acidic water is with hydrogen ion:

$$CaCO_3^{\ast} + H^+ = Ca^{++} + HCO_3^- \tag{1}$$

As pH increases, the hydrogen ion is consumed, and reaction with aqueous carbon dioxide occurs:

$$CaCO_3 + H_2CO_3^* = Ca^{++} + 2HCO_3^- \tag{2}$$

where $H_2CO_3^* = (CO_2)aq + H_2CO_3^o$.

When carbon dioxide is consumed, hydrolysis becomes the sole dissolution reaction:

$$CaCO_3 + H_2O = Ca^{++} + HCO_3^- + OH^- \tag{3}$$

Plummer and others (1978) define the net rate of dissolution (R) as:

$$R = k_1(H^+) + k_2(H_2CO_3) + k_3(H_2O) - k_4(Ca^{++})(CO_3^{--}) \tag{4}$$

where k_1, k_2, k_3 are first-order rate constants dependent on temperature and k_4 is dependent on temperature and pCO_2. Parentheses denote activity; most waters subject to acidification are dilute, and usually molarity can be assumed equal to activity.

The first reaction dominates at low pH because calcite reacts rapidly with hydrogen ion, the rate varying with the particle size of the calcite. The boundary layer around each particle quickly causes the reaction to become diffusion limited in quiescent lake waters (Sverdrup, 1986b). Turbulence in streams decreases the boundary layer and speeds up the dissolution reaction. The practical significance of hydrogen ion in lakes is probably minimal because at pH 4.5 about 3 mg $CaCO_3$/L will consume the hydrogen ion, and at pH 5.0 only 1 mg/L is required, resulting in an increase in ANC of only 40 and 6 μeq/L, respectively. The second reaction (Equation 2) then dominates until the $H_2CO_3^*$ is consumed. The third reaction (Equation 3) is zero order ($H_2O = 1.0$), occurs very slowly, and is essentially constant in practical applications. Thus, the dissolution of calcite can be written as (based on Sverdrup and Bjerle, 1982; Scheffe et al., 1986b):

$$-dM/dt = D/h\,(H^+ - HSAT) + k_2\,(H_2CO_3) + k_3 - k_4\,(Ca^{++})\,(HCO_3^-) \tag{5}$$

where M is the mass of undissolved calcite, D is the diffusion coefficient for H^+, h is the boundary layer thickness, HSAT is the surface H^+ on the calcite particles, and the other terms are as defined for Equation 4.

In streams there is a continuous mixing of calcite with low-pH water, so reaction 1 dominates. In most lakes reaction 2 is the most important; because the available aqueous carbon dioxide is consumed fairly rapidly, the rate of dissolution ultimately depends on the rate of aqueous carbon dioxide input. Sources of carbon dioxide include atmospheric influx, neutralization, groundwater influx, and respiration. Quantitatively, atmospheric influx of CO_2 affects the dissolution rate in the water column of lakes substantially, especially with fine particles with low effective settling rates. Young and others (1989) estimated the mass transfer coefficient for CO_2 at 2 to 3 m/day for two limed lakes in the Adirondacks. Correcting for chemical reaction gave enhancement factors of 3 to 6 and a "true" mass transfer coefficient of 0.55 m/day, which agrees with results of other gas-transfer studies in lakes.

In both streams and lakes, the amount of calcite added becomes important at greater dosages, due to feedback inhibition as would be predicted from Equation 5. Overdosing occurs in the top layers of lakes because the whole water column dose is added at a single time to the surface. Consequently, high pHs of more than 9.3 have been observed due to consumption of aqueous H^+ and CO_2 (Driscoll et al., 1987; Fordham and Driscoll, 1989; Young et al., 1989).

Sverdrup and others (1984) have indicated that calcite is inactivated by clay, metals, and humic acids. This reaction is especially important in streams draining boggy areas and in lakes fed by these streams. Laboratory experiments indicate that calcite particles that settle to the bottom in extreme "brown-water" lakes can be inactivated within 15 to 30 hours (Sverdrup et al., 1984). Consequently, Sverdrup and Bjerle (1983) estimated that the dissolution rate of undissolved limestone in the bottom sediments was close to zero after 2 years. However, Hongve and Abrahamsen (1984) indicated that in highly colored lake waters, fine-grained calcite dissolution is not inhibited, but after dissolution the cations exchange with sediment H^+; thus, a calcium mass balance would not account for dissolution. Fine particles dissolve rapidly, and exchange of Ca^{++} with H^+ can satisfy the base-neutralizing capacity of sediments (Molot et al., 1986; Young et al., 1989). The exchange reaction will continue for several years if the water column concentrations remain elevated (Boyd and Cuenco, 1980); sediment bulk chemistry responds rapidly to changing water column chemistry, with significant changes occurring within days to weeks (Young et al., 1989). Similar results were observed in the treatment of sediments with soda ash (Na_2CO_3) (Lindmark, 1982, 1984; Warfvinge et al., 1983), indicating the general nature of the sediment response to neutralization chemicals.

III. Treatment Criteria and Methods

The successful mitigation of acidic lake and stream waters with limestone involves (1) creating appropriate water quality throughout the water column and (2) providing a continuous supply of base cations to treat the water column effectively for a useful duration. Water quality must meet conditions safe for desirable aquatic communities, whereas the duration of treatment will depend (1) on the water residence time, which controls the washout rate (dilution) of ANC and calcium, and (2) on the consumption of ANC by the acidity in the inflowing waters.

The important water quality indicators of ecosystem quality include pH, ANC, calcium, and aluminum. These constituents interact to affect fish survival (Brown,

1981, 1983; Ingersoll et al., 1985) and reflect important components of pH buffering and calcite dissolution. Wright and Snekvik (1978) in a survey of 700 lakes in southern Norway showed that Ca^{++} and pH were the best correlates of the presence or absence of fisheries. Driscoll and others (1989a) showed that changes in Ca^{++} were stoichiometrically equivalent to ANC in limed lakes, accounting for ANC by covariance. Also, Schofield and others (1986) and Shortelle and Colburn (1987) suggest that Ca^{++} maintenance is more important than ANC for fish health. Similarly, Skogheim and others (1986, 1987) compared physiological responses of Atlantic salmon smolts to calcite and soda ash (Na_2CO_3) treatments and concluded that calcium protected the fish from aluminum toxicity but sodium did not.

Typically, fisheries managers want key variables to exceed the following concentrations: pH > 6, ANC > 50 μeq/L, and calcium > 100 μeq/L (2 mg/L) (also, see Brocksen et al., 1987, and Lessmark and Thornelof, 1986). When concentrations are substantially less, fish populations and other organisms may be affected directly, and inorganic aluminum may increase. With pH and Ca^{++} as the indicators most protective of fish, one can illustrate this concept graphically, defining safe conditions for fish where retreatment is unnecessary (Figure 6–2). The reacidification time is the interval between liming and when the concentration of calcium and pH decreases below safe conditions. In a practical sense, if no reaction with the dissolved calcite takes place, that is, inflow ANC and sediment exchange equal zero, and the lakes are completely mixed, three residence times will theoretically reduce by 95% the ANC measured immediately after liming. Depending on the initial dose, this may be adequate to maintain the biological

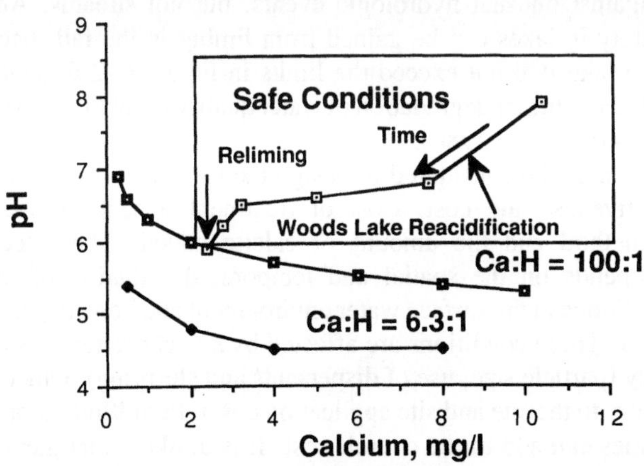

Figure 6–2. Guidelines for when to relime lakes. Woods Lake measured ratios are shown to illustrate how the ratio changes with time (arrow shows direction). The box outlines the safe conditions. The curves for calcium to hydrogen ionic ratios (μeq/L) suggest safe (ratio = 6.3) and conservatively safe (ratio = 100) levels for salmonid life stages. Woods Lake was relimed during the conservatively safe conditions.

integrity of the lake. However, reactions do take place as fresh acidic inputs are neutralized, incomplete mixing washes out stratified surface waters, and sediments and water exchange ANC. Lessmark and Thornelof (1986) show that at short residence times (less than 1 year), reacidification time is only two to three multiples of the residence time, whereas at longer residence time the multiple decreases to about one. The reacidification of Woods Lake in New York is shown in Figure 6–2, illustrating the process of washout and acid-base reactions in this Adirondack lake; the duration of treatment was 2.5 times the average residence time before retreatment became necessary (DePinto et al., 1989).

In streams the duration of treatment is more straightforward. Because stream flow is mixed vertically but not horizontally (plug flow), the duration of treatment or length of stream treated, depends on four factors: the dose, the dissolution efficiency, flow rate, and the ANC consuming reactions. Dilution of ANC by downstream tributaries and runoff inflows will lessen the length of stream treatment (see Rogers et al., 1986, for the DEACID model, which incorporates these additional acidic inputs).

Two lines for the ratio of calcium to hydrogen ion (μeq/L basis) are shown in Figure 6–2. The line with a ratio of 6.3 was based on laboratory studies of brown trout at concentrations showing 100% survival (Brown, 1987). Furthermore, a ratio of 4 would separate Wright and Snekvik's (1978) 700 Norwegian lakes into two groups: The lakes generally tended to be fishless if less than 4 and contained fish when greater than 4 (Brown, 1987). One would expect a similar ratio to apply to streams, although that work has not been performed. The line for a ratio of 100 shows the safe condition to have a large margin of safety. Such protection will likely be sufficient for most reproducing fisheries, accounting for both put-take and put-grow-take fisheries. Using the relationship in Figure 6–2 should protect most lakes against unusual hydrologic events, but not streams. An additional margin of safety in lakes can be gained from liming in the fall, because if one predicts that a lake will not exceed the limits in Figure 6–2 through the spring runoff period, then further degradation of water quality is not likely to occur during the summer (Booty et al., 1988).

Two factors must be considered to design the treatment of a water body with calcite: effectiveness and cost. Cost of treatment depends primarily on the application method and the amount of calcite added. The effectiveness of mitigation depends on the spatial and temporal distribution of the resulting chemical conditions in the surface water environment and how long the conditions are maintained. These conditions are affected by site characteristics and dissolution efficiency (particle size, use of dispersants and slurrying). Other cost factors include delivery to the site and site application costs. Permitting, monitoring, and related activities can add to the overall cost. It is unlikely that these latter costs would be the same magnitude as material and application costs in an operational liming program. There are examples of regional cost estimates for lakes (Menz and Driscoll, 1983), and cost optimization is included in some computer models (Sverdrup et al., 1984). However, most costs are estimated for individual sites, taking into account the different components in the labor, material, and equipment costs. When comparing costs for alternatives for lakes and streams, treatment should be calculated as a rate. Cost comparisons should be based on the rate, for example, metric tons per year for a particular lake or stream.

The design for applying the limestone will be least costly when the duration and extent of treated area are maximized for a given application. Effectiveness is maximized when the water quality necessary for an ecologically desirable community (Figure 6–2) is maintained throughout the ecosystem for the desired treatment duration. Balancing cost and effectiveness is interactive, but more cost-effective techniques can have the most benefit, ecologically. For example, stream bottom invertebrates near the treatment site can be smothered if poor dissolution results, due to overdosing or from treatment with powder that is too coarse. Generally, limestone particles on the bottom of streams do not effectively buffer acidic pulses or high flows (Pearson and McDonnell, 1978; Sverdrup, 1986a). Thus, poor dissolution efficiency can be both costly and ineffective.

The particle size of the calcitic limestone is the most important variable affecting the treatment design. Proper selection of powder sizes can allow a broad range of treatment alternatives as well as minimize dissolution problems. For accurate design of calcite treatment, particle size of the calcite must be known. Sverdrup (1986b) states that particle size must be directly measured with standard techniques, and rates of reaction must be determined from laboratory or field measurements, such as Scheffe and others (1986b). Moreover, he cautions that dry sieve techniques for determining particle size are inaccurate, and wet techniques should be used (Sverdrup, 1983). Surfaces of limestone particles are not regular and smooth; thus, surface area cannot be calculated accurately from geometric shapes and particle sizes directly. An empirical coefficient to correct for surface irregularities can be calculated from laboratory measurements.

Factors that modify how much limestone to add, where and when the treatment takes place, the properties of the limestone, and how to add it are listed in Table 6–1. The microcomputer program DEACID (based on Equations 1 through 5) has been used to assist in treatment design by rapid calculations of the dose, based on information provided by the user regarding site chemistry, mitigation method, and neutralizing agent (Rogers et al., 1986; Schreiber and Britt, 1987). This program is useful for comparing alternatives, including the effects of the factors listed in Table 6–1.

As indicated in Figure 6–3, treatment approaches are substantially different for lakes and streams. The treatment of lakes results in high initial concentrations of ANC and calcium; these gradually decrease due to washout and neutralization. In streams, flow-based lime dosages usually are designed to treat peak flows (Sverdrup et al., 1985; Sverdrup, 1986a) to control the low pH–high aluminum events (Nyberg et al., 1986). Liming of headwater lakes or of watersheds offers considerable promise to minimize difficulties in operating stream dosers, especially high in the watershed and in remote areas (Lessmark and Thornelof, 1986). When lakes have low residence times, lake liming requires more continuous treatment. For example, in the Swedish program, a lake with residence time less than 0.2 years is treated as a stream (Lessmark and Thornelof, 1986).

Table 6-1. Design variables for liming of lakes and streams.

Treatment dosage required
Volume (V, M^3)
Flowthrough rate (Q, M^3/time)
Bottom sediment characteristics (pH, BNC, organic content, bulk density, sedimentation)
Water chemical characteristics (pH, ANC, DIC, calcium, magnesium)
Organic constitutents (DOC, fulvic and humic acids)
Acidic loading to system
Treatment placement
Depth distribution
Ecological actors (spawning habitat, wetlands, macrophytes)
Tributaries (location, flow, chemistry)
Time pattern of treatment
Storm event and snowmelt flow rate (Q) as a function of time
Thermal stratification
Ecological factors (spawning season, seasonal habitat for life stages)
Calcite characteristics
Dissolution rate (particle size)
Equivalence (noncalcite minerals, water content)
Evenness of distribution (slurried)

BNC = base-neutralizing capacity; DIC = dissolved inorganic carbon; DOC = dissolved organic carbon.

Boat, truck, or aerial application can be selected for treating lakes; choices depend on site accessibility and cost. Powdered calcite either as a slurry or as dry powder has been sprayed from trucks and from boats (Sverdrup et al., 1984). In less accessible areas, both airplanes and helicopters have applied calcite in dry form (Booth et al., 1986) and as slurry (Scheffe et al., 1986b). Site remoteness does not preclude liming; however, the cost of liming will increase, in some cases substantially. Slurrying can take place at the site or prior to delivery (Scheffe et al., 1986b), in which case the shipping costs include the water. Boat application can slurry the calcite by mixing the powder with lake water pumped on board into a slurry box just prior to spraying (Sverdrup et al., 1984).

In general, the most important factor in lakes is the hydraulic residence time (T = volume/flowthrough rate). Small lakes with large flowthrough rates can have very short residence times on the order of weeks, whereas the residence time of large lakes can approach tens of years. Even in a single lake, residence time varies during the annual cycle due to seasonal precipitation and runoff patterns, and morphometric characteristics such as depth affect residence time. Furthermore, inflow from the watershed does not mix completely with lake waters during stratification. Episodic acidic events from storms and spring snowmelt may flow across the surface layers of a lake, short-circuiting the system and overestimating

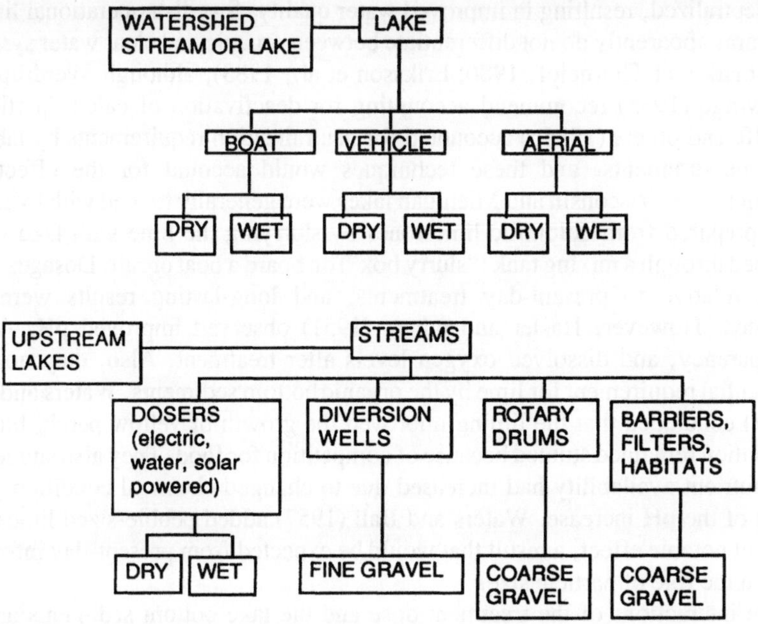

Figure 6–3. Schematic of alternative methods of adding limestone to surface waters.

the residence time and lakewide reacidification (DePinto et al., 1987; Hasselrot et al., 1987; Schofield et al., 1986). Consequently, the concept of residence time can give a poor estimate of washout and lead to inaccurate estimates of the reacidification time. Thus, temporal and spatial variation in water quality may reduce the expected duration of treatment, and computations that account for these differences may be necessary, for example, by using models (DePinto et al., 1989).

Because washout is the most significant loss process for lakes (DePinto et al., 1989), a mechanism for continuous dosing has been sought. Mechanical devices for lakes, such as stream dosers, are useful but a relatively expensive capital investment. Sediment dosing with calcite has been used to extend the treatment duration in lakes (e.g., Wright, 1985; Sverdrup et al., 1984; DePinto et al., 1987). The calcite dissolves slowly, exchanging with H^+ on sediment exchange sites. Later, these cations exchange with the H^+ in the overlying water, providing a more continuous treatment of acidic waters. The continued treatment may not be as effective in deep lakes because of a reduced sediment surface area to lake volume ratio. For the effectiveness of sediment dosing, the decisive parameter is the overflow rate, mean depth/hydraulic residence time (DePinto et al., 1987).

Base treatment of highly colored, soft water seepage lakes (bog lakes) in Wisconsin (Hasler et al., 1951; Stross and Hasler, 1960) and Michigan (Waters, 1956; Waters and Ball, 1957) indicates that highly organic systems can be treated and neutralized, resulting in improved water quality. Swedish operational liming programs apparently do not differentiate between organic and clear water systems (Lessmark and Thornelof, 1986; Eriksson et al., 1983), although Sverdrup and Warfvinge (1985) recommend accounting for deactivation of calcite particles. Scheffe and others (1986b) accounted for neutralization requirements by laboratory measurements, and these techniques would account for the effects of organics. The Wisconsin and Michigan lakes were generally treated with hydrated lime prepared from dolomitic limestones by slurrying the lime with lake water pumped through a mixing tank ("slurry box") on board a boat or raft. Dosages were small relative to present-day treatments, and long-lasting results were not obtained. However, Hasler and others (1951) observed improved pH, ANC, transparency, and dissolved oxygen levels after treatment. Also, they noted a substantial requirement for lime by the organic bottom sediments. Waters and Ball (1957) concluded that the liming improved the growth of yellow perch, but the population remained stunted because of competition for food. They also suggested that nutrient availability had increased due to changed chemical conditions as a result of the pH increase. Waters and Ball (1957) added pebble-sized limestone without notable effect, a result that would be expected from present-day information on the role of particle size.

The interactions of the treatment dose and the lake bottom sediments are an important part of the dose calculation (Young et al., 1989), especially in shallow lakes or deeper lakes where a substantial part of the limestone is applied to shallow zones. Interactions of particulate and dissolved calcite with sediments appear to be of several types. Particulate calcite can be directly buried (DePinto et al., 1989) or inactivated and buried (Sverdrup and Warfvinge, 1985). After calcite dissolution, exchange of Ca^{++} with H^{+} can lead to a marked increase in base saturation to depths of 7 cm (Gahnström, 1984), 10 cm (Young et al., 1989), and greater than 15 cm (Boyd and Cuenco, 1980), depending on sediment characteristics and treatment dose. Organic matter can contribute much of the ion exchange capacity of sediments. For example, sphagnum moss provides substantial ion exchange and can contribute acidity during active growth (Clymo, 1963). Similarly, insoluble or surface-sorbed humic acids in sediments can act as ion exchangers (Rhea and Young, 1987). Young and others (1986, 1989) concluded that in shallow lakes the base-neutralizing capacity (BNC) was an important part of the dose calculation; in deeper lakes, sediment BNC may not be a substantial factor (Molot, 1986).

The objective in streams is to obtain complete dissolution of the calcite as quickly as possible (Abrahamsen and Matzow, 1984; Sverdrup et al., 1985; White et al., 1984; Watt, 1986). Fine particle calcite that is slurried appears to be the most effective for complete dissolution (Sverdrup, 1986a). In streams, mechanical devices are used to provide more continuous input of calcite than in lakes (Sverdrup, 1986a). These can be classified into two groups: those that dispense a fine powder or slurry and those that are designed to use flowing water to cause large particles (gravel size) to abrade and grind and thereby enhance the rate of dissolution. The latter include diversion wells (Sverdrup, 1986a), which require constant hydraulic head, rotary drums (Zurbach, 1963, 1984), and limestone barriers (Pearson and McDonnell, 1975). Pearson and McDonnell (1978) performed a theoretical analysis of treatment with rotary drums that indicated that these systems are more suitable for complete neutralization of a relatively constant flow of mildly acidic (pH > 4) stream waters. In contrast they found barriers provided incomplete neutralization of highly acidic waters. By combining a barrier with a rotary drum, highly acidic (pH < 4) waters could be efficiently treated provided hydraulic design variables were adequate (Pearson and McDonnell, 1978). Habitat modification by the addition of limestone gravels to littoral areas of lakes has benefited spawning trout (Gunn and Keller, 1980, 1984a, 198b4) and presumably might be effective in streams; however, minimal neutralization is accomplished.

Dosers generally use very finely ground particles that are added by slurrying with upstream waters, although slurried calcite has been applied directly (Sverdrup, 1986a). All of these methods can be effective, but generally the mechanical dosers are better able to adjust dosage to high flows than are diversion wells, rotary drums, barriers, filters, and habitats (Sverdrup, 1986a). The water quality requirements of the fish species needing protection will affect the decision on which application method to select for streams.

Vertical mixing of constituents in the water column is complete in streams. However, episodic events from storms and spring snowmelt runoff can cause extensive pH depressions, dilution of calcium, and increased exposure to inorganic aluminum, with consequent biological effects. One alternative is to overdose with limestone so that treatment of these high flows is maintained. This is a costly alternative because dissolution of bed calcite is inefficient, and one of the major cost items in stream treatment is the supply of chemical. A feedback control is usually built into Swedish operational systems, so that dosage is adjusted to flow (e.g., Sverdrup, 1986a; Abrahamsen and Matzow, 1984). This feature is accomplished by coupling a microprocessor with a stage-flow algorithm to a stage-height recorder to set dosage to flow. An additional refinement is to add a pH monitor (e.g., Pearson and McDonnell, 1975) or to use a regression equation for the ANC dosing requirement as a function of flow (e.g., Sverdrup, 1986a).

Other considerations are timing of limestone additions to lakes and streams. These decisions are based on site hydrologic and biologic characteristics. For example, late summer or fall liming of lakes will minimize loss due to snowmelt runoff that flushes calcium and ANC from the lake. Also, liming during this period will usually protect adults and eggs of species that spawn in the fall. The pattern of thermal stratification of lakes needs to be considered because of its effects on treatment timing. The vertical density gradient can inhibit vertical transport of fine particles to deep waters and their sediments. This has two effects: (1) reduces or postpones treatment of deep layers and (2) enhances mixed-layer dissolution efficiency. Appropriate selection of calcite particle size lends flexibility to site-specific treatment designs. For example, large particle calcite would be necessary to treat deep waters in a stratified lake. Streams and nearshore areas of lakes usually require protection in the spring as well as the fall, when life stages of important fish species may be affected by snow runoff and storm-caused episodic events that have high acidity. Logistic problems are severe at remote sites and during the winter when silos for stream dosers may require replenishment. Normal access, electronic failures, and mechanical problems are further complicated by winter conditions.

Recently, interest has shifted to liming watersheds rather than surface waters, because application of basic cations directly to soils can potentially increase base saturation, replenishing the ANC of the watershed. Tributaries, headwater lakes, and soil systems are buffered by watershed liming. In addition, aluminum dissolution in the watershed is minimized, and the added calcium and other divalent cations are available for direct uptake by plants (Brown et al., 1987).

Following watershed liming, Brown and others (1987) showed that ANC and pH increase during larger flows, a result opposite of the situation with lakes receiving acidic runoff (Goldstein et al., 1985). However, soil liming with calcite in nutrient-poor soil systems may exacerbate nutrient problems, especially when magnesium is the limiting nutrient. Consequently, despite its lower dissolution rate, application of magnesium-rich dolomitic limestone may be more appropriate. Application of other highly soluble minerals [K_2O, MgO, CaO, $Mg(NO_3)_2$, $Ca(NO_3)_2$, $ZnSO_4$, $MnSO_4$] has benefited German forests (Zoettl and Huettl, 1986). Although watershed liming may be the most effective technique for whole-ecosystem mitigation of acidity (Howells and Brown, 1986; Brown et al., 1987; Warfvinge and Sverdrup, 1988), insufficient information on the practice is available at present.

IV. Modeling Reacidification of Surface Waters

Modeling of reacidification in lakes has several purposes (DePinto et al., 1989): (1) to assess the feasibility of restoring water quality necessary for a healthy biological community or preventing its deterioration; (2) to optimize treatment so as to minimize costs while protecting the aquatic community; and (3) based on predicted hydrologic conditions, to calculate when lake retreatment is necessary. In modeling lakes, the performance of selected water quality variables are calculated over time, the important variables being pH, ANC, Ca^{++}, and aluminum.

Although mass-balance calculations have been used to analyze the chemical effects of liming in lakes (e.g., see Dillon and Scheider, 1984; Wright, 1985), they do not offer the predictive capabilities of models. Accurate prediction of the reacidification time is performed with computer models that incorporate calcite dissolution and follow the time series for the key variables. The first of these was developed by Sverdrup and colleagues. DePinto and colleagues developed reacidification time estimates with models that used some of the concepts of Sverdrup. Also, the ILWAS (Integrated Lake-Watershed Acidification Study) model has been used for predicting reacidification time (Davis and Goldstein, 1988).

Two processes predominantly affect calcite particles: dissolution and settling

(Sverdrup, 1983). As the larger particles settle and dissolve, they simultaneously increase the pH, become smaller and enter the successively smaller size class, and begin to settle more slowly and dissolve more slowly. Beginning with Equations 1 through 5, Sverdrup and co-workers model H^+ and Ca^{++} and calculate ANC from a titration curve (Sverdrup and Warfvinge, 1985). Two versions of the model have been developed for lake reacidification (Sverdrup et al., 1984; Sverdrup and Warfvinge, 1985): (1) a single CSTR (continuously stirred tank reactor, i.e., one-box model) for completely mixed lakes, having time-variable tributary inflows and chemical reaction with the sediments; and (2) a two-box model to simulate stratification, otherwise similar to the one-box model. Note that graphical techniques are available also (Sverdrup et al., 1986). Coupled reactions for both model versions include (1) dissolution of calcite in the water column and increased pH from neutralization of acidity, (2) inhibition of calcite dissolution due to precipitation of metals and humates on particle surfaces, and (3) ion exchange of calcium with hydrogen in the sediments. The rate of water column dissolution of calcite is based on Equation 5 between pH of 5.5 to 8 but for pH < 5.5 simplifies to a linear function of H^+ concentration. Sverdrup and Warfvinge (1985) combine the CO_2 and hydrolysis terms as a single function, and in open systems approximate the function as a constant.

They calculate calcite dissolution in the water column by:

$$-dM/dt = R\,(A_l\,P)\,Y \tag{6}$$

where

M = calcite mass, kg
t = time
R = dissolution rate, $1/t$
A_l = surface area of the lake bottom, m^2
P = fraction of the bottom covered, unitless
Y = reactive fraction of the calcite, kmol/kg calcite.

Dissolution of calcite on the lake bottom is simulated with first-order kinetics, assuming that a constant fraction of the calcite is inactivated. Exchange of water column calcium with sediment hydrogen ions is computed assuming first-order kinetics.

DePinto and others (1987, 1989) and Scheffe and others (1986a) develop the modeling approach further by increasing the number of vertical layers and incorporating chemical equilibrium calculations for hydrogen ion. The equilibrium submodel is governed by a proton balance with dissolved inorganic carbon (DIC), DOC, and additional chemical species, including inorganic aluminum, fluoride, sulfate, and hydroxyl ions (DePinto et al., 1987). Also they assume that calcite deposited on the sediments rapidly mixes with the active sediment layer. The dissolution of this calcite is controlled by pore water chemistry. The typically larger pH of sediment pore water and greater sediment CO_2 concentrations are accounted for by relating dissolution to pore water chemistry. They account for burial of undissolved calcite by transporting calcite to the deeper, inactive layers of the sediment according to a first-order assumption.

Davis and Goldstein (1988) modeled the time series of ANC, pH, calcium, and aluminum as affected by liming of Woods Lake and its watershed using the ILWAS model. Aluminum species concentrations, important for protecting fish, are predicted from mechanistic formulations in the model; this capability is unique to this model. However, calcite dissolution was not modeled, and the time series for reacidification has not been compared to observations.

V. Ecological Effects of Base Treatment

A. Concerns about Base Treatment

The purpose of mitigation is to create and maintain water quality conditions appropriate to the development of a balanced and productive aquatic community. In a practical sense, the aquatic community of a treated system should be comparable to that of a circumneutral environment. Fish are the focus of treatment, owing to their ecological, recreational, and commercial importance, and treatment objectives can be directed at a balanced ecosystem or the more limited objectives of a self-maintaining fishery or one maintained by stocking. If the acidic lake is fishless, one of the greatest effects of mitigation results from the subsequent stocking of fish because the introduction of fish substantially affects the balance of predators, herbivores, and producers.

Liming causes a dramatic change in water quality in acidic lakes and streams, increasing pH, ANC, and divalent cations (most often Ca^{++}) and decreasing metal bioavailability. The consequences of these chemical changes have been reported in the literature and in a detailed case study of three Adirondacks lakes and are summarized in this section. Initial effects research focused on organism responses to the short-term chemical changes induced by limestone treatment. Long-term effects are still under investigation, although no evidence suggests that long-term effects of base treatment are deleterious. Two additional concerns are with changes in metal availability and toxicity and whether whole-lake treatment protects against localized or episodic acidic events.

B. Ecological Response

After an effective treatment, there is a shift from an acid-tolerant community to one associated with circumneutral conditions. One concern is whether the restored community has fewer species than would otherwise be present in a circumneutral system. Hultberg and Andersson (1982) suggested that succession in a limed lake proceeded toward a community composition close to that of the original preacidic state.

Eriksson and others (1983) reported on studies of humic and clear lakes that had been limed [$CaCO_3$, CaO, slag ($CaMgSiO_4$)] and concluded that phytoplankton, macrophytes, zooplankton, benthos, and fish communities changed markedly after liming, approximating circumneutral oligotrophic lakes. In the case of phytoplankton, Eriksson and others (1983) concluded that species composition became stable after 2 to 3 years. However, fish populations had not approached a steady state as quickly. Fish recruitment improved rapidly after treatment with the reappearance of absent species and early life stages; unusually large year classes for some species were observed in some lakes that accounted for 50% of the total several years after liming. Some consequences of treatment were slow-growing fish due to improved survival and increased competition for food. Eriksson and others (1983) report that neutralization methods for treating some streams and lakes with short residence times were inadequate to protect fish.

Similarly, Raddum and others (1986) reported no observance of direct deleterious effects on biota of limestone treatment of Lake Hovvatn in Norway. However, they noted that effects due to food competition and utilization were observed. They did not observe reinvasion of new species during their 3-year study and concluded that coming to a new steady state in the community took longer than 3 years. Introduction of species from different trophic levels or stocking of fish for different management purposes (e.g., put-take) can lessen the time to restore a fishery. However, it remains to be seen whether the typical community is restored.

It is difficult to assess the results of Dillon and others (1979) because the lakes that they limed were contaminated with toxic metals in addition to being acidic. They concluded that hydrated lime increased the pH, ANC, and Ca^{++} and reduced the concentration of metals but did not affect nutrient levels. Fertilization with low levels of phosphorus had pronounced effects on phytoplankton. Yan and Dillon (1984) summarized fish bioassay results that indicated that the lakes would not support fish immediately or several years after treatment. Whitby and others (1976) came to similar conclusions, making it logical to assume that the major factor affecting ecological response to lime treatment of the Sudbury Lake communities was determined by excessive concentrations of toxic metals.

Schofield and others (1986), Booth and others (1986), Wright (1985), and Hasselrot and others (1987) argue that the long-term maintenance of stocked fish depends on the capacity of a lake to support reproduction and growth. The low nutrient conditions typical of acidic lakes may limit productivity and prevent maintenance of a large, exploitable fish population. Recently, Schofield and others (1989) showed that fish productivity in limed lakes did not differ from unlimed lakes, given the food base present in the different types of lakes; in other words, nutrient concentrations appear to control fish productivity in circumneutral lakes. These results are consistent with previous suggestions that lake liming should be accompanied by fertilization if increased productivity is desired (Hasler et al., 1951; Arce and Boyd, 1975; Dillon et al., 1979; Boyd and Cuenco, 1980; Wilcox and DeCosta, 1984; Broberg, 1987, 1988).

C. Metal Toxicity

Rosseland and others (1986) could find no evidence of aluminum toxicity at high pH (due to formation of $Al(OH)_4^-$), as previously suggested as a problem immediately following liming (Dickson, 1983).

Other concerns focus on long-term consequences of treatment on metals. For example, a concern is whether repeated liming of a lake will lead to accumulation of trace metals (Al, Cd, Zn) in sediments, which will dissolve later during reacidification and cause more substantial exposure of fish or otherwise increase human exposure through bioaccumulation via the food web to fish. Although no evidence has indicated this to be a problem, there is ongoing research to measure long-term metal accumulation by fish in limed lakes.

Aluminum, in conjunction with increased hydrogen ion and low calcium ion, is a known toxicant (Driscoll et al., 1980; Brown, 1981), and reacidification does not seem to cause higher inorganic aluminum concentrations than previously observed in the same lake after liming (Driscoll et al., 1989b). Results from Bagatto and Alikhan (1987) indicate that metal levels in crayfish decreased after liming.

D. Episodic Acidification

Probably the most important concern about liming is whether or not it is protective against episodic acidification. Lake liming does not prevent low pH and high Al in surface waters of winter-stratified lakes during runoff periods (Driscoll et al., 1987). Consequently, nearshore areas may not be well protected. This problem is not as significant in lakes with long residence times, but many of the sensitive or acidic lakes in the United States (Linthurst et al., 1986) and Norway (Johannesson and Hindar, 1987) have rapid flushing rates. In lakes the vertical layers vary chemically because of differences in density and water flowpaths, acidic inputs, and processes that influence ANC. For example, during snowmelt lakes are typically ice covered. Stream inputs are less dense than the slightly warmer lake water and are restricted to shallow waters that extend to the lake bottom near shore. These littoral zones are important habitat for food organisms and for fish spawning and growth of young life stages. These nearshore, shallow zones are subject to pulses of acidic inflows during spring snowmelt and storm runoff events. Liming of the shoreline areas (Wright, 1985), placement of calcite habitats (Gunn and Keller, 1980, 1984a, 1984b; Gloss et al., 1987), or peat bog liming (Hasselrot and Hultberg, 1984) have all been used with good success, especially for protection of younger life stages of fish.

Models are not yet sophisticated enough to predict low pH and ANC associated with episodic events, although horizontally averaged ANC in the mixed layer is predictable (DePinto et al., 1989). There is considerable uncertainty about whether nearshore and shallow-water pH depressions actually lead to species losses from lakes because of water quality refuges and whether the timing of abrupt water chemistry changes coincide with organism sensitivity.

E. Case Study: Adirondack Lakes

The Lake Acidification Mitigation Project (LAMP) is sponsored by the Electric Power Research Institute as a collaborative study to examine the effects of liming on acidified Adirondack lakes (Porcella, 1989). The overall objective of LAMP is to describe the physical, chemical, and biological responses to calcite addition and subsequent reacidification. Reacidification following base application is of particular interest because LAMP investigators can examine the response of lakes to acidification processes.

Lakes represent a continuum of physical, chemical, and biological characteristics that vary in time and space. The experimental design was directed at how ecological responses to liming vary among three study lakes. Specific objectives included evaluation of changes in stocked and unstocked fish populations and other biota in relation to liming and reacidification, the risk of metal accumulation in fish during reacidification after repeated liming, and whether liming protects biota against episodic or littoral zone acidic events.

The Adirondack Mountains have high acidic deposition and contain a number of low-ionic-strength and low-ANC lakes, many of which have a relatively good baseline of data (Schofield and Driscoll, 1987). Based on accessibility, permission to lime, ability to perform hydrologic measurements, and physical, chemical, and biological characteristics, Cranberry Pond, Woods Lake, and Little Simon Pond were selected (Table 6–2). The first two lakes were fishless prior to liming, whereas Little Simon Pond contained *Salvelinus namaycush* (lake trout), which we concluded were under stress based on observations that only older-year classes occurred in the population. The fish community also included *Salvelinus fontinalis* (brook trout), *Catastomus commensoni* (white sucker), and *Servatilus atromaculatus* (creek chub). Woods Lake supported brook trout historically (Gloss et al., 1987), and our objective was to evaluate its potential for a self-maintaining fishery. These three clear water, drainage lakes provide a range of response times and biological conditions for study of physical and chemical responses to liming and reacidification. Mean hydraulic residence times of the three lakes ranged over a factor of eight.

LAMP began field sampling in January 1984. Seventeen months later (May 30–31, 1985), a helicopter limed the two short-residence-time lakes, Cranberry and Woods, with a water slurry (71% solids) of fine-particle (median 2-μm diameter) calcite (98% calcite) and 0.15% sodium polyacrylate dispersant to aid

Table 6–2. Adirondack lakes treated with limestone in lamp study.

Dimensions	Cranberry Pond	Woods Lake	Little Simon Lake
Area (ha)	7.2	23	63
Mean depth (m)	2.9	3.5	10
Maximum depth (m)	7.6	12	33
Lake-watershed ratio (%)	5	10	10
Residence time (months)	2	6	15
Calcite treatment			
Date (month/year)	5/85	5/85; 9/86	8/86
Calcite added (1,000 kg)	7	23; 35	86
Concentration (mg/L)	34	28; 44	13

suspension (Scheffe et al., 1986b). The helicopter sprayed the slurry onto the lake surface at a rate of 6.5 metric tons/hour; about 7 metric tons of dry calcite were added to Cranberry Pond, and 23 tons to Woods Lake; on a volumetric basis, the liming increased nominal water concentrations of calcite to 34 mg/L in Cranberry and 28 mg/L in Woods. The slurry was added to obtain uniform concentrations in the water column and uniform areal deposition of a portion of the calcite to the sediments. Four days after treatment, yearling and fingerling brook trout were introduced to each lake.

Cranberry Pond was allowed to reacidify, which occurred by the end of November 1985. Woods Lake was relimed in September 1986, when ANC had decreased to about 20 μeq/L. In the second liming, 35 tons of calcite were added, with nominal water concentrations of 44 mg/L of calcite. The first liming of Little Simon Pond occurred in August, 1986, when 86 tons of calcite were added, with nominal water concentrations of 13 mg/L of calcite. In both of the 1986 limings, several size ranges of powdered calcite were used so that coarser material would settle to the bottom sediments, on the hypothesis that this calcite would prolong the treatment duration. Retreatment of Woods Lake has not been necessary, and based on present predictions, water quality will meet criteria until mid-1989, an interval of 36 months compared to the 15 months obtained from the initial treatment. Only results from the 1985 liming of Cranberry Pond and Woods Lake are summarized here because the remaining studies are not completed.

The spatial and temporal distribution of water entering and exiting the lake, coupled with the chemical composition of those flows, will substantially affect the in-lake chemical concentrations and mass balances. Staubitz and Zarriello (1989) showed that runoff in the two treated lakes was highly variable, with high peak flows and poorly sustained low flows. Diffuse seepage from shallow groundwater responded rapidly to seasonal recharge and entered the lakes along the lake shore. Error analysis indicated the hydrologic balance was accurate within 10% on an annual basis.

Fordham and Driscoll (1989) described the transitory effects of limestone addition on water column acid-base chemistry. Within the first 24 hours after base application, the two limed lakes responded like systems closed to atmospheric CO_2 because the dissolution of the fine particles (median diameter 2 μm) exceeded the rate of atmospheric CO_2 invasion. Rapid dissolution of calcite, coupled with very low concentrations of DIC prior to treatment, resulted in pH increases in the upper 1 to 2 m of the mixed layer from 4.9 to 9.4 in Woods and from 4.6 to 9.1 in Cranberry Pond. Solutions were at saturation with respect to $CaCO_3$ solubility. Stoichiometric increases in pH, ANC, DIC, and dissolved Ca^{++} were observed. Over the next three weeks, the upper mixed water equilibrated with atmospheric CO_2 (Young et al., 1989), facilitating dissolution of suspended particles of calcite and releasing additional ANC and Ca^{++}. By the end of the 4 weeks, calcite dissolution efficiency attained 86% in Woods and 79% in Cranberry, according to budgets based on Ca^{++}. In effect, all of the added calcite dissolved in the two lakes by the end of the summer.

Driscoll and others (1989a) noted marked changes in the concentrations of the trace metals Al, Mn, and Zn following liming. There was a shift in the speciation of Al from acid-soluble to labile (inorganic) monomeric forms, which increased by about 20%. Gloss and others (1987) observed no increase in toxicity for in situ bioassays from exposures performed during the liming and the subsequent 12 days. Within 1 to 2 months of liming, the concentrations of the three metals were greatly reduced, typically by an order of magnitude. Because thermal stratification limited penetration of base into the 5% of lake volume contained in the hypolimnion, trace metal concentrations were considerably higher there than in the mixed layer.

The mechanism(s) for immobilization of trace metals following base treatment is not entirely clear. Chemical equilibrium calculations suggest that base addition results in pH increases and conditions of oversaturation with respect to the solubility of readily forming aluminum and manganese minerals. Throughout the study, waters were undersaturated with respect to the solubility of Zn-containing minerals. Therefore, retention of Al and Mn in the lakes may have been due to direct precipitation, whereas Zn retention probably occurred by adsorption or coprecipitation. Acidic Adirondack lakes are generally net sinks for Al and conservative with respect to transport of Mn and Zn (Driscoll et al., 1989a). Liming greatly increased retention of all three metals.

For the long-term chemical response to liming, Driscoll and others (1989b) noted that during the autumn after treatment, completely mixed water columns and atypically elevated runoff speeded reacidification, especially in Cranberry Pond. When pH values decreased below 5.5 in Cranberry, transport of elevated concentrations of inorganic Al through the lake was evident, although the concentrations were not higher than those prior to liming. Annual ANC budgets suggest that little $CaCO_3$ penetrated to the sediments, limiting prolonged release of ANC from sediment dissolution. Hydrolysis of aluminum, due to the elevated lake pH after liming, served to consume ANC, and there is evidence of exchange of water column Ca^{++} with sediments after treatment, followed by release during reacidification. However, these processes did not significantly accelerate or attenuate the reacidification rate. The rate of reacidification could largely be explained by the flushing of ANC by hydrologic inputs.

Young and others (1989) concluded that observed calcite dissolution in the two lakes exceeded predicted values significantly due to (1) thermocline inhibition of calcite settling, (2) hydrodynamic dispersion, and (3) atmospheric CO_2 influx. Immediately after treatment, sediment pH gradients changed from positive to negative downward, and amounts of total calcium in surficial sediments increased in parallel with decreases in sediment base-neutralizing capacity. This result suggested that ion exchange was important for sediment neutralization. In Woods Lake during the subsequent 15 months of reacidification, only partial return to preliming conditions occurred, typical of active exchange mechanisms in diffusion-limited systems. It is apparent that sediment interactions become quantitatively important errors if neglected when calculating mass balances of limed systems.

The physical and chemical variables in the LAMP study fit easily into a modeling framework. Using a lumped parameter dynamic model (ALaRM), DePinto and others (1989) evaluated the importance of inflow, sediment interactions with water, seasonal effects, and other variables that affect the duration of treatment. ALaRM was confirmed by its accurate simulation of the observed reacidification profiles of Woods Lake and Cranberry Pond, predicting the near zero ANC of Cranberry within 6 months and ANC concentrations of about 20 μeq/L for Woods at 15 months just prior to reliming. Calcium, ANC, and pH are the state variables followed in these profiles. Results showed that these profiles were controlled primarily by hydrologic flushing and secondarily by sediment-water ANC transfer. It is apparent that a prediction tool must consider these variables to provide a reasonable estimate of reacidification.

Bukaveckas (1989) analyzed phytoplankton dynamics during liming and reacidification in Cranberry and Woods as affected by light, temperature, and chemical variables. Corresponding changes of productivity and biomass observed in the mixed and lower layers of the water column suggests that the effect of liming on phytoplankton growth rates and loss rates was not restricted to the maximum depth of penetration of calcite particles. The neutralized (mixed) and acidic layers exhibited similar increases in phytoplankton abundance, but changes in species composition were limited to the mixed zone. Physiological capacities for using CO_2 versus bicarbonate forms of DIC appeared to play an important role in species composition, but other factors seem to regulate abundance. Post-treatment increases in phytoplankton abundance are attributed to (1) increased nutrient availability following precipitation of hydroxyl-aluminum flocs and (2) reduced zooplankton grazing pressure.

Without ecological modeling, the zooplankton responses to liming and species shifts of food organisms and predators cannot easily be deciphered. Schaffner (1989) attributes zooplankton responses to effects of phytoplankton changes and predation by stocked fish that cause changes in both zooplankton production and loss terms. Most rotifer taxa were greatly reduced in numbers within a week following base addition. The microcrustacean community also responded to liming. *Diaptomus minutus,* dominant in both lakes, declined after liming, but longer-term population responses appeared to be related to the dynamics of individual populations and changes in predation pressure. This caused some species to become abundant again in Cranberry Pond after reacidification while remaining rare in Woods Lake. Four additional crustacea became prominent in the lakes after liming: the cladocerans *Bosmina longirostris* and *Daphnia catawba* in Cranberry Pond, and *D. catawba* and the cyclopoid copepod *Cyclops scutifer* in Woods Lake.

Evans (1989) characterized the macroinvertebrate fauna prior to liming that forms the initial base of the food supply for the stocked trout. Within 2 to 3 months after liming and trout introductions, all limnetic insect populations (notonectids, corixids, dytiscids, and *Chaoborus americanus*) were near or below the detection limit of 0.02 individuals/$m.^3$ Available evidence strongly indicated that brook trout predation was the major factor reducing limnetic insect populations.

Prior to, during, and immediately after the liming, in situ bioassays were conducted with caged brook trout to assess the effects of the treatments on survival rates (Gloss et al., 1987). Significant improvements in survival of caged brook trout (equivalent to circumneutral control lakes) were observed in Cranberry Pond and the epilimnion of Woods Lake during and after liming. However, zero survival rates (100% mortality within 2 days), comparable to preliming, were observed for caged trout placed below the thermocline in Woods Lake, where acidic conditions persisted during and for several months after treatment. The incomplete neutralization of the deeper waters in Woods Lake resulted from the failure of very fine calcite particles to penetrate the denser hypolimnion. These results are consistent with the observation that no significant evidence of liming stress has been detected during liming. Schofield and others (1986) and Gloss and others (1987) compared survival of caged yearlings and fingerling brook trout in two separate experiments with ten and three lakes, respectively. The pattern was similar in all cases: limited or no survival prior to liming and survival equivalent to circumneutral lakes from bioassays performed either during the liming or several weeks later. Perhaps more significant, no indication of immediate fish mortality was observed during the liming of lakes having existing fish communities (C. L. Schofield, Cornell University, personal communication). No obvious stress or mortality was observed in fish stocked 4 days after liming of Woods Lake and nearby Cranberry Lake (Gloss et al., 1987).

The fingerling (0+) and yearling (1+) brook trout introduced into the two previously acidic and fishless lakes (4 days after liming) were monitored by semiannual trap-netting and continuous operation of outlet traps to measure emigration; stomach samples were collected during the ice-free season (Gloss et al., 1989). Gloss and others concluded that the growth and condition of both ages of fish were comparable to those of circumneutral lakes while water quality remained suitable. During reacidification, Cranberry Pond fish growth and survival declined rapidly. About 55% of the 0+-aged and 25% of the 1+-aged fish emigrated. Some limited survival of fish within the lake occurred, but practically speaking, in-lake mortality accounted for the remainder. Annual survival in Woods Lake fish was somewhat lower than other limed or circumneutral Adirondack lakes, averaging 35% for age 0+ and 25% for age 1+ fish. Occurrence of false annuli on scales during summer suggested that growing conditions were not optimal, although annual growth was similar to other waters. Initially, large conspicuous invertebrates made up most of the caloric intake by fish over 200 mm long. However, these taxa were replaced in importance by zooplankton late in the first summer and during the second year in Woods Lake. Gloss and others (1989) reported that reproduction was insignificant in Woods Lake, occurring only on man-made spawning areas, both calcitic and granitic substrates. Reproduction appeared limited by low-velocity groundwater flows in the lake and high acidity in tributary streams. The emergence period for fry from the man-made substrates peaked in mid-May, well after episodic snowmelt inflows produced toxic conditions in littoral areas.

Schofield and others (1989) used a bioenergetics model based on diet analysis to compare actual versus estimated production to test whether liming is a suitable management technique for maintaining adequate recruitment and growth. The zooplankton and macroinvertebrate data provided the initial conditions for parameterizing the bioenergetics model. They showed that the yearling and fingerling brook trout stocked in Woods and Cranberry after liming produced 10 to 14 kg/ha-yr at average standing crops of 8 to 10 kg/ha during the first year after stocking. These values were similar to levels observed in other limed and circumneutral Adirondack lakes, which collectively support a limited range of biomass (10 to 20 kg/ha) and production (7 to 16 kg/ha-yr) compared to more fertile waters in other regions. The estimated level of food consumption required to maintain initial production rates observed in the two limed lakes was 80 to 100 kg/ha-yr. This level of predation was sufficient to alter substantially the size structure and composition of the invertebrate prey community within a few months of stocking. However, mean production efficiency remained at levels of 20% to 25%, and the development of adverse water quality conditions in the first year after liming (reacidification of Cranberry Pond and temperature/pH gradient in Woods Lake; see Driscoll et al., 1989a) appeared to be of greater significance than food limitation in regulating production and biomass of these stocked brook trout populations. By 5 months after liming, reacidification in Cranberry Pond resulted in rapid declines in biomass and production as a result of both increased mortality and reduced growth.

VI. Conclusions

Limestone treatment is a practical and proven method for treating acidic surface waters. The major controlling factors have to do with the cost of treatment and the effectiveness of restoring ecological conditions in the surface waters. Two major uncertainties with regard to the latter factor concern the effect of episodic acidification on fish and invertebrates and the levels of fish production attainable with long-term maintenance liming.

Specific conclusions can be drawn from the results presented:

1. Calcitic limestone is favored and most commonly used for treating surface waters.
2. Proper selection of powder sizes can allow a broad range of treatment alternatives as well as minimize dissolution problems.
3. Models should incorporate CO_2 influx from the atmosphere and sediment interactions so as to accurately simulate calcite dissolution.
4. Sediment treatment can extend the treatment duration of lakes.
5. Highly organic waters do not require special evaluation for treatment with limestone.
6. Aluminum and trace metals are reduced effectively by liming, but the mechanisms that cause their removal appear to vary. Long-term effects of liming need evaluation.
7. Treatment with limestone benefits fish survival and reproduction. Fish production of limed lakes is typical of circumneutral lakes.

8. A major ecological effect of mitigation is caused by the stocking of fish.
9. Introduction of fish and other species may be necessary to shorten the time required to attain species composition similar to circumneutral lakes. Some estimates of the time required to meet this criterion without such introductions exceed 3 years.
10. The low nutrient conditions found in treated lakes apparently limit production. Increased production would require addition of nutrients.
11. No evidence of aluminum toxicity at high pH exists, and the hypothesis that initial toxicity to fish exists due to formation of $Al(OH)_4^-$ is untenable.
12. Nearshore areas of lakes are not protected by lake liming in all cases. However, the necessity for protection of littoral habitat by lake liming has not been proved. Because of the potential for harm to the aquatic community and the uncertainty about effects, efforts have shifted from whole-lake liming to nearshore liming (Booth et al., 1986; Wright 1985), riparian liming of peat bogs (Hasselrot et al., 1987), and watershed liming (Brown et al., 1987; Howells and Brown, 1986; Warfvinge and Sverdrup, 1988). Watershed liming offers considerable promise for restoration and management of surface water communities because low-order streams can be neutralized, as well as lakes and downstream waters.

References

Abrahamsen, H., and D. Matzow, 1984. Verh Internat Verein Limnol 22:1981–1985.

Arce, R. G., and C. E. Boyd. 1975. Trans Am Fish Soc 104:308–312.

Bagatto, G., and M. A. Alikhan. 1987. Bull Environ Contam Toxicol 30:401–405.

Barber, S. A. 1984. *In* F. Adams, ed. *Soil acidity and liming*, SSA, Inc., Madison WI. 171–209.

Barton, P., and T. Vatanatam. 1976. Environ Sci Technol 10:262–266.

Bengtsson, B., W. Dickson, and P. Nyberg. 1980. Ambio 9:34–36.

Björkqvist, D., and K. Weppling. 1987. *In Acidification and water pathways,* 365. Norwegian Water Resources and Electrical Board, Oslo.

Blake, L. M. 1981. New York Fish Game J 28:208–214.

Booth, G. M., J. G. Hamilton, and L. A. Molot. 1986. Water Air Soil Pollut 31:709–720.

Booty, W. G., J. V. DePinto, and R. D. Scheffe. 1988. Water Resourc Res 24:1024–1036.

Boyd, C. E., and M. L. Cuenco. 1980. Aquaculture 21:293–299.

Britt, D. L., J. E. Fraser, R. J. Fares, and J. D. Kinsman. 1984. *The proceedings of the lake acidification* mitigation workshop. U.S. Dept. of Energy, Office of Environmental Analysis. Washington, DC. 123 p.

Broberg, O. 1987. Hydrobiologia 150:11–24.

Broberg, O. 1988. Ambio 17:22–27.

Brocksen, R. W., G. J. Filbin, M. B. Bonoff, P. T. Gremillion, R. Danehy, and J. E. Fraser. 1987. *The effects of limestone application on water quality in fifteen acidified New York and Massachusetts lakes*. Paper No. 87-35.2, Air Pollution Control Association. 34 p.

Brown, D. J. A. 1981. Fish Biol 18:31–40.

Brown, D. J. A. 1983. Bull Environ Cont Tox 30:582–587.

Brown, D. J. A. 1987. CEGB Research No. 20:30–38.
Brown, D. J. A., G. D. Howells, and K. Paterson. 1987. *In Acidification and water pathways,* 349–363. Norwegian Water Resources and Electrical Board, Oslo.
Bukaveckas, P. A. 1989. Can J Fish Aquat Sci 46:352–359.
Clymo, R. S. 1963. Annals of Bot 27:309–324.
Davis, J. E., and R. A. Goldstein. 1988. Water Resourc Res 24:525–532.
DePinto, J. V., R. D. Scheffe, W. G. Booty, and T. C. Young. 1989. Can J Fish Aquatic Sci 46:323–332.
DePinto, J. V., T. C. Young, R. D. Scheffe, W. G. Booty, and J. R. Rhea. 1987. Lake Reserv Manage 3:421–429.
Dickson, W. 1983. Vatten 39:400–404.
Dillon, P. J., and W. A. Scheider. 1984. *In* J. L. Schnoor, ed. *Modeling of total acid precipitation impacts,* 121–154. Butterworth, Boston.
Dillon, P. J., N. D. Yan, W. A. Scheider, and N. Conroy. 1979. Arch Hydrobiol Beih 13:317–336.
Driscoll, C. T., W. A. Ayling, G. F. Fordham, and L. M. Oliver. 1989b. Can J Fish Aquat Sci 46:258–267.
Driscoll, C. T., J. P. Baker, J. J. Bisogni, and C. L. Schofield. 1980. Nature 284:161–164.
Driscoll, C. T., J. F. Fordham, W. A. Ayling, and L. M. Oliver. 1987. Lake Reserv Manage 3:404–411.
Driscoll, C. T., G. F. Fordham, W. A. Ayling, and L. M. Oliver. 1989a. Can J Fish Aquat Sci 46:249–257.
Driscoll, C. T., J. R. White, G. C. Schafran, and J. D. Rendall. 1982. J Environ Eng Div ASCE 108:1128–1145.
Eriksson, F., E. Hornstrom, P. Mossberg, and P. Nyberg. 1983. Hydrobiologia 101:145–164.
Evans, R. A. 1989. Can J Fish Aquat Sci 46:342–351.
Fordham, G. F., and C. T. Driscoll. 1989. Can J Fish Aquat Sci 46:306–314.
Gahnström, G. 1984. Verh Internat Verein Limnol 22:760–764.
Gloss, S. P., C. L. Schofield, and R. L. Spateholts. 1987. (*Salvelinus fontinalis*) populations in acidic lakes following base addition. Lake Reserv Manage 3:412–420.
Gloss, S. P., C. L. Schofield, R. L. Spateholts, and B. A. Plonski. 1989. Can J Fish Aquat Sci 46:277–286.
Goldstein, R. A., C. W. Chen, and S. A. Gherini. 1985. Water Air Soil Pollut 26:327–337.
Gunn, J., and W. Keller. 1984a. Fisheries 9:19–24.
Gunn, J. M., and W. Keller. 1980. Can J Fish Aquat Sci 37:1522–1530.
Gunn, J. M., and W. Keller. 1984b. Can J Fish Aquat Sci 41:319–329.
Hasler, A. D., O. M. Brynildson, and W. T. Helm. 1951. J Wildlife Manage 15:347–352.
Hasselrot, B., I. B. Andersson, I. Alenas, and H. Hultberg. 1987. Water Air Soil Pollut 32:341–362.
Hasselrot, B., and H. Hultberg. 1984. Fisheries 9:4–9.
Hongve, D., and H. Abrahamsen. 1984. Verh Internat Verein Limnol 22:700–703.
Howells, G., and D. J. A. Brown. 1986. Water Air Soil Pollut 31:817–825.
Hultberg, H., and I. B. Andersson. 1982. Water Air Soil Pollut 18:311–331.
Hultberg, H., and P. Grennfelt. 1986. Water Air Soil Pollut 30:31–46.
Ingersoll, C. G., T. W. LaPoint, H. L. Bergman, and J. Breck. 1985. *In* P. Rago and R. K. Schreiber, eds. *An early life stage brook trout experiment to determine the effects of pH, calcium, and aluminum in low conductivity water. Acid rain and fisheries: A debate of issues,* 42–48. Biol Rep. No. 80 (40.21). U.S. Fish and Wildlife Service, Washington, DC.

Johannessen, M., and V. Hindar. 1987. *In Acidification and water pathways,* 325–348. Norwegian Water Resources and Electrical Board, Oslo.
Kretser, W., and J Colquhuon. 1984. Fisheries 9:36–41.
Lessmark, O., and E. Thornelof. 1986. Water Air Soil Pollut 31:809–815.
Lindmark, G. K. 1982. Hydrobiologia 92:537–547.
Lindmark, G. K. 1984. Verh Internat Verein Limnol 22:772–779.
Linthurst, R. A., D. H. Landers, J. M. Eilers, D. F. Brakke, W. S. Overton, E. P. Meier, and R. E. Crowe. 1986. *Characteristics of lakes in the Eastern United States, vol. I. Population descriptions and physicochemical relationships*. EPA/600/4-86/007a, U.S. EPA. Washington, DC. p. 136.
Menz, F. C., and C. T. Driscoll. 1983. Water Resourc Res 19:1139–1149; 20:412.
Molot, L. A. 1986. Water Air Soil Pollut 27:297–304.
Molot, L. A., J. G. Hamilton, and G. M. Booth. 1986. Water Res 20:757–761.
Nyberg, P., M. Appelberg, and E. Degerman. 1986. Water Air Soil Pollut 31:669–688.
Ormerod, S. J., and R. W. Edwards. 1985. J Environ Manage 20:189–197.
Pearson, F., and A. J. McDonnell. 1975. J Environ Eng Div Am Soc Civ Eng 101:425–440.
Pearson, F. H., and A. J. McDonnell. 1978. J Wat Pollut Control Fed 50:723–733.
Plummer, L. N., D. L. Parkhurst, and T. M. L. Wigley. 1979. *In* E. A. Jenne, ed. *Chemical modeling in aqueous systems,* Paper 25, 537–573. ACS Symp. No. 93, Washington, DC.
Plummer, L. N., T. M. L. Wigley, and B. L. Parkhurst. 1978. Amer J Sci 278:179–216.
Porcella, D. B. 1987. Lake Reser Manage 3:401–403.
Porcella, D. B. 1989. Can J Fish Aquat Sci 46:246–248.
Raddum, G. G., P. Brettum, D. Matzow, J. P. Nilssen, A. Skov, T. Svealv, and R. F. Wright. 1986. Water Air Soil Pollut 31:721–764.
Rhea, J. R., and T. C. Young. 1987. Environ Geol Water Sci 10:169–173.
Rogers, P. W., J. V. DePinto, D. L. Britt, W. P. Saunders, H. U. Sverdrup, and P. G. Warfvinge. 1986. Lake Reser Manage 2:75–78.
Rosseland, B., and O. Skogheim. 1984. Fisheries 9:10–16.
Rosseland, B. O., O. K. Skogheim, H. Abrahamsen, and D. Matzow. 1986. Can J Fish Aquat Sci 43:1888–1893.
Schaffner, W. R. 1989. Can J Fish Aquat Sci 46:295–305.
Scheffe, R. D., W. G. Booty, and J. V. DePinto. 1986a. Water Air Soil Pollut 31:857–864.
Scheffe, R. D., J. V. DePinto, and K. R. Bilz. 1986b. Water Air Soil Pollut 31:799–807.
Scheider, W., and T. G. Brydges. 1984. Fisheries 9:17–18.
Schofield, C. L., and C. T. Driscoll. 1987. Biogeochemistry 3:63–85.
Schofield, C. L., S. P. Gloss, and D. Josephson. 1986. *Extensive evaluation of lake liming, restocking strategies, and fish population response in acidic lakes following neutralization by liming*. NEC-86/18. U.S. Fish and Wildlife Service, Washington, DC. 117 p.
Schofield, C. L., S. P. Gloss, B. Plonski, and R. Spateholts. 1989. Can J Fish Aquat Sci 46:333–341.
Schreiber, R., and P. Rago. 1984. Fisheries 9:31–35.
Schreiber, R. K., and D. L. Britt. 1987. Fisheries 12:2–6.
Shortelle, A. B., and E. A. Colburn. 1987. Lake Reserv Manage 3:436–443.
Skogheim, O. K., B. O. Rosseland, E. Hoel, and F. Krogland. 1986. Water Air Soil Pollut 30:587–592.
Skogheim, O. K., B. O. Rosseland, F. Kroglund, and G. Hagenlund. 1987. Water Res 21:435–443.
Staubitz, W. W., and P. J. Zarriello. 1989. Can J Fish Aquat Sci 46:268–276.

Stross, R. G., and A. D. Hasler. 1960. Limnol Oceanogr 5:265–272.
Sverdrup, H. 1983. Chemica Scripta 22:12–18.
Sverdrup, H., and I. Bjerle. 1982. Vatten 38:59–73.
Sverdrup, H., and I. Bjerle. 1983. Vatten 39:41–54.
Sverdrup, H., R. Rasmussen, and I. Bjerle. 1984. Chemica Scripta 24:53–66.
Sverdrup, H., and P. Warfvinge. 1985. Water Resourc Res 21:1374–1380.
Sverdrup, H., P. G. Warfvinge, and I. Bjerle, 1986. Vatten 42:10–15.
Sverdrup, H. U. 1986a. Water Air Soil Pollut 31:827–837.
Sverdrup, H. U. 1986b. Water Air Soil Pollut 31:689–707.
Sverdrup, H. U., P. G. Warfvinge, and J. Fraser. 1985. Vatten 41:155–163.
Underwood, J., A. P. Donald, and J. H. Stoner. 1987. J Environ Manage 24:29–40.
Warfvinge, P , and H. Sverdrup. 1988. Lake and Reserv Manage 4:99–106.
Warfvinge, P., H. Sverdrup, and I. Bjerle. 1983. Vatten 39:337–345.
Warfvinge, P., H. Sverdrup, and I. Bjerle. 1984. Chemica Scripta 24:67–72.
Waters, T. F. 1956. Trans Am Fish Soc 86:329–344.
Waters, T. F., and R. C. Ball. 1957. J Wildlife Manage 21:385–391.
Watt, W. D. 1986. Water Air Soil Pollut 31:775–790.
Whitby, L. M., P. M. Stokes, T. C. Hutchinson, and G. Myslik. 1976. Can Mineral 14:47–57.
White, W. J., W. D. Watt, and C. D. Scott. 1984. Fisheries 9:25–30.
Wilcox, G. R., and J. DeCosta. 1984. Int Revue ges Hydrobiol 69:173–199.
Wright, R. F. 1985. Can J Fish Aquat Sci 42:1103–1113.
Wright, R. F., and E. Snekvik. 1978. Verh Internat Verein Limnol 20:765–775.
Yan, N. D., and P. J. Dillon. 1984. *In* J. O. Nriagu, ed. *Environmental impacts of smelters*, 418–456. John Wiley and Sons. New York.
Young, T. C., J. V. DePinto, J. R. Rhea, and R. D. Scheffe. 1989. Can J Fish Aquat Sci 46:315–322.
Young, T. C., J. R. Rhea, and G. McLaughlin. 1986. Water Air Soil Pollut 31:839–846.
Zoettl, H. W., and R. F. Huettl. 1986. Water Air Soil Pollut 31:449–462.
Zurbach, P. 1984. Fisheries 9:42–47.
Zurbach, P. E. 1963. Trans Am Fish Soc 92:173–178.

Recovery of Acidified and Metal-Contaminated Lakes in Canada

Magda Havas*

Abstract

Three case studies that deal with the recovery of aquatic ecosystems from acidic stress are presented. These studies include (1) the experimental acidification and partial recovery of Lake 223 in the Experimental Lakes Area (ELA) of northwestern Ontario, (2) the recovery of two severely acidified and metal-contaminated lakes near the now-closed Coniston Smelter in Ontario, and (3) the recovery of moderately acidified lakes in the Sudbury region of Ontario following regional reductions of SO_2 emissions.

All of these studies indicate that lake water quality improves rapidly once acidic inputs are reduced. Improvements include an increase in pH, a decrease in SO_4 concentrations, and a decrease in concentrations of trace metals. Rapid recovery is attributed primarily to in situ alkalinity production (bacterial SO_4 and NO_3 reduction) and to the natural flushing of lakes. Geological weathering within the drainage basin in areas of base-poor bedrock is not an important source of lake alkalinity.

Improvements in water quality are followed by changes in biota. New species may invade unoccupied niches, and rare species may become more abundant when their natural predators and competitors are eliminated or when their food supply increases. Rapidly reproducing species, such as rotifers and phytoplankton, are among the first to become better established in lakes recovering from acidification. Amphibians and fish become more abundant once the pH is sufficiently high to allow them to reproduce successfully. However, information about biological recovery is still insufficient to enable us to determine to what extent and at what rate the biota do recover naturally following reductions of acidic and trace metal inputs.

---

*Institute for Environmental Studies and Faculty of Forestry, University of Toronto, Toronto, Ontario, M5S 1A4, Canada. Current address: Environmental and Resource Studies, Trent University, Peterborough, Ontario K9J 7B8, Canada.

I. Introduction

The terms *recovery, reversibility,* and *restoration* have all been used to describe improvements in ecosystems following an anthropogenic stress. These terms, however, are not synonymous, nor are they used consistently. Ecosystems may improve naturally once the stress is removed (reduced PO_4 loading, reduced SO_2 emissions), or they may be restored by direct human interventions (liming, stocking). The term *reversibility* implies a return to the original state (i.e., original functional and structural conditions with or without original species present), whereas *recovery* implies an improvement (biologically safe, aesthetically attractive) or a return to usefulness (drinking water, recreation). In this chapter deliberate attempts to induce change (such as liming) will be called *restoration,* and a return to prestress conditions will be called *reversibility*. Otherwise, the more general term *recovery* will be used.

In order to determine whether an ecosystem has returned to its original state (both structural and functional), one must know what that original state was prior to the stress and how that original state was altered. This information is seldom available, and thus true reversibility is difficult to determine. Instead, stressed ecosystems are compared with nearby unstressed ecosystems, with the underlying assumptions that they resembled each other prior to the stress and that any differences are attributable to the stress.

Several ecosystem characteristics relate to recovery. These include inertia, elasticity, resiliency, and vulnerability to irreversible damage (Cairns et al., 1977). *Inertia* refers to the ability of an ecosystem to resist change (e.g., acid-neutralizing capacity of lake ecosystems exposed to acidic rain). *Elasticity* refers to the ecosystems ability to rebound from damage (e.g., flushing time). *Resiliency* refers to the number of times and ecosystem is able to recover from stress (this often refers to pulsed stress such as acidic snowmelt during the spring). *Vulnerability to irreversible damage* refers to restocking capability and the time of response (e.g., feeder streams to a lake ecosystem).

Inertia is perhaps one of the single most important characteristics that determines the sensitivity of a particular ecosystem to stress. In general, aquatic ecosystems containing large volumes of water will be more inert to chemical and temperate changes than those with small volumes. Ecosystems with structural and functional redundancy will be able to adjust to stresses with fewer changes than those with little or no redundancy. Chemical inertia may refer to acid-neutralizing capacity, as mentioned above, and may include not only the water column alkalinity but also the buffering capacity of sediments, as well as in situ alkalinity production. For metals, *chemical inertia* may refer to presence of chelators that bind metals and thus make them less available to biota. This type of chemical inertia increases the apparent volume of a lake from its true volume by chelating, adsorbing, absorbing, or consuming the incoming chemical.

Another important characteristic is the inherent variability of the environment and thus the ability of biota within that environment to respond to change. Biota from highly variable environmentals (rivers) have a much broader ecological threshold for change and are thus better able to tolerate change than those from stable environments (oceans). The inertia of the environment must be distinguished from the inertia of its biota. An environment may be relatively resistant (inert) to a particular stress, but once the environment responds to that stress, the biota may be highly sensitive (e.g., temperature changes in marine ecosystems).

The ability of an ecosystem to recover also depends on how the physical, chemical, and biological characteristics have changed. Physical changes in substrate characteristics (e.g., erosion) may have a long-lasting influence on the type of benthic species able to survive in that environment. Similarly, residual toxic chemicals in the sediments that are returned seasonally to the water column during turnover are likely to have a long-lasting effect. The ability of biota to recover depends on their reproductive potential (elasticity) and their dispersal ability. Species that produce many young several times each year have a greater potential to recover more quickly than slower-reproducing species. Ready dispersal of species (spores, etc.) from nearby rivers and lakes is likely to accelerate the rate of recovery.

Much has been written on recovery of aquatic ecosystems from many different stresses (Cairns et al., 1977; Cairns, 1980). Some of the better-known studies include the recovery of Lake Washington in the United States from eutrophication (Edmondson, 1977); recovery of streams in Pennsylvania from acidic mine drainage (Herricks, 1977); recovery of the Thames Estuary in Great Britain from sewage effluent; recovery of Lake Hornborga in Sweden from lake level alterations (Bjork, 1977); and recovery of coastal and marine ecosystems from oil spills (Dicks, 1977; Foster and Holmes, 1977; Nelson-Smith, 1977). Recovery may occur naturally once the stress is reduced or eliminated, as in some of the examples above, or it may be induced as in the case of alum additions and artificial aeration to treat eutrophication (Fast, 1977). This chapter deals exclusively with the natural ability of aquatic ecosystems to rebound after acidic and trace metal inputs are reduced. Several of the examples offered represent lakes, near smelting operations, that were influenced by elevated inputs of both acid sulfate and trace metals. Although acidified lakes in background areas may not be subject to quite the same stresses as lakes near smelters, many of the same processes that influence lake recovery will apply. The major differences may be in rates of recovery and in the relative role of trace metals in the recovery process.

II. Case Study 1: Lake 223 in the Experimental Lakes Area

A. Background Information

Lake 223 is a small, poorly buffered lake (alkalinity = 100 μeq L^{-1}, pH 6.6) located on the Precambrian Shield in the Experimental Lakes Area (ELA) of northwestern Ontario. It has one inlet and one outlet with little input or loss of groundwater. It is surrounded by an uncut boreal forest of pine and spruce in a relatively pristine environment that receives less than 5 kg ha^{-1} y^{-1} of wet SO_4^{2-} deposition (Schindler et al., 1980b).

This lake was studied for 2 years (1974 and 1975) in its pristine state prior to any experimental manipulation. From 1976 until 1981, sulfuric acid was added to Lake 223, and the pH was lowered incrementally from 6.6 to 5.0. The annual input of acid was approximately four times the loading in the northeastern United States (Cook et al., 1986). During the period 1981 to 1983, the pH of Lake 223 was kept at pH 5.1 to 5.0. In subsequent years, less H_2SO_4 was added, and the lake was allowed to recover slowly to pH 5.4 in 1984 and to 5.6 in 1985.

Physical, chemical, and biological changes were reported during acidification (Malley, 1980; Schindler et al., 1980a, 1980b; Schindler and Turner, 1982) and during the subsequent recovery (Anderson and Schiff, 1987; Cook et al., 1986; Kelly et al., 1982; Schiff and Anderson, 1987; Schindler et al., 1985, 1986; Schindler, 1986). This is one of the few studies that allows us to determine whether a lake can return to its original state (reversibility) because preacidification measurements are available.

B. In Situ Alkalinity Production

Alkalinity, or the acid-neutralizing capacity, of lakes can be defined as:

$$\text{Alk} = [Ca^{2+}] + [Mg^{2+} + [Na^+] + [K^+] + [\text{other cations}] - [SO_4^{2-}] - [NO_3^-] - [CL^-] - [\text{other strong acid anions}] \text{ units are eq } L^{-1} \text{ (Anderson and Schiff, 1987).} \quad (1)$$

The other cations may include H^+, NH_4^+, Mn^{2+} Fe^{3+} (in oxygenated water), and Fe^{2+} (in anoxic water). Other anions may include organic acids and HS^- in anoxic waters.

Alkalinity can also be defined as:

$$\text{Alk} = [HCO^{3-}] + [CO_3^{2-}] + [OH^-] + [A^-] - [H^+] \text{ (all in } \mu\text{eq } L^{-1} \quad (2)$$

where A^- represents organic anions that are protolytic in the pH 4 to 7 range (Cook et al., 1986).

One of the most intriguing results that emerged from this study is that soft-water lakes can generate a substantial amount of their own alkalinity. This in situ alkalinity generation occurs rapidly, increases with inputs of SO_4^{2-} and NO_3^-, and can exceed the amount of alkalinity provided by the surrounding watershed in regions of base-poor geology (Schindler, 1986). Acidified lakes, therefore, may rebound more quickly from acidic stress than previously expected. In regions of calcareous bedrock and deep fertile soils, geological weathering provides the major source of alkalinity. However, in regions of base-poor bedrock and shallow soils—regions that are known to be acid-sensitive—in situ alkalinity production provides the major source of lake alkalinity.

In situ alkalinity production was discovered unexpectedly in Lake 223. Based on the static acid-neutralizing capacity of the lake (titratable alkalinity), its volume, and its flushing rate, H_2SO_4 was added to decrease the pH of the lake to a certain, predetermined level. During the first year (1976), alkalinity decreased but there was no change in pH. A mass balance for SO_4 revealed that the H_2SO_4

stimulated SO_4 reduction and thus alkalinity production in the anoxic hypolimnion (Schindler et al., 1985). During the first 2 years of the study, the acid-neutralizing capacity of the lake was underestimated, and the H_2SO_4 additions were only 31% to 38% efficient in decreasing pH and alkalinity in Lake 223 (Cook et al., 1986). In the years to follow, average annual pHs were always higher than the intended pHs because the lake continued to neutralize incoming acidity.

Studies were initiated to examine the source of alkalinity in Lake 223 because alkalinity in freshwater ecosystems can be consumed or generated by several processes. Processes that consume alkalinity include aerobic respiration, nitrification, ammonium uptake during photosynthesis, and oxidation of reduced sulfur compounds; those that generate alkalinity include denitrification, NO_3 uptake during photosynthesis, Mn (IV) and Fe (III) reduction, SO_4 and NO_3 reduction, methane fermentation, organic acid protonation, cation exchange, and mineral weathering (Table 7–1).

Hydrogen ions added to a lake, provided that they remain in the water column, decrease pH and thus increase acidity. However, H^+ can react with alkaline species present in the water, they can exchange with nonprotolytic ions in the sediments, or they can be taken up by the biota with accompanying anions such as SO_4 or NO_3 (Cook et al., 1986).

Under natural conditions, 85% of the alkalinity of Lake 223 was generated in situ, whereas the remaining 15% came from the watershed (Cook et al., 1986). Following experimental acidification, alkalinity production increased from 24 keq y^{-1} to 82 to 100 keq y^{-1}, with approximately 95% generated in situ (Cook et al., 1986). An alkalinity and ion budget for Lake 223 revealed that 66% to 81% of the added H_2SO_4 was neutralized by alkalinity generated within the lake. Most of the in situ alkalinity (85%) was produced by bacterial SO_4 reduction, Fe reduction, and FeS formation in littoral sediments (60%) and in the hypolimnion (25%) (Cook et al., 1986). Therefore, although an anoxic hypolimnion may encourage SO_4 reduction, it is not essential for SO_4 reduction because littoral sediments, or sediments beneath well-oxygenated waters, become anoxic a few cm below the water-sediment interface (Figure 7–1). An additional 19% of the alkalinity was generated by exchange of H^+ for Ca^{2+} and Mg^{2+} in the sediments, whereas the remaining 5% came from the watershed.

These results lead to several interesting conclusions: (1) that a large portion of the alkalinity can be generated within a lake (95% in Lake 223); (2) of the alkalinity generated in situ, most is due to bacterial SO_4 reduction (85% in Lake 223); and (3) SO_4 reduction occurs not only in anoxic hypolimnetic sediments but also in littoral sediments of the oxygenated epilimnion. Therefore, lakes that do not have an anoxic hypolimnion can generate their own alkalinity by bacterial SO_4 reduction within the sediments, and these lakes need not rely exclusively on geological weathering for alkalinity generation.

Additional studies demonstrated that the rate of SO_4 and NO_3 reduction are not identical, and thus the rates of recovery of lakes acidified with sulfuric and nitric acid differ. Sulfuric acid tends to be neutralized more slowly than HNO_3 (Schiff and Anderson, 1987), which suggests that it may be more harmful in the long term.

Table 7-1. Processes that either consume (−) or generate (+) alkalinity in aquatic ecosystems.

1. Aerobic respiration including nitrification:	
$(CH_2O)_{106}(NH_3)_{16}(H_3PO_4) + 138\ O_2 \rightarrow 106\ CO_2 + 16\ NO_3^- + HPO_4^{-2} + 122\ H_2O + 18\ H^+$	$\Delta Alk/O_2$(eq/mole = − 18/138
2. Denitrification:	
a. With oxidation of organic nitrogen[a]:	
$(CH_2O)_{106}(NH_3)_{16}(H_3PO_4) + 94.4\ NO_3^- + 92.4\ H^+ \rightarrow 106\ CO_2 + 55.2\ N_2 + HPO_4^{-2} + 177.2\ H_2O$	$\Delta Alk/\Delta NO_3^-$(eq/eq = +92.4/94.4
b. Without oxidation of organic nitrogen:	
$(CH_2O)_{106}(NH_3)_{16}(H_3PO_4) + 84.8\ NO_3^- + 98.8\ H^+ \rightarrow 106\ CO_2 + 42.4\ N_2 + 16\ NH_4^+ + HPO_4^{-2} + 148.4\ H_2O$	$\Delta Alk/\Delta NO_3^-$(eq/eq) = +98.8/84.8
c. Ammonia production	
$(CH_2O)_{106}(NH_3)_{16}(H_3PO_4) + 53\ NO_3^- + 120\ H^+ \rightarrow 106\ CO_2 + 69\ NH_4^+ + HPO_4^{2-} + 53\ H_2O$	$\Delta Alk/\Delta NO_3^-$ (eq/eq) = +120/53
3. MnO_2 reduction:	
a. With oxidation of organic nitrogen:	
$(CH_2O)_{106}(NH_3)_{16}(H_3PO_4) + 236\ MnO_2 + 470\ H^+ \rightarrow 236\ Mn^{+2} + 106\ CO_2 + 8\ N_2 + HPO_4^{-2} + 366\ H_2O$	$\Delta Alk/\Delta Mn^{2+}$(eq/eq) = + 470/472
b. Without organic nitrogen oxidation:	
$(CH_2O)_{106}(NH_3)_{16}(H_3PO_4) + 212\ MnO_2 + 438\ H^+ \rightarrow 212\ Mn^{+2} + 106\ CO_2 + 16\ NH_4^+ + HPO_4^{-2} + 318\ H_2O$	$\Delta Alk/\Delta Mn^{2+}$ (eq/eq) = +438/424
4. Fe (III) reduction:	
$(CH_2O)_{106}(NH_3)_{16}(H_3PO_4) + 424\ FeDOH + 862\ H^+ \rightarrow 424\ Fe^{+2} + 106\ CO_2 + 16\ NH_4^+ + HPO_4^{-2} + 742\ H_2O$	$\Delta Alk/\Delta Fe^{2+}$ (eq/eq) = +862/848
5. SO_4^{2-} reduction:	
$(CH_2O)_{106}(NH_3)_{16}(H_3PO_4)\ 53\ SO_4^{-2}\ 67\ H^+ \rightarrow 106\ CO_2 + 16\ NH_4^+ + 53\ HS^- + HPO_4^{-2} + 106\ H_2O$	$\Delta Alk/\Delta SO_4^{-2}$ (eq/eq) = +120/106
6. CH_4 fermentation:	
$(CH_2O)_{106}(NH_3)_{16}(H_3PO_4) + 14\ H^+ \rightarrow 53\ CO_2 + 53\ CH_4 + 16\ NH_4^+ + HPO_4^{-2}$	$\Delta Alk/\Delta CH_4$ (eq/mole) = +14/53
7. Cation exchange:	
$X - M^{+a} + nH^+ \rightarrow X - n(H^+) + {}^{+ab}$	$\Delta Alk/\Delta M^{R+}$ (eq/eq) = + 1/1
8. Organic acid protonation:	
$R - COO^- + H + \rightarrow R - COOH$	
9. Mineral weathering:	
e.g., $CaA1_2Si_2O_8 + H_2O + 2H^+ \rightarrow Ca^{+2} + A1_2Si_2O_5(OH)_4$	$\Delta Alk/\Delta Ca^{2+}$ (eq/eq) − + 1/1

[a] Note: Alk is consumed due to nitrification.
[b] X = Mineral or organic substrate; M = Ca^{+2}, Mg^{+2}, Na^+, K^+, NH_4^+, etc.
From Schiff and Anderson (1987).

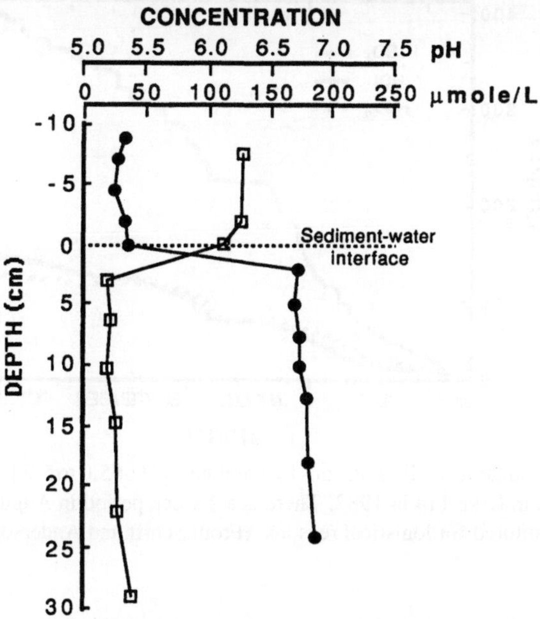

Figure 7–1. Porewater pH (●) and sulfate (□) at 4 m in Lake 223. Sulfate was measured on August 25, 1981, and pH on October 12, 1981. (From Cook et al., 1986.)

Considerably more equivalents of HNO_3 had to be added than equivalents of either H_2SO_4 or HCl to keep the water in the tubes in Lake 114 between pH 5.0 and pH 5.2 (Figure 7–2). The tubes separated a column of water in the lake that could be manipulated independently of the lake.

C. Conclusions

Poorly buffered soft water lakes, therefore, need not rely on the slow process of mineral weathering within their drainage basins as the primary source of alkalinity. These acid-sensitive lakes can generate most of their alkalinity internally by SO_4 and Fe reduction in sediments, whether or not they have an anoxic hypolimnion. Thus the potential for recovery seems to be much faster than originally assumed, based on an overestimate of the importance of external sources of alkalinity to those generated internally. Additions of sulfuric and nitric acid accelerate SO_4 and NO_3 reduction and thus the recovery process, although not at the same rate (Cook et al., 1986). Experimental data indicated that nitric acid is reduced more rapidly than sulfuric acid and thus may be less damaging to aquatic ecosystems in the long term.

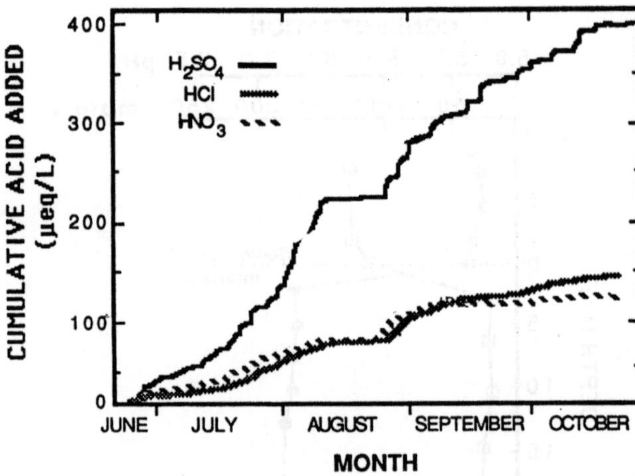

Figure 7–2. Cumulative acidic additions to maintain a pH of 5.0 to 5.2 in the HCl, H_2SO_4, and HNO_3 tubes in Lake 114 in 1982. There is a 2-week period in August when the tubes could not be monitored for logistical reasons. (From Schiff and Anderson, 1987.)

D. Physical and Chemical Changes Associated with Acidification and Recovery

Acidification of Lake 223 induced several changes, including increased transparency, increased rates of hypolimnetic heating and thermocline deepening, increased concentrations of Mn, Ca, Zn, Al, and chlorophyll, and decreased concentrations of suspended carbon, total dissolved silica N, Fe, and Cl. Approximately a quarter to a third of the added SO_4 was consumed and thus reduced the efficiency of acidification under both anoxic and oxic conditions (Cook et al., 1986).

During the initial stages of recovery of Lake 223 (i.e., fall turnover 1983 and 1985), SO_4 concentration decreased by 13% (from 228 to 202 μeq L^{-1}; pH increased by 67% (from 5.18 to 5.67); alkalinity increased from −4 to 10; and the concentration of Mn decreased by 40% (from 155 to 93 μg L^{-1}) (Schindler, 1986). Data for other chemical parameters are not available. These changes suggest a reversal of the effects of acidification.

E. Biological Changes Associated with Acidification and Recovery

Biological changes associated with acidification of Lake 223 were as follows (Schindler et al., 1985):

1. Phytoplankton: Chlorophytes, cyanophytes, and peridineae appeared at the expense of the original diatoms and chrysophycean species. Species were replaced by larger forms covered by mucilage. Biomass increased to a peak in 1980 and then decreased slightly. Chlorophyll production peaked in 1982.

2. Periphyton: *Mougeotia* appeared in the littoral zone and interfered with spawning of lake trout.
3. Benthic invertebrates: The opossum shrimp (*Mysis relicta*), the crayfish (*Orconectes virilis*), leaches, and the mayfly (*Hexagenia*) disappeared. Emergence of dipteran larvae increased and remained elevated during acidification.
4. Planktonic crustaceans: Some species diappeared (*Diatomus sicilis* and *Epischura lacustris*); others increased in number (*Holopedium gibberum*, and *Bosmina longirostris*); still others (*Daphnia catawba*) appeared for the first time in Lake 223 after it was acidified. Copepods were replaced by Cladocera, and there was a shift from larger to smaller organisms.
5. Fishes: Fathead minnow (*Pimephales promelas*) was the most sensitive and stopped reproducing. As the fathead minnow populations decreased, populations of pearl dace (*Sematilus margarita*) increased. Lake trout (*Salvelinus namaycush*) stopped reproducing and lost fitness.
6. Decomposers: There were no changes in the rate of decomposition.

Some of these changes were initiated not by the decline in pH but rather by the loss of one or more species. For example, the disappearance of *Mysis* (which preys on large daphnids like *D. catawba*) and the decline of *D. galeata mendotae* (a potential competitor) enabled *D. catawba* to invade and become established in Lake 223. Similarly, the increase in the pearl dace population corresponded to the decline in the fathead minnow population. Finally, the loss at pH 5.8 of two key organisms in the food web leading to lake trout (*Mysis* and *Pimephales*) probably contributed to the reduced fitness of lake trout and accelerated its decline in Lake 223.

Once the amount of acidic inputs was reduced and Lake 223 started to recovery chemically, the biota also started to respond to the change. Some groups were elastic (phytoplankton) and rebounded rapidly; others showed no improvement (crayfish, opossum shrimp).

The biological changes associated with recovery of Lake 223 were as follows (Schindler, 1986):

1. Phytoplankton: During the first stages of recovery, phytoplankton diversity increased. The species of phytoplankton that returned during recovery were similar to the ones eliminated during acidification.
2. Planktonic crustaceans: Two rather large, acid-tolerant invertebrates, *Holopedium gibberum* and *Daphnia catawba*, decreased in number during recovery, which may be due to the return of planktivorous fishes rather than to the change in pH, because both *H. gibberum* and *D. catawba* are found in lakes with pHs above 5.5 (Havas and Likens, 1985).
3. Fishes: Once the pH of Lake 223 increased from 5.0 to 5.4, white sucker (*Catostomus commersoni*) and pearl dace (*Semotilus margarita*) spawned successfully. The fitness of lake trout (*Salvelinus namaycush*) improved, although as of 1986 there was still no evidence of successful spawning. An unexpected result was the invasion of Lake 223 by brook sticklebacks (*Culaea inconstans*), which were rare before and during acidification. Presumably, with

slow biological recovery, new species have a greater opportunity to invade vacated niches.

4. Other species: As of 1987, several of the species that disappeared from the lake had not returned. These include crayfish (*Orconectes virilis*), fathead minnows (*Pimephales promelas*), sculpins (*Cotus cognatus*), and opossum shrimp (*Mysis relicta*).

Biological recovery is affected not only by changes in water chemistry but also by changes in biota. Some species may recover rapidly (phytoplankton), others slowly (lake trout), and some may not return unless they are restocked. However, it is too soon to tell how much, how rapidly, and how completely Lake 223 will recover.

III. Case Study 2: Alice and Baby Lakes at Coniston, Ontario

A. Background Information

There were three smelters in the Sudbury region, at Falconbridge, Coniston, and Copper Cliff. The Coniston and Copper Cliff smelters were both owned by the International Nickel Company (INCO). In 1972, INCO permanently closed its Coniston smelter and built the world's highest stack (381-m superstack) at Copper Cliff. These two activities greatly reduced the amount of sulfur emitted and the amount deposited in the Sudbury region, especially near the town of Coniston (Figure 7–3).

The Falconbridge smelter emitted 3.5×10^5 tons SO_2 y^{-1} in the early 1970s, which dropped steadily to or below their legal limit of 1.54×10^5 tons y^{-1} imposed in 1985. Prior to 1972, emissions of SO_2 from the two INCO smelters was in excess of 20×10^5 tons y^{-1} with approximately 10% coming from the Coniston smelter (Warner, 1984). Starting in 1973, SO_2 emissions from INCO decreased to 12×10^5 tons y^{-1}. A prolonged strike at INCO in 1977 and 1978 further reduced emission of SO_2 to 6×10^5 tons y^{-1}, and in 1985 legislation limited SO_2 emissions to 7.28×10^5 tons y^{-1} (Brydges, 1985). Since that time, INCO has been operating at or below its legal limit.

Copper, Ni, and at least a dozen trace metals are also emitted from the stacks and were consequently reduced with lower S emissions. At Coniston, bulk deposition of Cu decreased from 70 mg m^{-2} 30 d^{-1} (summer mean) in 1970 to 6.6 in 1977, and to 1.1 during the Copper Cliff strike in 1977 and 1978. Bulk deposition of Ni decreased from 102 to 7.1 and to 1.4 mg m^{-2} 30 d^{-1} during the same period (Hutchinson and Havas, 1986). Closure of the Coniston smelter, building the superstack, the strike, and new air pollution legislation have greatly improved air quality in the Sudbury region. The result is that nearby lakes have begun to improve, as has the roadside vegetation.

Two lakes in Coniston, Alice and Baby, have been studied since 1968. These lakes are located between 600 and 1,000 m southwest of the now closed Coniston

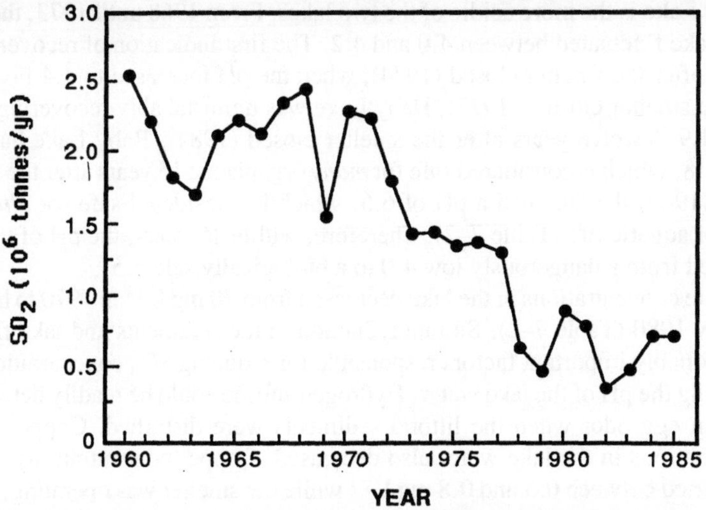

Figure 7–3. SO_2 emmisions (10^6 tons/yr) in the Sudbury basin (1960 to 1985). (From Dillon et al., 1986.)

smelter. Detailed descriptions of these sites can be found in Hutchinson and Havas (1986). Despite their close proximity, the geomorphology and the chemistry of these two lakes are quite different.

Baby Lake is an oligotrophic lake (11.7 ha surface area) in a granitic and gneissic bedrock depression that is devoid of soil and vegetation. It is a deep and relatively clear lake (Secchi depth of 2 m in October 1977). It has one outflow into Alice Lake that consists of a slow, meandering stream that flows through glacial till deposits that are well vegetated with cattails and other marsh plants.

Alice Lake has a larger surface area (26.7 ha) than Baby Lake but is not as deep. It sits in a depression in glacial till that contributes to the turbidity of the water and gives it a milky appearance (Secchi reading 0.6 m in October 1977). This lake has a much greater resiliency to acidification and metal contamination because of the neutralization capacity and cation exchange capacity of the till. Its drainage basin was virtually devoid of vegetation prior to the early 1970s but after the Coniston smelter closed, the grass *Deschampsia cespitosa* invaded drier areas and the cattail *Typha latifolia* spread along the lake edge. Two streams flow into Alice Lake, one from Baby Lake and the other from the slag pond that drains the Coniston Valley. This latter stream dries up during the summer and leaves behind bluish green crystals of Cu and Ni sulfate.

B. Chemical Changes

Chemical composition of the waters in Baby and Alice Lake have been studied since 1968. Water samples collected during fall turnover (October) each year give good integrated samples for the whole lake.

Baby Lake is the more acidic of the two lakes. From 1968 until 1973, the pH of Baby Lake fluctuated between 4.0 and 4.2. The first indication of recovery came 2 years after the smelter closed (1974), when the pH increased to 4.4 Five years after the smelter closure (1977), Baby Lake was unmistakably recovering with a pH of 4.9. Twelve years after the smelter closed (1984), Baby Lake had a pH above 5.8, which is considered safe for *many* organisms; 15 years after the smelter closed (1987) the lake had a pH of 6.5, which is considered safe for *almost all* forms of aquatic life (Table 7–2). Therefore, within 15 years, the pH of this lake increased from a dangerously low 4.0 to a biologically safe 6.5.

Sulfate concentrations in the lake decreased from 70 mg L^{-1} in 1970 to half that value by 1980 (Table 7–2). Sulfate reduction in the sediments and lake flushing were probably important factors responsible for reducing SO_4 concentrations and increasing the pH of the lake water. Hydrogen sulfide could be readily detected by its rotten egg odor when the littoral sediments were disturbed. Copper and Ni concentrations in the lake water also decreased. Copper concentrations in Baby Lake varied between 0.5 and 0.8 mg L^{-1} while the smelter was operating, and by 1987 they had dropped 98% to 0.02 mg L^{-1}, a level equal to normal background concentrations in non-copper-contaminated areas. Nickel concentrations in Baby Lake showed a similar pattern, with a maximum concentration of 3.2 mg L^{-1} in 1970 dropping by 93% to 0.2 mg L^{-1} in 1987 (Table 7–2).

Although Zn is not smelted at the Sudbury smelters, it is found in stack dust and was elevated in the lake water. Zinc concentrations in Baby Lake decreased by

Table 7-2. Chemical composition of Baby Lake and Alice Lake water from 1968 to 1987.

	Baby Lake				Alice Lake			
Year	pH	SO_4	Cu	Ni	pH	SO_4	Cu	Ni
1968	4.2	60	0.64	3.0	6.1	266	0.25	6.9
1969	4.1	nd	0.78	3.0	5.9	nd[a]	0.15	6.8
1970	4.2	72	0.76	3.2	6.3	nd	0.06	6.4
1971	4.0	nd	0.52	2.7	6.3	nd	0.06	6.4
1972	4.1	nd	0.66	2.6	6.1	nd	0.95	nd
1973	4.1	nd	0.50	2.8	6.9	nd	0.05	3.5
1974	4.4	nd	0.35	3.0	6.9	nd	0.01	5.0
1977	4.9	40	0.34	1.5	6.5	176	0.06	3.7
1978	4.5	nd	0.28	1.3	6.2	153	0.05	2.0
1980	4.9	31	0.15	1.7	6.8	76	0.15	3.6
1983	5.4	30	0.06	0.6	6.7	40	0.07	1.7
1984	5.8	nd	0.06	0.4	6.3	nd	0.01	1.4
1985	6.1	27	<0.01	0.4	7.4	108	<0.01	1.3
1986	6.8	23	0.03	0.4	6.9	116	<0.01	1.2
1987	6.5	nd	0.02	0.2	nd	nd	nd	nd

From Hutchinson and Havas (1986) and Hutchinson (1987).

nd = no data. Values are in mg L^{-1} except for pH.

83% from 0.18 mg L^{-1} in 1973 to 0.03 mg L^{-1} in 1987. Iron and Mn concentrations were much more variable but also decreased with time. In contrast, Ca, Mg, Na, and K concentrations did not change significantly after the smelter closed (Hutchinson and Havas, 1986).

Alice Lake was never very acidic, with the lowest recorded pH of 5.9 in 1969. However, it was contaminated with Cu and Ni (Table 7–2). Both Cu and Ni concentrations have decreased significantly since the Coniston smelter closed in 1972, although Ni concentrations may still be too high for successful colonization of some species (see next section on biological recovery). Manganese concentrations have also decreased from 1.0 mg L^{-1} in 1968 to 0.2 mg L^{-1} by 1977 (Hutchinson and Havas, 1986).

C. Biological Changes

Algal samples were first collected from Baby and Alice Lakes during the summers of 1969 and 1970 (Myslik, 1970). Three and four species were found in Baby and Alice Lakes, respectively. This was significantly lower than the 46 species found during the same period in Richard Lake, which is approximately 6 km from the Coniston smelter. Algal density was recorded as negligible (less than 1 cell mL^{-1}) in both Baby and Alice.

By the summer of 1985 and 1986, there were 19 species in Baby Lake and 37 species in Alice Lake (Yung, 1987). Some of the more abundant species in Baby Lake were *Dinobryon accuminatum, D. Bavaricum,* and *Chromulina glacialis,* and in Alice Lake were *Rhizosolenia eriensis* and *Chrysochromulina breviturri*. The diatom *R. eriensis* was also the dominant species in an oligotrophic lake recovering from metal contamination on Vancouver Island (Deniseger et al., 1986). This species has long spines that probably make it inedible for zooplankton grazers. Algal densities had increased but were still low for oligotrophic lakes.

Bioassays of the water gave some interesting results regarding water quality (Hutchinson and Havas, 1987). The growth of a laboratory strain of algae, *Chlorella vulgaris,* was tested in 1970, 1972, 1977, 1984, and 1986. Growth of *C. vulgaris* in both lakes water increased steadily with time.

Prior to 1980, growth was less than 5% in Alice Lake water as compared with a nutrient medium (100%). Growth in Alice Lake water increased to 34% of that in the growth medium by 1984 and to 55% by 1986, which demonstrates a significant improvement in water quality, although growth was still poor. Neither pH adjustments nor nutrient additions stimulated growth of *C. vulgaris* in the 1986 bioassay, which may be due to the elevated Ni concentrations in the lake (Table 7–2).

In Baby lake, growth of *C. vulgaris* increased from 10% (of a reference nutrient medium) in 1972, to 20% in 1977, to 35% in 1984, and to 60% by 1986. Therefore, by 1986, although the water quality had improved significantly, it was still not ideal for the growth of this species (compared to an ideal growth medium). Raising the pH of Baby Lake water in the bioassays significantly improved growth of *C. vulgaris* in the mid-1970s when the pH was below 5 but had no beneficial effect in either 1984 or 1986, by which time the pH had increased naturally to 6. Addition of nutrients did stimulate growth of *C. vulgaris,* which suggests that poor nutrient status may be the main factor limiting primary production in this lake.

Just as phytoplankton are returning to Baby and Alice Lake, zooplankton are also returning (MacIsaac, 1987). Prior to the mid-1970s, planktonic life-forms in either of these lakes were rare. In 1985 and 1986, rotifers found in Baby Lake included *Keratella taurocephala, Lecane tenuisela,* and *Monostyla lunaris, Chydorus sphaericus, Bosmina,* and *Eubosmina* were also found, as were copepod nauplii and cyclopoid copepods.

In Alice Lake a total of six zooplankton species have been found, including *Bosmina longirostris, Chydorus sphaericus,* and the rotifers *Keratella cochleans* var. *cochlearis, Polyarthra vulgaris,* and *P. renata.* The last three species are good indicators of nonacidic waters. Although this does not comprise a healthy zooplankton assemblage, it does indicate that some species are now able to survive in these two lakes, which were virtually devoid of zooplankton 10 years ago.

Water boatmen and dragonflies have also been seen quite recently at the edges of these two lakes, as have frogs, minnows, and seagulls. The presence of frogs, which are known to be extremely acid sensitive, is a definite sign of improvement.

D. Conclusions

Within a relatively short period of time, even extremely acidic, metal-contaminated lakes can improve after the acidic and metal emissions are reduced. Lakes with no apparant buffering provided by their drainage basins (Baby Lake has exposed granitic bedrock, no soil, and no vegetation) can rely on internal sources of buffering. Concurrent with an increase in pH is a decrease in the concentrations of SO_4 and trace metals. The biota respond more slowly, but changes in phytoplankton, zooplankton, insect larvae, amphibians, fishes, and waterfowl can be detected 10 to 15 years after acidic and metal inputs to the lake are reduced.

IV. Case Study 3: Recovery of Moderately Acidic Lakes in the Sudbury Region

A. Background

The reduction in the emissions of SO_2, Cu, and Ni from the Sudbury smelters affected not only the lakes close to the smelters but also those many kilometers away. The Ontario Ministry of the Environment (OMOE) conducted several studies from the mid-1970s until the early 1980s to determine changes in water quality in both acidic and circumneutral lakes and streams located up- and downwind of the Sudbury smelters. Their studies ranged from once-only sampling of over 200 lakes covering an area of 250 km^2 to weekly sampling of four streams.

B. Lake Water Surveys

During 1974 to 76, Conroy and others (1978) sampled a total of 209 lakes within a 250-km radius of Sudbury. They found that many of the lakes near the smelters were acidic, with pHs below 5.5. The acidic lakes were found primarily downwind of the smelters in a northeast and southwest direction. Acidic lakes also had slightly elevated concentrations of Cu and Ni and were depauperate of fish. These same lakes were resampled in 1981 and 1983.

All but one of the 45 acidic lakes (pH less than 5.5) sampled in 1974 and 1976 had a higher pH in 1981 and 1983. (Keller and Pitblado, 1986). The pH in the circumneutral lakes showed no obvious trends, with some values below, some above, and some the same as in the earlier study (Figure 7–4). Sulfate concentrations decreased during the same period in most of the lakes, with the greatest decreases in the high-SO_4 lakes closer to the smelters.

In a separate but related study, the OMOE sampled 37 lakes from 1981 to 1985. These lakes had pHs below 5.5 in the early Conroy and others (1978) study. They found that the average annual pH of these lakes increased from 4.74 to 4.93 and that the average annual SO_4 concentrations decreased from 13.9 to 11.9 mg L^{-1} during the 5-year study (Keller and Pitblado, 1986).

In a more intensive survey of 29 lakes sampled 5 to 6 times each year from 1979 until 1982, 10 lakes showed small but significant increases in pH (Keller and Pitblado, 1986).

Dillon and others (1986) have also documented substantial improvements in water quality (increased pH and decreased SO_4 and metal concentrations) of two lakes, Clearwater and Swan, which are approximately 15 km southwest of

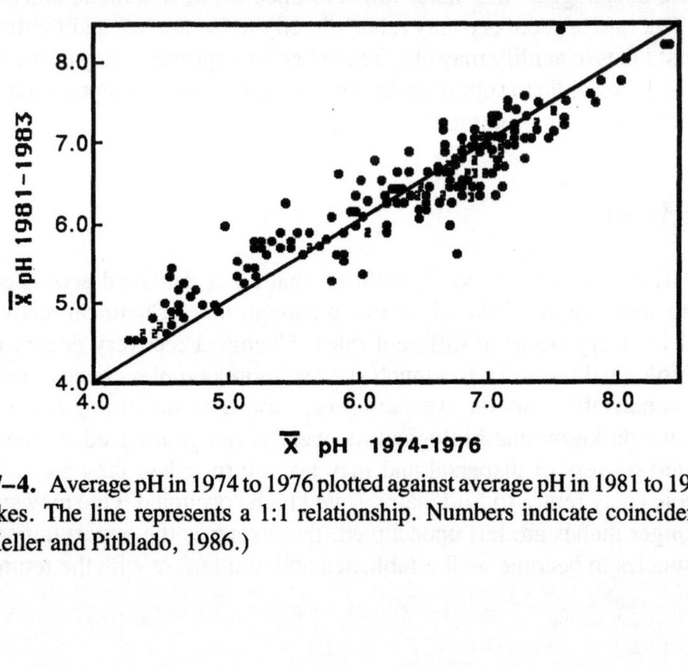

Figure 7–4. Average pH in 1974 to 1976 plotted against average pH in 1981 to 1983 for the study lakes. The line represents a 1:1 relationship. Numbers indicate coincident points. (From Keller and Pitblado, 1986.)

Sudbury. The average pH (arithmetic mean H^+ concentration) of Clearwater Lake between 1973 and 1977 was 4.23, and by 1984 it had increased to 4.61 (60% change in H^+ activity). Concentrations of metals in the water column during the same period decreased by 24% for Ni, 48% for Cu, and 52% for Zn.

Similarly, the pH of Swan Lake increased from 4.0 (range 3.8 to 4.1) in 1977 to 5.1 (range 4.6–5.7) in 1984 (Table 7–3) (MacIsaac et al., 1986). During the same period, conductivity decreased by 50%, SO_4 by 70%, and the Secchi depth by 2.6 m (i.e., light penetration decreased). Concentrations of Cu, Ni, Zn, Al, and Mn were also lower in 1984 than in 1977. This is one of the few lakes for which changes in the rotifer populations have been reported. *Keratella taurocephals,* an acidophile, decreased in abundance and was replaced by *Polyarthra* spp., *Chromogaster ovalis, Conochiloides natans,* and *Trichercera similis*.

C. Stream Water Survey

Streams are much more variable than lakes, and for that reason they are less useful for identifying subtle changes in water quality. OMOE sampled four streams weekly during 1979, 1980, and 1981 in the Killarney area (50 km southwest of Sudbury) (Keller and Pitblado, 1986). No obvious trend in pH was identified in any of the streams due to inherent variability and to the large fluctuations in the hydrologic cycle. The only difference was that the pH decline during the spring snowmelt was greater in 1979 than in either 1980 or 1981, although this difference may relate to the flow of water as much as to the reduced emissions of SO_2.

D. Conclusions

All of these data suggest that lakes improve once acidic and metal emissions are reduced. The rate of recovery may relate directly to the rate of acidification, such that the first lakes to acidify may also be the first to improve. Many of the Sudbury lakes are still too acidic to support a diverse population and, except for a few, have not yet shown signs of biological recovery.

V. Conclusions

Case histories of recovery clearly indicate that most damaged ecosystems will improve to some degree naturally if given enough time. Chemical recovery and biological recovery occur at different rates. Chemical recovery occurs rapidly, whereas biological recovery is a much slower, more complex process and thus is also less predictable. So far we have very few data on biological recovery; however, we do know that biological recovery is not guaranteed. Some species have limited powers of dispersal and may not return unless they are restocked. Other species may return to find their vacated niche occupied. One may speculate that the longer niches are left unoccupied, the greater is the opportunity for new and rare species to become well established and to interfere with the return of the

Table 7-3. Selected chemical characteristics of Swan Lake between 1977 and 1984.

Parameter	1977 $\overline{X}$ + SD	1977 *n*	1982 $\overline{X}$ + SD	1982 *n*	1983 $\overline{X}$ + SD	1983 *n*	1984 $\overline{X}$ + SD	1984 *n*
pH	4.0 (3.8–4.1)[a]	19	4.8 (4.7–5.0)	12	5.0 (4.7–5.6)	13	5.1 (4.6–5.7)	12
Conductivity (μS/cm)	98 + 4	18	51 + 2	12	46 + 2	13	46 + 2	12
Sulfate (μeq/L)	583 + 17	10	223 + 17	11	202 + 16	13	182 + 16	11
Total phosphorus (μg/L)	11 + 3	16	10 + 5	11	12 + 8	13	11 + 4	11
Color (Hazen)	—	—	8.1 + 4.5	11	6.0 + 4.9	13	16.5 + 8.5	11
Secchi disk (m)	6.7 + 1.3	18	4.5 + 0.4	12	5.3 + 0.9	13	4.1 + 0.8	12
Cu (μg/L)	64 + 12	17	17 + 8	11	12 + 5	12	9 + 3	11
Ni (μg/L)	301 + 30	18	82 + 11	11	80 + 12	12	58 + 22	11
Zn (μg/L)	36 + 5	17	17 + 12	11	11 + 3	13	12 + 11	11
Al (μg/L)	289 + 51	11	76 + 17	11	67 + 38	13	47 + 17	11
Mn (μg/L)	213 + 10	11	116 + 6	11	100 + 3	13	99 + 3	11

From MacIsaac et al. (1986).

[a]Range given for pH.

original species. Species with the greatest reproductive potential are the most likely to succeed once favorable conditions return.

Scientific intervention in the form of liming, fertilization, or restocking programs may accelerate recovery, but this type of intervention may not be essential for recovery to occur. In some cases, methods used to eliminate a stress may be worse than the stress itself. Liming of metals-contaminated acidic lakes, for example, may kill the few remaining organisms by suddenly increasing the pH and increasing metal toxicity (Havas, 1986). Liming programs are mitigative measures that cannot substitute for a cleaner environment.

But how much do acidic emissions need to be reduced to enable lakes to recover? According to Schindler (1988), SO_4 deposition between 9 and 14 kg ha^{-1} y^{-1} would protect some of our most acid-sensitive ecosystems and would enable many acidified ecosystems to recover. This value is much lower than the 20 to 50 kg ha^{-1} y^{-1} currently measured in eastern North America and in western Europe. Results from the Coniston region of Ontario suggest that lakes can improve even if the 14 kg ha^{-1} y^{-1} target is not met, provided that SO_4 deposition is significantly reduced (i.e., greater than 50% reduction in this region).

References

Anderson, R. F., and S. L. Schiff. 1987. Can J Fish Aquat Sci 44:188–193.

Bjork, S. 1977. *In* J. Carns, Jr., K. L. Dickson, and E. E. Herricks, eds. *Recovery and restoration of damaged ecosystems,* 110–133. University Press of Virginia, Charlottesville.

Brydges, T. 1985. Atmospheric Environment Service, Downsview, Ontario (personal communication).

Cairns, J., Jr. 1980. *The recovery process in damaged ecosystems,* Ann Arbor Science Publ., Ann Arbor, MI. 167 p.

Cairns, J., Jr. K. L. Dickson, and E. E. Herricks, eds. 1977. *Recovery and restoration of damaged ecosystems.* University Press of Virginia, Charlottesville. 531 p.

Conroy, N., K. Hawley, and W. Keller. 1978. *Extensive monitoring of lakes in the Greater Sudbury area, 1974–1976.* Ontario Ministry of the Environment Tech. Rep., Rexdale. 40 p plus appendices.

Cook, R. B., C. A. Kelly, D. W. Schindler, and M. A. Turner. 1986. Limnol Oceanogr 31:134–148.

Deniseger, J., A. Austin, M. Roch, and M. J. R. Clarks. 1986. Environ Exp Bot 26:217–226.

Dicks, G. 1977. *In* J. Cairns, Jr., K. L. Dickson, and E. E. Herricks, eds. *Recovery and restoration of damaged ecosystems,* 208–240. University Press of Virginia, Charlottesville.

Dillon, P. J., R. A. Reid, and R. Girard. 1986. Water Air Soil Pollut 31:59–65.

Edmondson, W. T. 1977. *In* J. Cairns, Jr., K. L. Dickson, and E. E. Herricks, eds. *Recovery and restoration of damaged ecosystems,* 102–109. University Press of Virginia, Charlottesville.

Fast, A. W. 1977. *In* J. Cairns, Jr., K. L. Dickson, and E. E. Herricks, eds. *Recovery and restoration of damaged ecosystems,* 134–165. University Press of Virginia, Charlottesville.

Foster, M. S., and R. W. Holmes. 1977. *In* J. Cairns, Jr., K. L. Dickson, and E. E. Herricks, eds. *Recovery and restoration of damaged ecosystems,* 166–190. University Press of Virginia, Charlottesville.

Havas, M. 1986. *In* A. Stern, ed. *Air pollution,* vol. VI, 351–389. Academic Press, Orlando, FL.

Havas, M., and G. E. Likens. 1985. Can J Zool 63:1114–1119.

Herricks, E. E. 1977. *In* J. Cairns, Jr., K. L. Dickson, and E. E. Herricks, eds. *Recovery and restoration of damaged ecosystems,* 43–71. University Press of Virginia, Charlottesville.

Hutchinson, T. C. 1987. Unpublished data, University of Toronto, Toronto, Ontario, Canada.

Hutchinson, T. C., and M. Havas. 1986. Water Air Soil Pollut 28:319–333.

Keller, W., and J. R. Pitblado. 1986. Water Air Soil Pollution 29:285–296.

Kelly, C. A., J. W. M. Rudd, R. B. Cook, and D. W. Schindler. 1982. Limnol Oceanogr 27:868–882.

MacIsaac, H. J. 1987. Unpublished data, University of Toronto, Toronto, Ontario, Canada.

MacIsaac, H. J., W. Keller, T. C. Hutchinson, and N. D. Yan. 1986. Water Air Soil Pollut 31:791–797.

Myslik, G. 1970. Unpublished data, University of Toronto, Toronto, Ontario, Canada.

Nelson-Smith, A. 1977. *In* J. Cairns, Jr., K. L. Dickson, and E. E. Herricks, eds. *Recovery and restoration of damaged ecosystems,* 191–207. University Press of Virginia, Charlottesville.

Schiff, R. F., and S. L. Anderson. 1987 Can J Fish Aquat Sci 44:173–187.

Schindler, D. W. 1986. Recovery of Canadian lakes from acidification. Proc. Effects of Air Pollution on Terrestrial and Aquatic Ecosystems, Reversibility of Acidification, Grimstad, Norway 9–11, June 1986. Royal Norwegian Council for Scientific and Industrial Research and the Commission of the European Communities.

Schindler, D. W. 1988. Science 239:149–157.

Schindler, D. W., K. H. Mills, D. F. Malley, D. L. Findlay, J. A. Shearer, I J. Davies, M. A. Turner, G. A. Linsey, and D. R. Cruikshank. 1985. Science 228:1395–1401.

Schindler, D. W., M. A. Turner, M. P. Stainton, and G. A. Linsey. 1986. Science 232:844–847.

Schindler, D. W., R. Wageman, R. B. Cook, T. Ruszcynski, and J. Prokopowich. 1980. Can J Fish Aquat Sci 37:342–354.

Warner, S. 1984. INCO, Sudbury, Ontario. Personal communication.

Yung, K. 1987. Unpublished data, University of Toronto, Toronto, Ontario, Canada.

Paleoecological Analysis of Lake Acidification Trends in North America and Europe Using Diatoms and Chrysophytes

Donald F. Charles,* Richard W. Battarbee,† Ingemar Renberg,‡ Herman van Dam,§ and John P. Smol||

Abstract

Analysis of sediment diatom and chrysophyte assemblages is the best technique currently available for inferring past lake water pH trends, and use of this approach is increasing rapidly. Sediment-core-inferred pH data exist for at least 100 lakes in both North America and Europe. This number will approximately double within the next 2 years. The pH inference equations are based on at least 15 calibration data sets for North America and 10 for Europe, involving totals of at least 500 and 300 lakes, respectively.

Paleoecological studies indicate that recent acidification has been caused by acidic deposition in the Adirondack Mountains (New York), northern New England, Ontario, Quebec, and Canadian Atlantic provinces in North America; England, Scotland, and Wales in the United Kingdom; and Norway, Sweden, Finland, The Netherlands, and West Germany in Europe. Inferred pH decreases are commonly as much as 0.5 to 1.5 pH units. No acidification trends were observed in regions currently receiving low deposition of strong acids (e.g., Rocky Mountains and Sierra Nevada in the western United States, northwestern Norway, and northwestern Scotland). Slight or no trends toward decreasing pH were observed in study lakes receiving moderately acidic deposition (upper Midwest and northern Florida, United States). The magnitude of pH decline in lakes studied is greater, on the average, in Europe than in North America. The amount of inferred acidification (increase in H^+ concentration) correlates with the

*Department of Biology, Indiana University, Bloomington, IN, 47405 USA. Current address: U.S. EPA, Environmental Research Laboratory, Corvallis, OR 97333, USA.

†Palaeoecology Research Unit, Department of Geography, University College London, 26 Bedford Way, London WC1H 0AP, UK.

‡Department of Ecological Botany, Umeå University, S-90187 Umeå, Sweden.

§Research Institute for Nature Management, P.O. Box 46, 3956 ZR Leersum, The Netherlands.

||Department of Biology, Queen's University, Kingston, Ontario K7L 3N6, Canada.

amount of S and N loading and the ability of watersheds and lakes to neutralize acidic inputs and is generally consistent with current lake acidification models.

In most cases, the primary cause of recent acidification trends (post-1850) is atmospheric deposition of acidic material, as opposed to land use changes or natural processes, although these may be contributing factors. Acidic loading has decreased in some regions since 1970 (e.g., northeastern United States, United Kingdom). Some lakes have become less acidic in response, but others continue to lose buffering capacity and are becoming more acidic. Many currently acidic lakes were naturally acidic (pH < 5.5) prior to the onset of anthropogenic acidification. These lakes are typically small (<10 ha), located at moderately high elevations, have thin or peaty soils, or are located in glacial outwash deposits. Many of these have acidified further recently.

I. Introduction

Acidification of lakes by atmospheric deposition of strong acids is a major environmental issue, and causes and effects have been studied intensively. However, there are still major uncertainties concerning the acidity status of lakes prior to the onset of acidic deposition and how these lakes have changed in response to inputs of acids. More information on these topics is necessary to assess fully the effects on aquatic resources and to predict accurately the nature and extent of response to future changes in atmospheric deposition. Historical changes have been assessed by (1) comparison of recent and historical water chemistry, fish, and diatom data; (2) use of empirical and dynamic models; and (3) analysis of paleoecological data (National Research Council, 1986; Davis and Stokes, 1986). Of these, paleoecological studies have provided some of the best information. Other than the sparse chemistry and fisheries information, which is typically difficult to interpret, it provides the most direct indication of trends in acidity status. Also, new paleoecological studies of lakes can be made at any time. They do not depend on the existence of historical records. Paleolimnological studies can provide information on past lake pH and acid-neutralizing capacity (ANC) and how it has changed through time, the rates and magnitude of the change, possible causes of the change, and the relative importance of processes and mechanisms (Battarbee, 1984; Charles and Norton, 1986; Smol et al., 1986).

Because of the usefulness and importance of information on lake acidification that can be derived from paleoecological investigations, there has been a rapid increase in the number of such studies in Europe and North America (Battarbee and Charles, 1986, 1987). This chapter provides an up-to-date assessment of the studies, first by describing the underlying rationale, assumptions, and methods, and then by summarizing, comparing, and contrasting the results of studies on the two continents.

Much of the background information on method and approach is derived from Charles and Norton (1986). Summary and assessment of studies in the different geographic regions are treated in this chapter as follows: Charles, United States;

Smol, eastern Canada; Battarbee, United Kingdom; Renberg, Norway, Sweden, and Finland; and Van Dam, Austria, Belgium, Czechoslovakia, Denmark, The Netherlands, and West Germany.

The sediments of a lake contain information on the lake's past: its biota, water chemistry, and watershed characteristics, and material it has received directly from the atmosphere (Frey, 1969; Pennington, 1981). The information is provided by the organic and inorganic substances, in dissolved or particulate form, that entered or were formed within a lake and deposited in its sediments. The primary materials are controlled by watershed geology, climate, and biological processes. Once deposited at the bottom of a lake, sediments can be affected by secondary processes, such as transport (horizontal and vertical) and a variety of chemical and biological activities.

Sediments can be sampled by taking cores; the core samples represent the history of sediment deposition. The times when particular intervals of sediment were deposited can be determined either by analyzing the radioactive content (^{210}Pb, ^{137}Cs, ^{14}C) or by examining changes in sediment characteristics, such as pollen and charcoal as indicators of logging and forest fires, that correspond to well-dated local events.

Several components of sediments provide information on factors related to lake acidification. Concentrations of polycyclic aromatic hydrocarbons (PAH), spherical carbonaceous particles (soot), lead, sulfur, vanadium, sulfur isotope ratios, and magnetic particles can be interpreted to indicate trends in atmospheric deposition of substances derived from combustion of fossil fuels. The pH of lake water during the past can be inferred by analyzing assemblages of diatoms and chrysophytes. Remains of chydorids (littoral crustaceans) and chironomids (midge larvae) may also provide insight into changes in aquatic biota related to acidification.

II. Inferring Past Lakewater pH: Rationale and Methods

Analyzing and interpreting sediment diatom and chrysophyte assemblages is the best paleolimnological technique available for inferring past lakewater pH. Investigators are using this approach increasingly to assess changes that may have been caused by atmospheric deposition of strong acids because patterns of change in lakewater pH can be used to help determine trends in acidic deposition.

A. Diatoms and Chrysophytes as Indicators of Lake Chemistry

Diatoms make up a large group of single-celled freshwater and marine algae (class Bacillariophyceae). They have siliceous cell walls and are formed of two halves or valves. Chrysophtes (Chrysophyceae) are primarily planktonic. In this chapter, the term *chrysophyte* refers to only one family, the Mallomonadaceae, also known as the *scaled chrysophytes*. Its members have flagella and an external cell covering of overlapping siliceous scales and bristles. The scales are used for paleoecological reconstructions.

The distributions of diatom taxa among lakes are closely related to water chemistry (Cleve, 1891; Kolbe, 1932; Hustedt, 1927, 1939; Jørgensen, 1948; Cholnoky, 1968; Patrick and Reimer, 1966, 1975; Patrick, 1977). For this reason, diatoms are commonly used as indicators of pH, nutrient status, salinity, and other water quality characteristics (e.g., Lowe, 1974). Stratigraphic analysis of fossil diatom assemblages can be used to investigate changes in lakes resulting, for example, from shifts in climate, development of watershed soils and vegetation, local human disturbance of watersheds, and acidic deposition (Battarbee, 1984, 1986; Fritz and Carlson, 1982; Brugam, 1983).

Diatom assemblages in sediment are good indicators of past lake pH because (1) diatoms are common in nearly all freshwater habitats; (2) distributions of diatom taxa are strongly correlated with lake water pH (Hustedt, 1927, 1939; Meriläinen, 1967; Battarbee, 1984; Gasse and Tekaia, 1983; Gasse, 1986; Huttunen and Meriläinen, 1983; Dixit, 1983; Davis & Anderson, 1985; Charles, 1985; Anderson et al., 1986a; Dixit, 1986b; Scruton and Elner, 1986; Flower, 1987; Taylor, 1986); (3) diatom remains are preserved well in sediment and can usually be identified to the lowest taxonomic level; (4) their remains are usually abundant in sediment (10^4 to 10^8 valves/cm^3 of sediment) so that rigorous statistical analyses are possible, and (5) many taxa are typically represented in sediment assemblages (20 to 100 taxa per count of 500 valves is typical), so that inferences are based on the ecological characteristics of many taxa.

Some disadvantages in using diatoms as pH indicators are that (1) diatom identification requires considerable taxonomic expertise, (2) diatoms are not well preserved in some cores because of dissolution (particularly in some peaty and some calcareous sediments), (3) sometimes the number of taxa is low (e.g., in some peatland pools), (4) calibration data sets (the current relationship between water chemistry and surface sediment diatom assemblages) are not available for all the lake regions studied, and (5) good ecological data are not always available for all dominant taxa. Other problems associated with interpretation of diatom data are discussed in Charles and Norton (1986), Battarbee (1986), and Davis and Smol (1986). Except for the dissolution problem, all the above disadvantages are equal to or less than disadvantages for other groups of aquatic organisms.

In general, the use of chrysophyte scales for pH reconstructions involves the same advantages and disadvantages as those for diatoms (Smol, 1980, 1987, 1988; Smol et al, 1984a, 1984b; Kristiansen, 1986; Steinberg and Hartmann, 1986; Hartmann and Steinberg, 1987), except that the number of chrysophyte taxa in a sediment assemblage is typically a tenth the number of diatom taxa, and most chrysophyte taxa are euplanktonic (normally suspended in the water). The latter characteristic provides an advantage over diatoms in the study of acidic lakes because euplanktonic diatoms are normally rare or nonexistent in lakes with pH below about 5.5 to 5.8 (Battarbee, 1984; Charles, 1985; Renberg et al., 1985). In these cases, chrysophytes may be more sensitive indicators of water chemistry changes than diatoms because they live in direct contact with the open water. Most diatoms grow in the shallower water of the littoral zone, which may be chemically different from the open water.

B. Techniques for Inferring Past pH

Several techniques based on diatom assemblages have been used to assess trends in acidification and to derive equations for inferring lake water pH. These are described briefly below. Further descriptions and discussion of details and uncertainties associated with the reconstruction of lake water pH are covered in depth by Gasse and Tekaia (1983), Battarbee (1984), Davis and Anderson (1985), Charles (1985), Smol et al., (1986), Oehlert (1984), Taylor et al., (1986), Birks (1987), Ter Braak and Van Dam (1988), and Stevenson et al., (1988).

The simplest and most straightforward approach is to count sediment core diatom and chrysophyte assemblages and prepare depth profiles of percentages of the dominant taxa. Changes in the profiles are then interpreted in light of the ecological data available for those taxa. Until very recently, this was the only technique used to analyze chrysophyte scale data. Now, however, more rigorous techniques have been developed (Charles and Smol, 1988; Dixit et al., 1988a).

Hustedt (1939; see also Battarbee et al., 1986) made one of the first significant steps toward establishing a more quantitative approach for using diatoms as pH indicators. He recognized the strong relationship between diatom distributions and lake water pH and defined the following pH occurrence categories (from Battarbee et al., 1986):

1. Alkalibiontic: occurring at pH values > 7
2. Alkaliphilous: occurring at pH about 7, with widest distribution at pH > 7
3. Indifferent: equal occurrence on both sides of pH 7
4. Acidophilous: occurring at pH about 7, with widest distribution at pH < 7
5. Acidobiontic: occurring at pH values < 7 with optimum distribution at pH 5.5 and under

Assignments of diatom taxa to these categories can be based on literature references and on the distribution of taxa within waters of particular geographic regions. Changes in the percentages of diatom valves in each pH category in a sediment core can be used to estimate trends in lake water pH. This method makes use of data on most of the taxa within a core, not just the most common.

Nygaard (1956) took the next major step with the development of a set of indexes. These indexes are based on ratios of the percentages of diatom valves in Hustedt's pH categories. Acid units and alkaline units are calculated as an intermediate step:

$$\text{Acid units} = 5\ (\%\ \text{acidobiontic}) + (\%\ \text{acidophilous})$$
$$\text{Alkaline units} = 5\ (\%\ \text{alkalibiontic}) + (\%\ \text{alkaliphilous})$$

The percentages of valves in the two extreme pH categories, acidobiontic and alkalibiontic, are weighted by a factor of 5 because diatoms in these categories are presumably stronger indicators of pH.

The formulas for Nygaard's indexes are:

$$\alpha = \frac{\text{acid units}}{\text{alkaline units}}$$

$$\omega = \frac{\text{acid units}}{\text{number of acid taxa}}$$

$$\epsilon = \frac{\text{alkaline units}}{\text{number of alkaline taxa}}$$

Of these, index α is the best predictor of current pH in acidic lakes (Meriläinen, 1967; Renberg, 1976; Davis and Anderson, 1985; Charles, 1985; Beeson, 1984) and has been used most frequently. Application of index α to acidic lakes is limited, however, if diatom assemblages have no alkaliphilous or alkalibiontic taxa. In these cases there are zero alkaline units, and the denominator of the index is zero. It is possible to circumvent this problem by arbitrarily setting the denominator equal to one (e.g., Charles, 1985) or some lower number. The index may still work reasonably well.

Meriläinen (1967) refined quantitative techniques even further by developing an approach for predicting lake water pH from the index values. The relationship between the $\log_{10}$ of index values and measurements of lake water pH for several lakes is determined by using simple regression analysis. Predictive equations are then derived directly from the slope and intercept of the regression equations.

Renberg and Hellberg (1982) derived a new index (Index B), also based on pH categories, that can be used to quantitatively infer pH:

$$\text{Index B} = \frac{(\%\ \text{indifferent}) + 5\ (\%\ \text{acidophilous}) + 40\ (\%\ \text{acidobiontic})}{(\%\ \text{indifferent}) + 3.5\ (\%\ \text{alkaliphilous}) + 108\ (\%\ \text{alkalibiontic})}$$

Index B has advantages over index α, including the use of more information and less reliance on alkaline taxa, which are typically rare or absent in acidic lakes. Other indexes incorporating Hustedt's (1939) pH categories have been developed (e.g., Watanabe and Yasuda, 1982).

Predictive equations can also be developed from multiple linear regression analysis of measured lake water pH with the percentages of diatoms in each pH category (e.g., Davis and Anderson, 1985; Charles, 1985; Figure 8–1). Standard errors and confidence intervals for prediction of new points are normally smaller than those for the indexes discussed above (e.g., Charles, 1985; Huttunen and Meriläinen, 1986b). Other multiple regression approaches for inferring pH involve the use of selected taxa or principal components of taxa data sets as independent variables (Davis and Anderson, 1985; Gasse and Tekaia, 1983). The standard error for inferred pH ranges between 0.25 and 0.5 pH units (Battarbee, 1984; Charles and Norton, 1986). Ter Braak and Van Dam (1988) have developed new methods for inferring pH and other chemical characteristics from diatoms using maximum-likelihood calibration based on Gaussian logic response curves of taxa against pH and weighted averaging. Oksanen and others (1988) used weighted averaging, least squares, and maximum likelihood to calculate pH

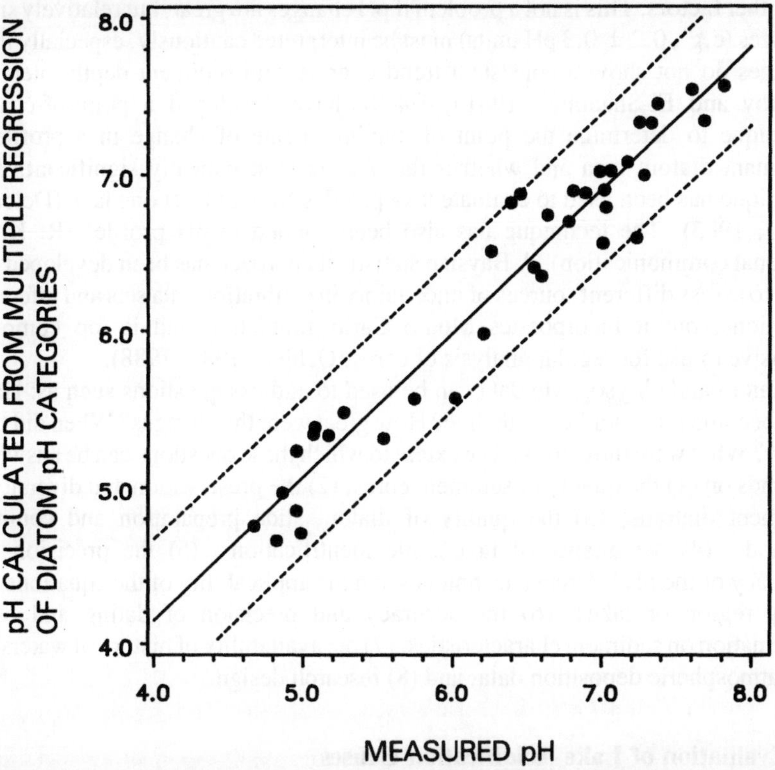

Figure 8–1. Inferred lake water pH calculated from multiple linear regression of diatom pH categories versus measured surface water pH for 37 Adirondack lakes, and 95% confidence intervals for an individual prediction of pH from diatom data. (From Charles and Smol, 1988.)

optima and tolerance of diatom taxa and then used these estimates to predict the pH of other lakes, using weighted averaging. Detrended correspondence analysis (DCA) has been used to infer pH trends (Huttunen and Meriläinen, 1986). A promising new technique incorporating DCA and canonical correspondence analysis (CANOCO) (Ter Braak, 1986) can also be used to infer pH and other water chemistry characteristics as well (Stevenson, et al., 1988). A multiple regression technique, similar to that for pH, has been developed to infer alkalinity (Charles et al., 1989). Also, there are new techniques to infer pH using both diatoms and chrysophytes (Charles and Smol, 1988). These are particularly useful in acidic lakes (pH < 5.0), because chrysophytes appear more sensitive to pH changes in these ranges than do diatoms (e.g., Gibson et al., 1987).

Trends in pH curves usually are not analyzed statistically. Instead, subjective interpretations are made that account for the nature of the diatom assemblages, the error associated with the predictive techniques, the evidence of sediment mixing, and other factors. This is not a problem if pH changes are great, but relatively small changes (e.g., 0.2 ± 0.3 pH units) must be interpreted cautiously, especially if the changes do not show a consistent trend over several sediment depth intervals. Esterby and El-Shaarawi (1981a, 1981b) have developed a point-of-change technique to determine the point of maximum rate of change in a profile of dominant diatom taxa and whether the change is statistically significant. This technique has been used to evaluate taxa profiles from at least one lake (Delorme et al., 1983). The technique has also been applied to pH profiles (R. Kreis, personal communication). A Baysian statistical approach has been developed and used to assess different sources of uncertainty in calibration data sets and inference equations, but it incorporates Monto Carlo simulation and is too computer intensive to use for regular analysis of cores (Oehlert, 1984, 1988).

Diatom and chrysophyte data can be used to address questions such as: Has a lake become more acidic or alkaline? How great were the changes? When did they occur? What were the causes? The extent to which these questions can be answered depends on (1) the quality of sediment cores, (2) the preservation and diversity of sediment diatoms, (3) the quality of diatom slide preparation and counting methods, (4) the quality of taxonomic identifications, (5) the precision and accuracy of the pH inference techniques and the applicability of the equations to a study region or lakes, (6) the accuracy and precision of dating and other information on sediment characteristics, (7) the availability of historical watershed and atmospheric deposition data, and (8) research design.

C. Evaluation of Lake Acidification Causes

Diatom and chrysophyte data can be used not only to infer the past pH trend of a lake but also, in many cases, to suggest the causes of the changes. There are three major potential causes of acidification of low-alkalinity lakes: (1) long-term natural acidification; (2) watershed disturbances, such as fires, blowdown, logging, complete deforestation, reforestation and decrease in agricultural use of land (primarily in Europe), and ensuing responses of vegetation and soils; and (3) atmospheric deposition of strong acids. Other factors may affect lake acidification, but not normally on a regional scale. These include nearby emission sources, discharge of factory effluent, cultural development (e.g., roads and houses), land clearance for agriculture, acidic mine drainage, mixing with seawater, liming, paludification, drainage of wetlands, and water-level changes such as those resulting from small dams, beaver activity, or changes in climate. The last factor is probably important only when a significant proportion of a watershed is wetland and net sulfur reduction-oxidation and cation exchange processes are affected. In shallow pools (e.g., in the Netherlands), low pH values may occur during dry summers because of oxidation of sulfides in the exposed sediments. Because these other factors have not, with few exceptions, affected the lakes evaluated in this chapter, they are not considered further.

The three primary potential causes of lake acidification are addressed below. Diatom studies of long-term acidification in both Europe and North America are also briefly reviewed. Following this is a summary of patterns of diatom changes to be expected in response to each major acidification cause.

1. Long-Term Natural Acidification

Long-term trends in the postglacial development of temperate zone European and North American lakes have been studied using sediment diatom, chrysophyte, pollen, and chemical analyses. These studies indicate that most acidic to weakly alkaline lakes in areas having bedrock that is relatively resistant to chemical weathering have undergone a gradual long-term acidification process. This process has been recognized at least since the studies of Lundqvist (1925) and has been observed in many geographic areas. In Europe, these areas include Sweden (Digerfeldt, 1972, 1975, 1977; Florin, 1977; Renberg, 1976, 1978; Renberg and Hellberg, 1982; Tolonen, 1972), Finland (Alhonen, 1967; Tynni, 1972; Tolonen, 1967, 1980; Salomaa and Alhonen, 1983; Tolonen et al., 1986), Denmark (Nygaard, 1956; Foged, 1969), northwestern England (Evans, 1970; Haworth, 1969; Round, 1957, 1961; Pennington, 1984), Scotland (Alhonen, 1968; Pennington et al., 1972), Wales (Crabtree, 1969; Evans and Walker, 1979; Walker, 1978), Czechoslovakia (Rehakova, 1983), and Greenland (Foged, 1972). Fewer regions in North America have been investigated, but diatom, chrysophyte, and sediment chemistry data indicate that long-term acidification has occurred in Mirror Lake, New Hampshire (Sherman, 1976, 1985); Cone Pond, New Hampshire (Ford, 1984, 1989); Compton, Quincy, and Bottomless Pit Bogs and Bear and George Ponds in New Hampshire (Ryan and Kahler, 1987); Bethany Bog, Connecticut (Patrick, 1954); Berry Pond, Massachesetts (Rochester, 1978); Duck Pond, Cape Cod, Massachusetts (Winkler, 1988); Holmes Lake (Rhodes, 1985), Heart Lake, Upper Wallface Pond, and Lake Arnold in the Adirondack Mountains, New York (Reed, 1982; Whitehead et al., 1986; Christie and Smol, 1986); Crystal Lake, Wisconsin (Conger, 1939); Lake Mary, Wisconsin (J.C. Kingston, personal communication); Vestaberg Bog, Michigan (Colingsworth et al., 1967); Red Rock Lake, Colorado (Norton and Herrmann, 1980); Green Lakes 5, Niwot Ridge, Colorado (Kerstin Williams, personal communication); Splan Lake, New Brunswick (Rawlence, 1988); and on Ellesmere Island, above the arctic circle (Smol, 1983). Some lakes have apparently been naturally acidic from the time they first formed (e.g., Round Loch of Glenhead, Jones et al., 1986).

We can make some generalizations based on these studies. The magnitude and the rate of long-term acidification vary among lakes and within lakes. Diatom-inferred pH (DI), when it has been calculated, indicates declines from 0.5 pH unit or less to as much as 2.5 pH units; for example, from pH 7.5 to pH 5.0 for Upper Wallface Pond in the Adirondacks (Whitehead et al., 1986) and from pH 7.0 to 6.0 for Lake Gårdsjön, southwestern Sweden (Renberg and Hellberg, 1982). Rates of change are gradual; declines of 1 pH unit took hundreds to thousands of years. In general, lakes with the highest current pH have acidified the least, and data for lakes with pH currently above about 7.5 indicate little or no long-term acidification, for example, Linsley Pond (Patrick, 1943), Pickerel Lake (Haworth, 1972),

Sunfish Lake (Sreenivasa and Duthie, 1973), Kirchner Marsh (Brugam, 1980), South King Pond (Ford, 1984, 1989), and Little Round Lake (Smol and Boucherle, 1985).

The most rapid acidification generally occurred during the early postglacial period. Following deglaciation, the lakes were commonly alkaline (pH 7.5 or above), with the major reason for the high pH being the presence of unweathered till and outwash deposits with large supplies of easily leachable base cations. Alkalinity declined after easily leaches bases were removed and as soil and vegetation developed. These processes have been shown to be important in soil and lake development (Crocker and Major, 1955; Livingstone et al., 1958; Andersen, 1966; Berglund and Malmer, 1971) and seem to recur during each interglacial period (Iverson, 1958; Anderson, 1966, 1967; Miller 1986). The extent of the decline in pH and alkalinity depended largely on the characteristics of the soil, bedrock, climate, and vegetation. Acidification was greater in regions where geologic materials are more resistant to chemical weathering, where soils and glacial deposits are thinner, where rainfall is greater, and where vegetation results in greater output of organic acids. The early period of most rapid acidification was typically followed by long periods of relative stability of pH or very gradual decline (typically 1.0 pH unit or less). Some fluctuations in pH occurred and are attributed to changes in factors such as climate, watershed vegetation, and lake level changes. For European lakes especially, other events such as seawater intrusion (connection with saltwater environments), and land use changes caused by early inhabitants also affected lake water pH.

2. Watershed Disturbance

Logging, fire, blowdown, and similar sudden disturbances can change lake pH and alkalinity, although there is evidence that these changes are often relatively minor and short-lived (Gorham et al., 1979; Wright, 1981; Nilsson et al., 1982; Martin et al., 1984). Several factors can affect the response of a lake to watershed disturbance and determine whether pH increases or decreases and by how much and how fast. These include (1) the nature, extent, and severity of the disturbance; (2) the past frequency of similar disturbances; (3) the influence of drainage pattern and density, geology, soil type and texture, steepness of watershed slopes, and hydrology; and (4) taxonomic composition and rate of vegetation regrowth.

One well-documented response to disturbance is the initial decrease in pH following logging of watersheds at Hubbard Brook. This decrease was caused by increased stream water nitrate concentrations resulting from mineralization of soil organic matter and reduced plant uptake (Bormann and Likens, 1979). Another possible pattern of pH change would begin with an initial, rapid increase in pH as the flux of cations released from a watershed increased. This would be followed by a gradual decline to below-predisturbance pH as the cation flux decreased because of landscape stabilization and increased uptake by rapidly regrowing vegetation (Nilsson et al., 1982; Gorham et al., 1979). The loss of a forest canopy could change the type and amount of dry deposition, thereby altering the depositional flux of acidifying compounds. Alternatively, reforestation could lead to increased scavenging of aerosols and therefore increased S loading (Harriman and Wells, 1985). More gradual land use change, such as the decline in certain agricultural practices in Scandinavia in the past 100 to 200 years, might also cause lake acidification. According to Rosenqvist (1978), the regeneration of natural vegetation would lead to an increase in soil humus content and ion exchange capacity, removing more cations from soil water, which would result in surface water acidification. Over an even longer period of hundreds to thousands of years, the change of forest to heathland, particularly in western Europe, might have affected the proton flux to lakes and pools.

Detecting the effects of watershed disturbance in the sediment record can be difficult if water chemistry changes are short-lived. This is likely if the sediment interval(s) representing the time of disturbance was not analyzed or if altered diatom and chrysophyte assemblage composition was not discernible because of sediment mixing or some other factor.

3. Atmospheric Deposition of Strong Acids

Precipitation containing excess strong acids is falling on large areas of eastern North America (Barrie, 1984) and western Europe (EMEP/CCC, 1984). Available evidence suggests that the quantities of strong acids now being deposited are significantly greater than they were before the year 1850 and that this precipitation can cause acidification of lakes (e.g., Asman et al., 1988; National Research Council, 1986; Altshuller and Linthurst, 1984; Schwela, 1983; Brimblecombe and Stedman, 1982; Ministry of Agriculture, 1982; Odén, 1968). Some authors suggest, for example, that the concentration of sulfate in many Adirondack lakes may be five times higher than before 1800 (Galloway et al., 1983; Holdren et al., 1984).

4. Evaluating Causes of Recent Acidification

Causes of lake acidification can be evaluated using paleoecological data to test alternative hypotheses. Studies can involve single lakes, comparisons of sites that have, for example, high and low acidic deposition, or comparisons of areas that are forested or not. First, the logical consequences of each hypothesis are specified in terms of trends that can be observed in sediment cores. Then, these are compared with measured trends in sediment cores in attempts to demonstrate that the logical consequences do not occur, and therefore that the hypothesis is false (Charles and Whitehead, 1986a; Jones et al., 1986; Battarbee et al., 1985).

a. Natural Long-Term Acidification

Recent declines in lake pH caused by natural, long-term acidification processes should be continual throughout the period studied. Based on studies of long-term lake histories, the overall decline should be small, no more than about 0.1 pH unit in a 20-year period. There should be no major changes in taxa composition or at least no major shifts in the percentage of diatom valves assigned to each pH category. There should be no correlation between DI pH changes and indicators of watershed disturbance or acidic deposition.

b. Watershed Disturbance

If a pH decline was caused only by watershed disturbance and processes associated with vegetation recovery, the period of pH changes should logically be related to the time of the disturbance; the larger and more rapid the pH changes, the closer they should have been to the time of disturbance. A pattern that might be expected is a sudden change in pH followed by a gradual return to previous conditions, although there are few data to substantiate this contention. In the case of fire by itself, or in combination with logging, a few studies suggest there may be an increase in pH after the disturbance, followed by a gradual decrease (Dickman and Fortescue, 1984; Tolonen et al., 1986; Davis et al., 1989). In Scotland, Flower and Battarbee (1983) and Flower and others (1987) were able to falsify the watershed change hypothesis by showing that acidification of some lakes began well before their watersheds were planted with coniferous trees. In the Adirondacks of New York, lake acidification began before a major blowdown event (Johnson et al., 1987).

In addition to changing the pH, watershed disturbance may affect the nutrient input to lakes. This effect alone can cause changes in the composition of diatom assemblages. In the Netherlands, changes in human impact (e.g., discontinuation of sheep washing in moorland pools) play a role in acidification (Van Dam et al., 1988; Van Dam, 1989a).

c. Atmospheric Deposition of Strong Acids

Lake acidification can be attributed to acidic deposition if (1) the lake is in a region receiving deposition of excess strong acids and there is evidence of industrial pollution in the upper levels of the sediment; (2) DI pH declines occur after an increase in acidic deposition could reasonably be expected to have taken place, based on knowledge of current water chemistry and watershed acidification processes; and (3) changes are not correlated with, and cannot be attributed to, watershed changes or other local factors.

When these criteria are used to evaluate causes of acidification, several factors must be emphasized. First, more than one of the above processes may be affecting a lake simultaneously. Second, the history of each lake watershed system and nearby emission sources should be investigated thoroughly, so that all possible causes of acidification can be evaluated. Third, interpretations based on diatom and chrysophyte data should be consistent with interpretations of other sediment data and current models of lake acidification processes. Fourth, assessments of regional acidification trends should be based on studies of several lakes within a region.

III. Assessment of Recent Lake Acidification Trends in Eastern North America and Europe

A. Selection and Analysis of Data Sets

Diatom and chrysophyte assemblage data were chosen using the following criteria as guidelines: (1) present lake water alkalinity less than 200 μeq/L (acid-neutralizing capacity low enough so that lake pH and diatom assemblage composition could have been altered as a result of significant change in watershed characteristics or atmospheric deposition of acids; some lakes with higher alkalinity chosen as reference lakes that should have had stable pH in recent years); (2) minimal or, if not minimal, well-known watershed disturbance or influence from local emission sources, industrial effluents, or other factors; (3) sufficient quality and quantity of diatom or chrysophyte data, with taxonomic identification to lowest possible level, an adequate number of time intervals analyzed, and access to primary data including counts; (4) sufficient ecological information on at least the most common diatom taxa, including distribution with lake water pH within the region; (5) quality of dating of sediment; and (6) the availability of data for other lake, watershed, and sediment characteristics relevant to acidic deposition and lake acidification.

The steps in analyzing the data were as follows: (1) assemble all relevant data sets, (2) evaluate pH and acidification trends by type of lake and by geographic region, (3) evaluate potential causes of the trends, and (4) draw conclusions concerning effects of acidic deposition. Most of the lakes discussed in this chapter are listed in Tables 8–1 and 8–2. Many of the data are as yet unpublished because investigators have produced the data only recently.

Summaries of research results for the major regions are presented in varying formats. This largely reflects differences in the amount, quality, and consistency of data for making regional assessments, the extent to which the data have already been synthesized, and the most relevant issues that need to be addressed.

Inferred pH values are based on calibration sets developed for regions in which the study lakes are located. The calibration data and ecological analyses of diatom and chrysophyte taxa are summarized or referred to in the various papers cited in the sections.

B. North America

1. United States

To evaluate recent acidification trends in North America, diatom data sets from several regions (Figures 8–2 and 8–3; Table 8–1), meeting the minimal criteria listed in the section above, were assembled and reviewed. In the United States, most of the lakes in the Adirondack Mountains of New York (12), Upper Midwest (9), and northern Florida (6) and some in New England (3) were studied as part of

Table 8-1. Summary of the extent and timing of pH changes for lakes in North America and Europe. North American regions are ordered east to west; European regions west to east. See Figures 8-3, 8-7, 8-10, and 8-13 for maps showing lake locations.

Map number	Lake	Year of core	Current lake water pH	Surface sediment inferred pH	Inferred pH (pre-1850)	Inferred pH change[a]	Approximate date of first pH change	Primary cause of pH change	Reference
North America									
(Figure 8-3)									
USA, Adirondack Park, NY									
1	Honnedaga Lake	1976	4.7–4.8	5.2	6.1	−1.1	na	A	1, 2
2	Seventh Lake	1976	6.4–7.0	6.5	6.6	ns	na	na	1
3	Woodhull Lake	1976	5.1–5.3	6.0	6.1	ns	na	na	1
4	Panther Lake	1980	7.2	6.9	7.5	−0.6	post-1960	A	3
5	Sagamore Lake	1978	6.0	6.3	6.8	−0.5	1940s	A	3
6	Woods Lake	1980	4.9	4.8	5.2	−0.4	1930s	A	3
7	Barnes Lake	1983	4.7	4.7	5.2	−0.5	post-1900	A	4
8	Bear Pond	1983	5.0	5.5	5.9	−0.4	1920–1940	A	4
9	Big Moose Lake	1982	4.6–5.1	5.0	6.0	−1.0	1920–1940	A	4, 5
10	Clear Pond (E.L.)	1982	7.0	7.0	7.0	ns	na	na	4
11	Deep Lake	1983	4.7	4.8	5.4	−0.6	1940–1950	A	4
12	Lake Arnold	1982	4.8	4.7	5.1	−0.4	du	A	4
13	Little Echo Pond	1982	4.8	4.5	4.6	ns	na	na	4
14	Merriam Lake	1982	4.9	5.1	5.4	−0.3	1910–1930	A	4
15	Queer Lake	1983	5.3	5.4	6.4	−1.0	1930–1950	A	4
16	Upper Wallface Pond	1983	5.0	4.9	5.3	−0.4	1910–1930	A	4
17	West Pond	1982	5.5	5.7	5.6	ns	na	na	4
18	Windfall Pond	1982	6.5	6.9	6.8	ns	na	na	4
19	Holmes Lake	1984	5.0 (1980)	Limed	5.3	?	na	na	6

USA, New England									
20	Branch Pond (VT)	1980	4.7	4.5	4.8	−0.3	1930	A	7, 8
21	Cone Pond (NH)	1980	4.5–4.8	4.6–4.7	4.5–4.8	ns	na	na	10
22	E. Chairback Pond (ME)	1979	5.2	5.0–5.1	4.8–4.9	+0.2	1920–1930	A	9
23	Haystack Pond (VT)	1983	4.8	5.0	5.3	−0.3	1920	A	11
24	Klondike Pond (ME)	1978	6.2	5.5	5.7	ns	na	na	9
25	Ledge Pond (ME)	1978	5.6	4.4–4.7	5.2	−0.6	1880s	A	9
26	Little Long Pond (ME)	1983	5.6	5.8	6.0	−0.2	1950	A	11
27	Mountain Pond (ME)	1979	6.1	5.0	5.1	ns	na	na	7, 8
28	Mud Pond (ME)	1983	4.5	4.7	5.2	−0.5	1950	A	11
29	Solitude Pond (NH)	1979	4.8	5.0	4.9–5.2	ns	na	na	9
30	Speck Pond (ME)	1978	5.2	4.9	5.1	ns	na	na	9
31	Tumbledown Pond (ME)	1978	5.2	5.1	5.0	ns	na	na	9
32	Unnamed Pond (ME)	1979	4.8	5.1	5.1	ns	na	na	9
33	Duck Pond (MA)	1983	4.6	5.0	5.2	−0.2	du	?	12
Canada, Ontario									
34	Lake B	1981	5.2	4.8	6.4–6.6	−1.6	1955	S	13, 14
35	Lake CS	1981	5.2	6.4	7.2	−1.0	1955	S	14
36	Batchawana Lake	1981	6.1	6.2	6.0	ns	na	na	15
37	Clearwater Lake	1984	4.5	4.3	6.3	−2.0	1930	S	16
38	Hannah Lake	1984	4.3 (1970)	4.6 (1970)	6.0	−1.4	1880	S	16
39	Lohi Lake	1984	4.7	nd	nd	na	na	na	16
40	Quirke Lake	1984	5.5–6.4	5.5	6.3	−0.8	1955–1959	S	17
Canada, Quebec									
41	C-22	1983	5.8	5.4	5.5	ns	na	na	18
42	Key Lake	1983	5.5	5.2	5.3	ns	na	na	18
43	Blais Lake	1986	7.1	7.2	7.3	ns	du	na	19
44	Bonneville Lake	1986	5.4	5.4	5.6	−0.2	1940	A	19

Table 8-1. *(Continued)*

Map number	Lake	Year of core	Current lake water pH	Surface sediment inferred pH	Inferred pH (pre-1850)	Inferred pH change[a]	Approximate date of first pH change	Primary cause of pH change	Reference
45	Chevreuil Lake	1986	7.0	7.1	7.0	ns	na	na	19
46	Chomeur	1986	6.7	6.9	7.0	ns	na	na	19
47	David Lake	1986	7.3	7.1	7.0	ns	du	na	19
48	Kidney Lake	1986	6.8	6.8	6.8	ns	na	na	19
49	Lagou Lake	1986	5.8	5.6	5.7	ns	na	na	19
50	Lemaine Lake	1986	6.0	6.2	6.1	ns	na	na	19
51	Nolette Lake	1986	6.5	6.4	6.5	ns	na	na	19
52	Thomas Lake	1986	6.8	6.4	6.5	ns	na	na	19
53	Truite Rouge Lake	1986	6.1	5.7	5.8	ns	1943	A	19
Canada, New Brunswick and Nova Scotia									
54	Big Indian Lake (NS)	1981	nd	5.3	6.1	−0.8	post-1939	A	20
55	Emigrant Lake (NB)	1981?	6.1	6.1	6.5	−0.4	pre-1948	A	20
56	Kinsac Lake (NS)	1981	5.5	6.1	6.3	−0.2	post-1917	A	20
57	Lilly Lake (NB)	1981?	5.6	5.7	6.2	−0.5	du	A	20
58	Round Lake (NS)	1981	4.6	6.2	6.2	ns	na	na	20
59	St. Patricks Lake (NB)	1981?	6.7	6.8	6.8	ns	na	na	20
60	Tomoowa Lake (NB)	1981?	5.3	6.4	6.4	ns	na	na	20
61	Tupper Lake (NS)	1981	nd	nd	nd	nd	na	na	20
62	8	1984	5.3	5.3	5.3–5.6	−0.2–0.3	du	na	21
63	12	1984	5.1	5.1	5.3–5.4	−0.2–0.3	du	A?	21
64	204	1984	5.1	5.2	5.5	−0.3	du	A?	21
65	219	1984	5.6	5.5	5.5	ns	na	na	21
66	660	1984	5.5	5.6	5.7	ns	na	na	21
67	817	1984	5.2	5.3	5.7	−0.4	1930	A?	21

Canada, Newfoundland									
68	Aides Pond	1984	6.0	6.4	6.0	+0.4	1930–1940	na	22
69	Lake #660	1984	5.5	5.5	5.7	−0.2	1946		21
70	Stephensons Pond (206)	1984	5.6	5.9	6.0	ns	na	na	22
Canada, NW Ontario, Experimental Lakes Area									
71	Lake 223	1984	5.3–5.5	5.4	6.7–7.0	−1.4	1976?	E	23
USA, Northern Minnesota, Wisconsin, and Michigan									
72	Andrus Lake (MI)	1984	5.6	6.0	6.0	ns	na	na	24
73	Brown Lake (WI)	1984	5.7	5.9	6.4	−0.3	1900	?	24
74	Camp 12 Lake (WI)	1984	5.4	5.6	5.6	ns	na	na	24
75	Denton Lake (WI)	1984	5.2	5.5	5.7	−0.3	1870	?	24
76	Dunnigan Lake (MN)	1984	6.7	6.4	6.6	ns	na	na	24
77	Hustler Lake (MN)	1984	6.8	6.9	6.6	+0.3	na	na	24
78	McNearney Lake (MI)	1984	4.4	4.9	4.9	ns	na	na	24
79	Nels Lake (MN)	1984	6.4	6.5	6.6	ns	na	na	24
80	Otto Mielke Lake (WI)	1984	5.3	5.8	5.4	+0.4	na	na	24
81	Bastile Lake (WI)	1986	5.0	5.6	5.6	ns	na	na	25
82	Hillis Lake (WI)	1986	5.6	5.5	5.8	−0.3	du	A?	25
83	McGrath Lake (WI)	1986	5.2	5.4	5.6	−0.2	du	na	25
84	Morgan Lake (WI)	1986	4.6	5.4	5.5	ns	du	na	25
85	Sugar Camp Lake (WI)	1986	5.4	5.7	5.0	ns	na	na	25
USA, Northern Florida									
	Fore Lake	1983	5.1	5.1	5.0	ns	na	na	26
	Lake Barco	1983	4.5	4.2	4.9	−0.7	1958	A?, F?	26
	Lake Lou	1983	6.3	6.8	6.8	ns	na	na	26
	Lake Mary	1983	4.6	4.8	4.9	ns	na	na	26
	Lake Suggs	1983	4.9	5.4	6.3	−0.9	Late 1800s	A?, F?	26
	Mirrow Lake	1983	4.9	5.0	5.1	ns	na	na	26

Table 8-1. *(Continued)*

Map number	Lake	Year of core	Current lake water pH	Surface sediment inferred pH	Inferred pH (pre-1850)	Inferred pH change[a]	Approximate date of first pH change	Primary cause of pH change	Reference
USA, Rocky Mountains									
	Emerald Lake (CO)	1982	6.7	6.7	6.8	ns	na	na	27
	Lake Haiyaha (CO)	1982	6.3	6.8	6.7	ns	na	na	27
	Lake Husted (CO)	1982	6.8	6.7	6.5	ns	na	na	27
	Lake Louise (CO)	1982	6.8	6.9	6.8	ns	na	na	27
	Black Joe Lake (WY)	1984	6.6	6.8	6.8	ns	na	na	28
	Deep Lake (WY)	1984	nd	6.7	6.7	ns	na	na	28
	Hobbs Lake (WY)	1984	6.4	6.8	6.8	ns	na	na	28
USA, California, Sierra Nevada									
	Emerald Lake	1986	6.3	6.6	6.4	+0.2	na	na	29, 30
	Eastern Brook Lake	1986	7.1	7.1	6.9	+0.2	na	na	30
	Harriett Lake	1986	6.5	6.4	6.4	ns	na	na	30
	Lake 45	1986	5.2	5.7	5.9	−0.2	na	na	30
Europe (Figure 8-7)									
Wales									
1	Llyn Hir	1984	4.8[d]	nd	6.1	−1.3[e]	1850	A	31
2	Llyn Gynon	1985	5.2	5.5	6.0	−0.5	1945	A	31
3	Llyn y Bi	1985	4.9	4.7	6.0	−1.3	1900	A	31
4	Llyn Dulyn	1985	4.9	4.7	6.0	−1.3	1850	A	31
5	Llyn Eiddew Bach	1985	4.8	nd	6.6	−1.8	1940?	A	31
6	Llyn cwm Mynach	1985	5.6	5.6	6.1	−0.5	1850	A	31
7	Llyn Llagi	1985	4.9	4.6	6.1	−1.5	1850	A	31

Scotland									
8	Loch Enoch	1982	4.5	4.4	5.3	−0.9	1840	A	31
9	Loch Valley	1981	4.7	4.6	5.6	−1.0	1860	A	31
10	Round Lake of Glenhead	1981	4.7	4.9	5.9	−1.0	1860	A	31
11	Loch Dee	1980	5.3	5.5	6.2	−0.7	1890	A	31
12	Loch Grannoch	1980	4.7	4.7	5.7	−1.0	1930	A	31
13	Loch Fleet	1985	4.5	4.6	5.8	−1.2	1975	A, R	31
14	Loch Skerrow	1980	5.3	5.8	5.9	ns	na	na	31
15	Loch Urr	1984	6.8	6.8	6.5	ns	na	na	31
16	Loch Tanna	1986	5.0	4.6	5.0	−0.4	1850	A	31
17	Loch Tinker	1985	5.7	6.0	6.4	−0.4	1850	A	31
18	Loch Chon	1986	5.4	5.5	6.2	−0.7	1900[c]	A, R	31
19	Loch Laidon	1985	5.4	5.3	5.8	−0.5	1860	A	31
20	Lochnagar	1986	5.0	4.8	5.7	−0.9	1850	A	31
21	Dubh Loch	1986	5.3	5.2	5.7[b]	−0.5	du	A	31
22	Loch nan Eun	1986	5.0	4.8	6.0[b]	−1.2	du	A	31
23	Lochan Uaine	1986	5.8	5.7	5.8	ns	na	na	31
24	Lochan Dubh	1986	5.6	5.2	5.5	−0.3	du	A	31
England									
25	Scoat Tarn	1984	5.0	4.6	6.0	−1.4	1850	A	31
26	Low Tarn	1984	5.0	4.6	6.0[b]	−1.4	du	A	31
27	Greendale Tarn	1984	5.2	4.9	6.0	−1.1	1900	A	31
28	Burnmoor Tarn	1979	6.4	6.2	6.3	ns	na	na	31
(Figure 8-10)									
Norway									
1	Langtjern	1975	4.7–5.2	<5.5	4.3–6.2	ns	na	na	32
2	Övre Målmesvatn	1977	4.5	<5.2	4.4–6.5	−0.5	1930	A	33
3	Blåvatn	1978	5.1	5.1	5.2	−0.1	1930	A	34, 35
4	Hovvatn	1978	4.4	3.9–4.4	4.8–5.4	−0.7	1918	A	34, 35

Table 8-1. *(Continued)*

Map number	Lake	Year of core	Current lake water pH	Surface sediment inferred pH	Inferred pH (pre-1850)	Inferred pH change[a]	Approximate date of first pH change	Primary cause of pH change	Reference
5	Brårvatn	1979	5.2	5.2–5.3	5.6–6.3	−0.7	1900	A	35, 36, 37
6	Botnavatn	1978	5.7	5.7	5.9	−0.2	1920	A	35
7	Dorsvatn	1978	5.0	5.0	5.0	ns	na	na	35
8	Grönlivatn	1978	6.5	6.5	6.5	ns	na	na	35
9	Holmvatn	1978	4.7	4.6	4.9–5.2	−0.5	1930	A	35, 37
10	Nedre Målmesvatn	1979	4.6	4.4–4.5	4.9–5.3	−0.6	1890	A	35
11	Oppljosvatn	1979	5.8	5.8	5.8	ns	na	na	35
12	Gulspettvatn	1985	4.9	4.8	5.6	−0.8	du	A	38
13	Röyrtjörna	1987	6.5	6.0	6.0	ns	na	na	38
14	Verevatn	1986	4.4	4.4	5.1	−0.7	du	A	38
15	Holetjern	1986	4.6	4.6	5.1	−0.5	du	A	39
16	Ljosvatn	1986	4.5	4.6	5.2	−0.6	du	A	39
17	Skomakarvatn	1987	4.6	4.5	5.1	−0.6	du	A	39
Sweden									
18	Stora Skarsjön	1971	4.5	4.5	6.0	−1.5	du	A	40
19	Gårdsjön	1979	4.6	4.5	6.0	−1.5	1950	A	41
20	Härsvatten	1979	4.4	4.2	5.9	−1.7	du	A	41
21	Lysevatten	1979	5.9	5.2[f]	6.1	−0.9[f]	du	A	41
22	Lilla Öresjön	1986	4.5	4.6	6.1	−1.5	du	A	42
Finland									
23	Hauklampi	1982	4.8–4.9	5.1–5.4	6.0–6.4	−1.0	1962	A	43
24	Häkläjärvi	1982	4.8–5.2	5.6	6.3	−0.7	du	A	43

25	Orajärvi	1982	4.7–4.8	4.9	6.3	−1.4	1930	A	43
26	Hirvilampi	1983	5.6	5.5	6.0–6.4	−0.7	1960	A	44
27	Matalajärvi	1983	4.7	4.9	5.8–6.2	−1.0	1950	A	44
28	Munajärvi	1983	4.7	5.2	5.5–6.0	−0.6	1950	A	44
29	Pappilanlampi	1966	5.3	5.0–5.5	5.0–5.5	ns	na	na	44
30	Työtjärvi	1975	5.1–6.7	6.6–7.0	6.6–7.0	ns	na	na	44
31	Valkeislampi	1983	5.1	5.2	5.2	ns	na	na	44
32	Valkjärvi	1983	5.1	5.5–5.7	6.1–6.4	−0.7	1930	A	44
33	Vitsjön I	1979	5.9	5.9–6.4	6.8	−1.0	du	W	44
34	Vitsjön II	1983	4.3	5.6	6.0–6.2	−0.5	1930	A	44
35	Vuorilampi	1983	5.9	5.3–5.5	6.3–6.5	−1.0	1960	W	44
36	Iso-Hanhijärvi	1984	4.9	5.0	5.0–5.5	−0.3[g]	du	A[g]	45
37	Kakkisenlampi	1984	4.8	5.0	4.5–5.1	ns	na	na	45
38	Kangasjärvi	1984	5.3	5.2	5.2–5.5	−0.3[g]	1960	A[g]	45
39	Kiiskilampi	1984	4.8	4.3	4.7	−0.4	du	A	45
40	Pieni-Kalliojärvi	1984	5.2	5.0	nd	—	na	na	45
41	Pieni-Löytönen	1984	5.1	6.0	6.5–7.0	−0.8[g]	du	A[g]	45
42	Siikajärvi	1984	5.3	5.1	5.9	−0.8	1960	A	45
43	Sonnanen	1984	6.5	6.5	6.5	ns	na	na	45
44	Suo-Valkeinen	1984	4.9	4.7	5.0	−0.3	1965	A	45
45	Valkealampi	1984	4.9[h]	4.7	4.5–4.8	ns	na	na	45
46	Iso-Tiilijarvi	1983	5.1	5.2–5.7	5.6–6.0	−0.2	du	A	46
47	Kankareenjärvi	1984	5.4	5.4–5.9	5.7–6.2	−0.3	1965	A	46
48	Koukjärvi	1984	5.6	5.5–6.0	5.7–6.4	−0.3	1965	A	46
49–72	Work in progress; see Figure 8-10 for pH change								47
73–78	Work in progress; see Figure 8-10 for pH change								48
(Figure 8-13)									
The Netherlands									
1	Kliplo	1985	5.3	5.0	4.5	−1.2[i,g]	1910	A, W	49
2	Gerritsfles	1985	4.3	4.2	4.4	−1.7[i,g]	1900	A, W	49
3	Achterste Goorven	1985	3.9	4.5	4.2	−1.7[i,g]	1920	A, W	50

Table 8–1. *(Continued)*

Map number	Lake	Year of core	Current lake water pH	Surface sediment inferred pH	Inferred pH (pre-1850)	Inferred pH change[a]	Approximate date of first pH change	Primary cause of pH change	Reference
West Germany									
4	Pinnsee	1983	—	4.7	6.0	−1.3	1930	A, W	51
5	Plötschersee	1984	5.8	6.0	6.5	−0.5	After 1900	A, W?	52
6	Grosser Bullensee	1983	3.7	3.7	4.5	−0.8	After 1930	A, W	52
7	Kleiner Bullensee	1983	3.7	3.7	4.5	−0.8	After 1930	A, W	52
8	Oderteich	1984	4.5	4.8	5.1	−0.3	du	A, W, F	52
9	Herrenwiesersee	1983	3.6–4.1	3.7–4.3	4.5	−0.2–0.7	1880	A, W, F	51
10	Huzenbachersee	1984	4.3	4.3	5.0	−0.7	c. 1850	A, W, F	52
11	Schurmsee	1984	4.1	4.0	4.7	−0.7	c. 1850	A, W, F	52
12	Wildsee (Ruhestein)	1984	4.3	4.4	4.4	ns	na	na	52
13	Glaswaldsee	1984	4.3	4.8	4.8	ns	na	na	52
14	Feldsee	1983	6.3	6.0	6.1	ns	na	na	52
15	Grosser Arbersee	1983	4.5–5.1	4.7	5.5	−0.8	1960	A, S, F	51
16	Kleiner Arbersee	1981	4.5–5.1	4.7	5.5	−0.8	1930	A, F	51
17	Rachelsee	1983	4.3–4.9	4.4–4.6	5.1–5.2	−0.5	1950	A, S, F	53
Austria									
18	Mutterbergsee	1985	5.2	5.2	4.8	+0.4	du	na	52
19	Oberer Plenderlesee	1984	7.0	7.3	6.9–7.1	+0.3	du	na	54
20	Schwarzsee ob Solden	1984	5.2–5.4	5.1	5.1–5.5	ns	na	na	54
21	Goasselesee	1984	6.2	6.0	6.0–6.1	ns	na	na	54
22	Gippersee	1984	6.2	6.9	6.7–6.8	ns	na	na	54

Czechoslovakia									
23	Teufelssee	1986	4.3	4.3	4.8	−0.5	After 1900	A, F, W	52

References are as follows: 1. Del Prete and Schofield (1981); 2. Charles, unpublished data; 3. Davis et al. (1988); 4. Charles et al. (1989); 5. Charles et al. (1989); 6. Rhodes (1985); 7. Norton et al. (1981); 8. R. B. Davis and D. S. Anderson, pers. comm.; 9. Davis et al. (1983); 10. Ford (1986); 11. R. B. Davis, M. C. Whiting, and D. S. Anderson, pers. comm.; 12. Winkler (1985, 1988); 13. Fortescue (1984); 14. Dickman et al. (1984); 15. Delorme et al. (1986); 16. Dixit (1986b), Dixit and Evans (1986), Dixit et al. (1987a); 17. McKee et al. (1987); 18. Hudon et al. (1986); 19. Dixit et al. 1987b, 1988c, 1988d), S. Dixit, Queen's University, pers. comm.; 20. Elner and Ray (1987); 21. Scruton and Elner (1986), Scruton et al. (1987a); 22. Scruton et al. (1987b); 23. Davidson (1984), Dickman et al. (1988); 24. Kingston et al. (1989); 25. Kingston and Kreis (1987); 26. Sweets et al. (1989); 27. Baron et al. (1986); 28. Stuart (1984), Norton and Kahl (1987), Beeson (1988); 29. Holmes (1986); 30. M. C. Whiting and D. R. Whitehead, pers. comm.; 31. Battarbee et al (1988); 32. Berge (1975); 33. Berge (1979); 34. Davis and Berge (1980); 35. Davis et al (1983); 36. Davis et al (1980); 37. Davis (1987); 38. Berge, pers. comm.; 39. Berge and Birks, pers. comm.; 40. Miller (1973); 41. Renberg and Hellberg (1982); 42. Renberg, unpublished; 43. Tolonen and Jaakkola (1983); 44. Tolonen et al. (1986); 45. Simola et al. (1985); 46. Tolonen and Suksi, pers. comm.; 47. Liukkonen, pers. comm.; 48. Turkia, pers. comm.; 49. Van Dam et al (1988); 50. Dickman et al. (1987); 51. Arzet et al. (1986a); 52. Arzet (1987); 53. Steinberg et al (1984); 54. Arzet et al. (1986b).

E. L. = Elk Lake; nd = no data; ns = not significant (pH decrease ≤ 0.2 units); na = not applicable; du = ^{210}Pb dating uncertain or core not dated.

Primary cause of pH change (other factors may also contribute): A = acid deposition (long-range transport); S = local acidic deposition source; L = liming; E = experimental acidification; W = watershed disturbance (e.g., logging, fire); H = sheep washing; R = reforestation?; F = water-level changes.

[a] Surface sediment minus pre-1850 DI pH.

[b] Calculated from sample at base of the core.

[c] Main change after reforestation (post-1970).

[d] Preliming data.

[e] Calculated using modern measured pH.

[f] Affected by liming in 1974.

[g] Reacidified after cultural eutrophication.

[h] Before liming.

[i] Maximal DI pH (1900–1920) minus current diatom-inferred pH.

Table 8-2. Changes in diatom-inferred pH between old and new samples from pools and lakes in western and central Europe.

Number	Name	Depth (m)	Area (ha)	Altitude (m)	Old sampling (diatoms)	Inferred old pH	New sampling (diatoms)	Inferred new pH	Inferred pH change	New sampling (pH measurement)	New pH (measurement)	Apparent cause of pH change
Denmark												
31	Lille Helmiskaer-W.	0.1	0.69	9	1945	6.9	1982	5.6	−1.3	1982	5.1	A
32	Birkemøse	0.5	2.69	7	1945	6.0	1982	5.0	−1.0	1982	4.7	A
33	Mosso	6.0	3.63	70	1941	5.1	1982	4.3	−0.8	1982	4.2	A
34	Lille Økssø	5.5	0.34	62	1943	4.3	1982	4.2	ns	1982	3.7	A
35	Sokland	1.0?	11.50	16	1959	5.0	1982	5.1	ns	1977–1982(23)	5.2	na
36	Tormål	0.5	7.71	16	1950	6.2	1982	6.6	+0.4	1982	7.3	na
37	Praestekaer 2	1.0?	1.34	16	1967	5.1	1982	5.0	−0.1	1977–1982(23)	5.0	A
38	Hykaer	1.0	13.00	16	1950	4.5	1982	5.2	+0.7	1977–1982(27)	4.6	na
39	Grønbakke	0.1	3.38	5	1943	7.0	1982	5.1	−1.9	1982	4.9	A
40	Hundsø	7.0	1.81	75	1944	4.9	1982	4.8	−0.1	1977–1982(27)	4.4	A
41	Raevsø	8.0	5.00	74	1949	6.7	1977–1982(2)	6.3	−0.4	1977–1982(27)	6.0	A
42	Grane Langsø	11.5	11.40	74	1949	6.5	1977–1982(2)	4.9	−1.6	1977–1982(27)	5.3	A
43	Kalgaard Sø	11.0	10.50	74	1944	6.5	1982	6.5	ns	1976–1982(37)	6.7	na
44	Kongsø	9.3	3.00	74	1949	7.2	1982	4.8	−2.4	1977–1982(27)	5.0	A
45	Lille Gribsø	6.0	0.39	50	1943	4.7	1982	4.8	ns	1982	5.0	A
46	Hjortesøle	3.5	0.05	10	1943(2)	5.1	1982(2)	5.5	+0.4	1982	6.1	na
47	Sorte Sø	4.8	1.10	66	1947	5.7	1982	5.2	−0.5	1982	4.8	A
The Netherlands												
48	Van Hunenplak	0.8	0.69	4	1963	5.6	1983	5.0	−0.6	1979–1983(16)	4.5	A
49	Reeënveen	1.0?	0.19	15	1929	4.2	1978	4.2	ns	1978	4.3	na
50	Diepveen	1.2	0.81	10	1924	4.9	1978–1986(3)	4.2	−0.7	1978–1986(3)	4.5	A
51	Poort 2	0.5	0.01	10	1924(2)	4.3	1978–1986(4)	4.2	ns	1978–1986(3)	4.5	na
52	Kliplo	1.1	0.62	13	1924–1929(2)	3.0	1972–1984(11)	4.6	−0.4	1970–1984(34)	5.3	A
53	Schurenberg	1.3	1.00	13	1924	4.5	1978	4.6	+0.1	1978	3.8	A
54	Pool in Echtenerzand	0.5	0.25	10	1933	4.4	1978–1986(3)	4.1	−0.3	1978–1982(3)	3.7	A
55	Gerritsfles	1.2	6.78	40	1916–1918(7)	4.8	1964–1984(21)	4.1	−0.7	1974–1984(18)	4.3	A, W
56	Kempesfles	0.5	0.28	25	1918	4.8	1973–1986(4)	4.1	−0.7	1973–1986(4)	3.6	A, W
57	Deelense Was	0.6	0.50	45	1917	6.4	1983	4.2	−2.4	1983(7)	4.1	A, W
58	Bloempjesven	1.0	0.63	17	1918	4.4	1982	4.1	−0.3	1982	3.8	A
59	Ganzenven	0.2	0.28	7	1950	4.1	1982	4.1	0.0	1982	4.3	A
60	Goudbergven	1.0	1.45	13	1929	4.2	1982	4.0	−0.2	1982	3.7	A
61	Zwarte Goor	0.4	5.25	12	1929	4.4	1982	4.0	−0.4	1982	3.7	A, W
62	Schaapsven	1.0	1.56	11	1919–1925(2)	4.6	1975–1982(3)	4.1	−0.5	1975–1986(5)	4.5	A, W
63	Galgeven	3.6	11.30	11	1919	4.9	1978	4.2	−0.7	1978	4.0	A
64	Diaconieven	1.6	1.45	10	1916	4.4	1978	4.0	−0.4	1978	3.9	A

65	Brandven	1.0	1.40	10	1926	4.3	1978	4.0	−0.3	1978	3.8	A
66A	Achterste Goorven A	1.8	2.35	10	1919–1929(3)	6.8	1978–1984(11)	4.3	−2.5	1975–1984(18)	4.1	A, W
66B	Achterste Goorven B	1.8	2.35	10	1925–1929(5)	5.8	1975–1984(12)	4.1	−1.7	1975–1984(20)	3.9	A, W
66C	Achterste Goorven E	1.8	2.35	10	1919–1928(12)	4.7	1975–1982(12)	4.0	−0.7	1978–1984(19)	3.9	A, W
67	Klein Aderven	1.3	1.10	10	1919–1920(2)	4.5	1978	4.0	−0.5	1978	4.2	A
68	Middelste Wolfsputven	0.6	0.72	9	1921(2)	4.4	1978–1982(3)	4.3	−0.1	1978–1986(5)	4.6	A
69	Tongbersven-W.	1.0	0.09	8	1919	4.4	1984	4.6	ns	1984	4.9	na
70	Groot Huisven	1.7	3.36	8	1929(2)	5.0	1976–1986(9)	4.2	−0.8	1978–1986(25)	4.7	A
71	Diepmeerven	1.5	2.80	25	1932	4.8	1982	4.0	−0.8	1981–1982(2)	3.5	A, W
72	Pikmeeuwenwater-SE	1.0	0.31	20	1952	4.5	1982	4.4	−0.1	1982	4.2	A
73	Ravenven-NW	0.6	0.40	25	1952	4.4	1982	4.0	−0.4	1982	3.7	A
74	Rouwkuilenven	1.0	2.13	30	1957	4.3	1983	4.4	+0.1	1982–1983(45)	4.9	A
75	Onderste Rolven-E	0.5	0.25	43	1956	5.4	1983	4.5	−0.9	1983	4.4	A
Belgium												
76	Pool W. of Zwartwater	0.5	0.22	29	1935	4.5	1982(2)	4.1	−0.4	1975–1982(18)	3.9	A
77	Pool N.W. of Zwartwater	0.5	0.16	29	1935	4.5	1982	4.1	−0.4	1975–1982(17)	3.7	A
78	2 pools in Fagne d'Ardenne	2.0	0.005–.01	650	1952–1953(2)	4.3	1982(2)	4.3	0.0	1982(2)	3.9	A, W
79	2 pools in Fagne de Massehottéé	0.2	0.02	590	1949(2)	4.3	1982(2)	4.6	+0.3	1982	4.6	na
80	Pool in Grande Fagne	0.2	0.003	560	1952	4.5	1982	4.4	ns	1982(2)	4.8	na
West Germany												
81	Plotschersee	13.3	8.68	42	1924	5.2	1983	6.0	+0.8	1983(1)	5.8	na
82	Grosser Bullensee	11.0	9.80	32	1928	4.4	1983	4.0	−0.4	1983(20)	3.8	A
83	Erdfallsee	11.3	4.50	43	1928	7.1	1983	5.9	−1.2	1978–1983(14)	6.1	A, W
84	Heideweiher	1.0	1.92	44	1928	6.1	1983	4.6	−1.5	1983	4.9	A, W
85	Weinfelder Maar	52.0	15.90	484	1942	7.6	1983	7.4	ns	1977–1983(3)	7.3	na
86	Wildsee (Kaltenbronn)	3.0	2.30	908	1961	4.1	1983(2)	4.1	0.0	1983(2)	3.6	A, F
87	Hohlohsee	3.0	0.70	980	1961	4.2	1983	4.1	−0.1	1983(2)	3.5	A, F
88	Feldsee	34.5	9.15	1109	1839–1909(4)	6.7	1983(2)	6.9	ns	1983(3)	6.7	na
89	Fichtsee	6.5	1.21	598	1916	4.2	1983(2)	4.0	ns	1983	3.9	na
90	Grosser Arbersee	15.9	7.72	935	1916–1949(5)	5.1	1983	4.7	−0.4	1979–1983(5)	4.9	A, S, F
91	Kleiner Arbersee	9.7	9.40	918	1900–1913(6)	5.1	1983	5.0	−0.1	1982–1983(8)	4.4	A, F

When more than one sample is available, the number of samples is given between parentheses and the values are medians. Inferred pH values are calculated by weighted averaging (Ter Braak and Van Dam, 1988). Diatom counts for localities 19, 23, 33, 34, and 37 from Van Dam et al. (1981); for localities 22, 25, and 36 from Van Dam (1988); other diatom counts original. New pH measurements were done simultaneously with diatom sampling and taken from various sources (Van Dam, in prep.). Inferred pH changes less than 0.3 units are not considered to be significant, unless coinciding with important changes in the relative abundance of the acidobiontic diatom *Eunotia exigua*.

Apparent causes of pH change: A = acid deposition (long-range transport); S = local acidic deposition source; W = watershed disturbance (e.g., logging, sheep washing); F = changes of water level; ns = not significant; na = not applicable.

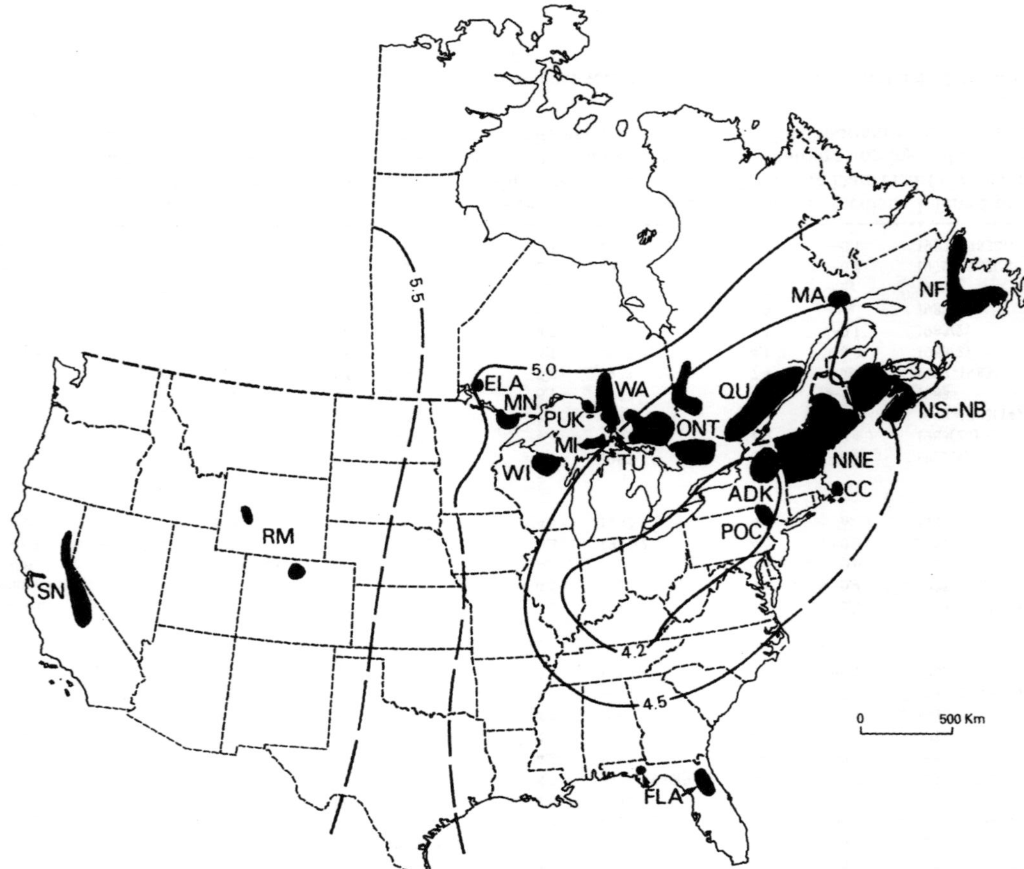

Figure 8–2. Areas of North America where paleolimnological studies to infer past pH using diatom and chrysophytes are completed or in progress. Area boundaries include lakes in calibration data sets and those analyzed stratigraphically (see Table 8–1). Contours of precipitation pH are from Barrie and Hales (1984).

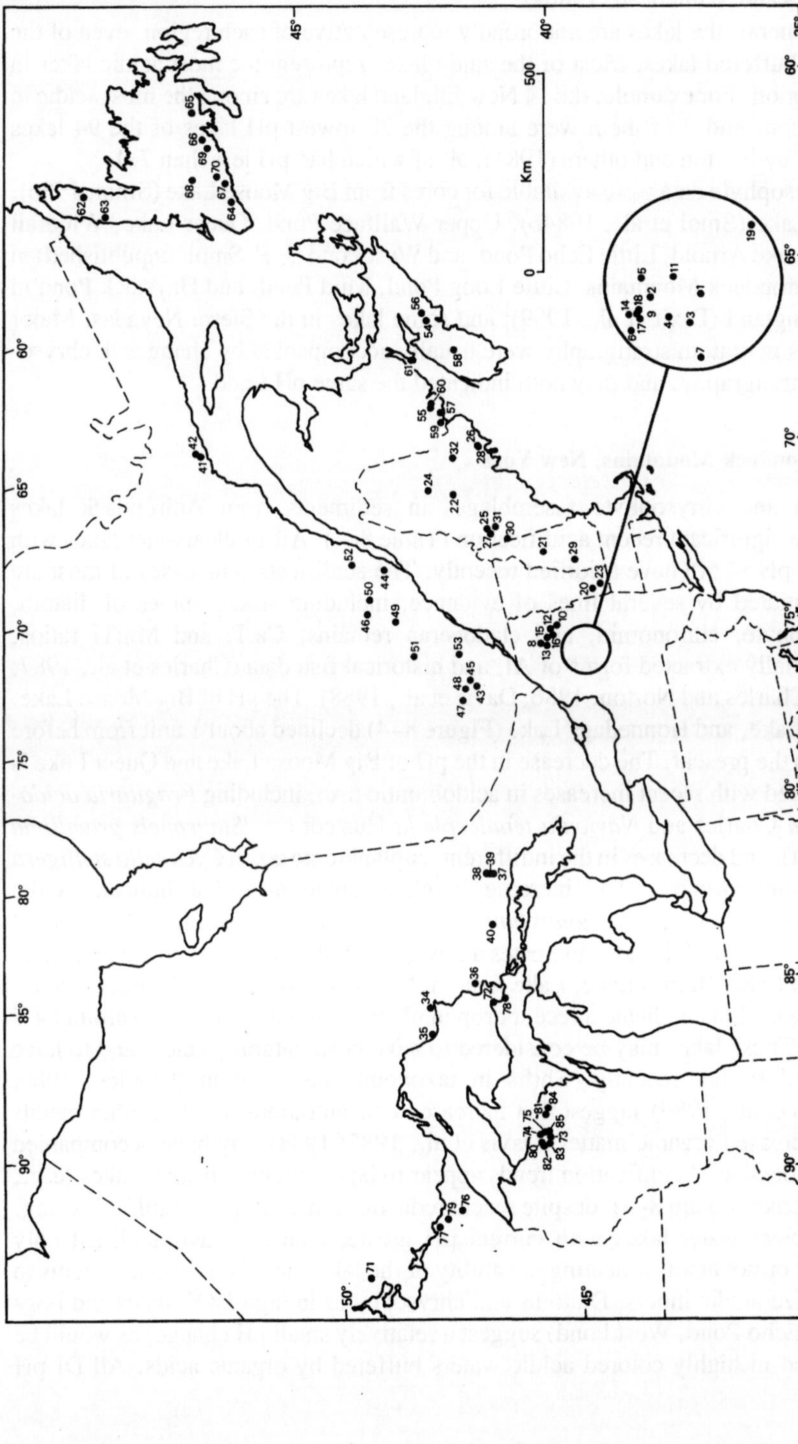

Figure 8–3. Lakes in the northeastern United States and southeastern Canada studied to infer historical diatom and chrysophyte pH. Lake names corresponding to map numbers, current pH, and inferred pH data are listed in Table 8–1.

the Paleoecological Investigation of Recent Lake Acidification project (PIRLA) (Charles and Whitehead, 1986a).

In general, the lakes are not broadly representative of each region, even of the poorly buffered lakes. Most of the study lakes represent the most acidic lakes in each region. For example, the 14 New England lakes are among the most acidic in the region, and 9 of them were among the 20 lowest-pH lakes of the 94 lakes studied by Norton and others (1981), all of which had pH less than 7.0.

Chrysophyte data were available for cores from Big Moose Lake (Smol, 1986), Deep Lake (Smol et al., 1984b), Upper Wallface Pond, Queer Lake, Windfall Pond, Lake Arnold, Little Echo Pond, and West Pond (J. P. Smol, unpublished) in the Adirondack Mountains; Little Long Pond, Mud Pond, and Haystack Pond in New England (Dixit et al., 1989); and some lakes in the Sierra Nevadas. Major changes in diatom stratigraphy were usually accompanied by changes in chrysophyte stratigraphy, and they both indicated the same pH trend.

a. Adirondack Mountains, New York

Diatom and chrysophyte assemblages in sediments from Adirondack lakes indicate significant recent acidification (Table 8–1). All 11 clearwater lakes with current pH < 5.5 have acidified recently. The acidification histories of most are substantiated by several lines of evidence, including stratigraphies of diatom, chrysophyte, chironomid, and cladoceran remains, Ca:Ti and Mn:Ti ratios, sequentially extracted forms of Al, and historical fish data (Charles et al., 1987, 1989; Charles and Norton, 1986; Davis et al., 1988). The pH of Big Moose Lake, Queer Lake, and Honnedaga Lake (Figure 8–4) declined about 1 unit from before 1800 to the present. The decrease in the pH of Big Moose Lake and Queer Lake is associated with recent increases in acidobiontic taxa, including *Fragilaria acidobiontica* Charles and *Navicula tenuicephala* Hustedt (= ?*Stauroneis gracillima* Hustedt), and decreases in the indifferent, euplanktonic taxon *Cyclotella stelligera* Cleve and Grunow. The increase in chrysophyte taxa that indicate acidic conditions, such as *Mallomonas hamata* Asmund and *M. hindonii* Nicholls (Smol et al., 1984a, 1984b), also indicates a lowering of pH. Diatoms and chrysophytes in Deep Lake, Barnes Lake, Lake Arnold, Merriam Lake, Upper Wallface Pond, and Woods Lake indicate a recent drop in pH from about 5.0 to 5.5 to around 4.7 to 4.9. These lakes may be considered to have been naturally acidic and to have acidified further recently. Shifts in taxonomic composition (Charles, 1986; Charles et al., 1989) suggest that increased concentrations of Al or other metals and decreased organic matter (Davis et al., 1985, 1988) may have accompanied the pH decline. Acidification trends appear to be continuing in some lakes (e.g., Queer Lake; Figure 8–4), despite recent reductions in atmospheric sulfur loading.

All clear water lakes with current pH greater than 6.0 have acidified only slightly or not at all, reflecting the ability of the lake and/or watershed systems to neutralize acidic inputs. Diatoms and chrysophytes in high DOC lakes and bogs (Little Echo Pond, West Pond) suggest a relatively small pH change, as would be expected in highly colored acidic waters buffered by organic acids. All DI pH

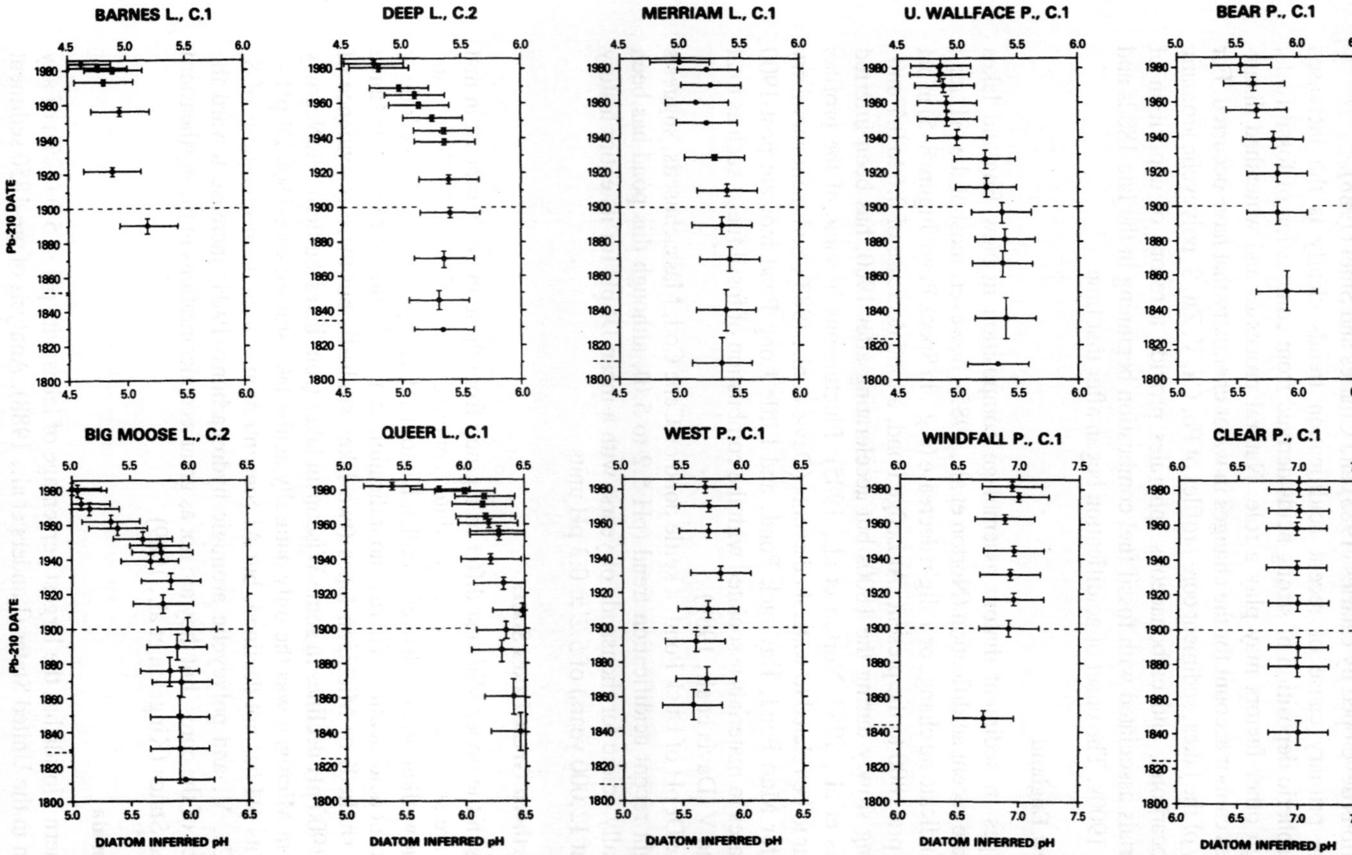

Figure 8–4. Diatom inferred pH versus ^{210}Pb date for ten Adirondack study lakes, calculated using multiple regression of Hustedt pH groups (Figure 8–1, Charles and Smol, 1988). Error bars for DI pH values are the standard error of the regression (± 0.26 pH units). Error bars for ^{210}Pb dates are standard deviations. (Binford, 1989.)

values (Table 8–1), except for Seventh Lake, were calculated using inference equations developed by Charles (1985) and Charles and Smol (1988).

The primary cause of recent acidification trends clearly is the increased atmospheric deposition of strong acids derived from combustion of fossil fuels, though other factors may play a role. Natural processes and watershed disturbances cannot account for the changes in water chemistry that have occurred. For many of the lakes, sediment core profiles of Pb, Ca, V, Zn, S, polycyclic aromatic hydrocarbons, and carbonaceous spherules provide a record of deposition of materials associated with fossil fuel combustion beginning in the late 1800s and early 1900s. The onset of acidification began after that time.

b. New England

Changes in sediment diatom assemblage composition in New England lakes indicate recent acidification (Norton et al., 1985); however, most of the pH trend data indicate no change or a slight decrease (e.g., in Speck Pond; Figure 8–5) in pH from pre-1800 to the present. At Ledge Pond, a total decrease of 0.6 to 0.7 unit, starting slowly during the 1800s but accelerating after 1960, has been inferred (Davis et al., 1983; Norton et al., 1985). Fluctuations in some of the profiles appear to be related to watershed events (Davis et al., 1983). Sediment chemistry data for Mud Pond, Haystack Pond, and Little Long Pond indicate post-1900 increases in materials associated with the combustion of fossil fuels, such as total Pb and V (Davis et al., 1989).

The DI pH of Duck Pond, a kettle pond on Cape Cod, Massachusetts, suggests a slight recent acidification trend (pH 5.2 to 5.0), although this pond has been naturally acidic for thousands of years, with a mean DI pH for its entire history (about 12,000 years) of 5.2 ± 0.3 pH units.

c. Northern Great Lakes States

Lakes in the upper Midwest differ as a group from those in the Northeast in that most are seepage lakes, are very dilute (conductivities as low as 9 μS/cm), and have low alkalinity (-38 to 80 μeq/L). Diatom-inferred pH trends indicate little or no recent acidification. There is no indication of acidification of Minnesota (three lakes studied) or Michigan lakes (two lakes studied), but there is evidence of post-1900 pH declines in some Wisconsin lakes (four lakes studied). McNearney Lake in Michigan was the only naturally acidic lake studied (pre-1800 DI pH = 4.9); its pH has not declined, but Al concentration is high. Concentrations of S, Pb, Cu, V, and polycyclic aromatic hydrocarbons (PAH) increase toward the surface of the cores, but they are not as great as concentrations in the northeastern United States (Kingston et al., 1989).

d. Florida

Northern Florida has the largest percentage of lakes with pH < 5.0 (12%) of any region in the United States (Landers et al., 1988). Analysis of pre-1850 sediment

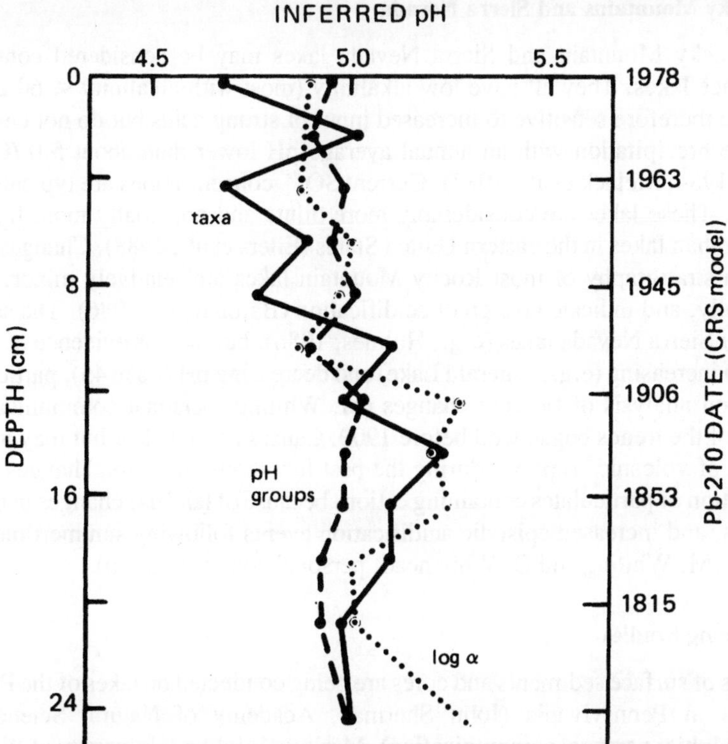

Figure 8–5. Diatom-inferred pH for Speck Pond, Maine, calculated from multiple-regression analysis of taxa and pH groups and from log 10 index α (From Battarbee, 1984; Charles and Norton, 1986.) Lead-210 dates have been added to published figure. (Data from R. B. Davis and S. A. Norton, University of Maine, Orono, personal communication.) CRS refers to constant rate of supply model.

core diatom assemblages indicates that four of six lakes had low pH values due to natural causes (Table 8–1; Sweets et al., 1989). Low cation concentrations are probably attributable to the low cation exchange capacity and base saturation of the deep, coarse sands in which the lakes are situated and to isolation from the local groundwater table (Floridan aquifer) by the clay-rich Hawthorne formation. The high SO_4^{2-} concentrations, which contribute most to the acidity, apparently result from high evapoconcentration rates (Pollman and Canfield, 1989).

Lake Barco and Lake Suggs have become more acidic recently, with DI pH decreases of 0.7 to 0.9 pH units. There are two possible causes: (1) acidic deposition based on the timing of pH changes and the analysis of sediment chemistry and (2) a decrease in the regional groundwater table, as shown by dynamic lake models (Sweets et al., 1989; Pollman & Canfield, 1989).

e. Rocky Mountains and Sierra Nevada

The Rocky Mountain and Sierra Nevada lakes may be considered control or reference lakes. They all have low alkalinity (most with alkalinity < 60 μeq/L) and are therefore sensitive to increased input of strong acids but do not currently receive precipitation with an annual average pH lower than about 5.0 (Gibson et al., 1984; Melack et al., 1987). Current SO_4^{2-} concentrations are typically < 5 μeq/L. These lakes are considerably more dilute and potentially more likely to acidify than lakes in the eastern United States (Eilers et al., 1988). Changes in the diatom stratigraphy of most Rocky Mountain lakes are relatively minor, occur gradually, and indicate no sign of acidification (Baron et al., 1986). The same is true for Sierra Nevada lakes (e.g., Holmes, 1986), but there is evidence of trends of both increasing (e.g., Emerald Lake) and decreasing pH (Lake 45), particularly based on analysis of floristic changes (M. Whiting, personal communication). Some of the trends began well before 1900. Causes are not clear but may include effects of volcanic eruptions during the past few thousand years, changes in dry deposition of particulates containing cations because of land use changes in nearby valleys, and increased episodic acidification events following summertime rainstorms (M. Whiting and D. Whitehead, personal communication).

f. Ongoing Studies

Studies of surface sediments and cores are being conducted on lakes of the Pocono Plateau in Pennsylvania (John Sherman, Academy of Natural Sciences of Philadelphia, personal communication). Marjorie Winkler (University of Wisconsin) is analyzing several short (0.5 to 1.0 m) and long cores from lakes on Cape Cod, Massachusetts. Mark Whiting, Donald Whitehead, Aruna Dixit, and John Smol are studying several more lakes in the Sierra Nevadas. John Smol, Donald Charles, John Kingston, Sushil Dixit, Keith Camburn, and Brian Cumming are analyzing tops and bottoms of cores from about 40 randomly selected Adirondack lakes (studied as part of the EPA Eastern Lake Survey, Phase I, Landers et al., 1988). Kerstin Williams (Institute of Arctic and Alpine Research) is studying the long-term trends of Blue Lakes and Green Lakes 4 and 5 at the Niwot Ridge Long-Term Ecological Research (LTER) site in Colorado.

2. Eastern Canada

A number of investigators, usually working independently, have used paleoecological techniques to study acidification trends in a wide variety of lakes in eastern Canada. In contrast to the United States, where the goal of most studies has been to assess the effects of acidic substances transported long distances, a large percentage of the Canadian studies focuses on lakes that are near large point sources of SO_x and metal emissions (e.g., Sudbury and Wawa), near mining effluents, or on lakes that have been manipulated (e.g., limed or acidified). In addition, several studies deal with lakes that are strongly influenced by organic rich soils and bogs. Nonetheless, in all lake regions studied thus far where acidic precipitation is falling, diatoms and chrysophytes have indicated recent pH decreases in at least some lakes. The taxonomic changes are similar to those recorded in the northern United States and some areas of Europe.

a. Ontario

Using Index α, calibrated with data for 28 lakes located north of Lake Superior in the Algoma District near Wawa, Ontario (Dickman et al., 1984; Dixit, 1986a; Dixit and Dickman, 1986), Dixit (1983) inferred that the pH in Lake CS had markedly declined, that Lake X4 has been rather acidic for the last 200 years, that Lake U3 has had a relatively stable pH over recent history, and that Lake W1 has had a fluctuating pH during its recent history. The acidification of Lake CS and another lake, B, in the past 30 to 50 years is apparently attributable to local emission of S near Wawa, Ontario, in the 1940s and 1950s (Dickman et al., 1984; Dickman and Fortescue, 1984; Somers and Harvey, 1984). Forest fires and logging may have affected diatom stratigraphy and lake water pH. Dickman and Thode (1985) provided profiles of DI pH for four additional lakes in the Wawa area: Logger, Beaver, WW1, and WW2. Beaver Lake has acidified considerably; WW1 and WW2 are located in a carbonate-rich greenstone area and did not acidify, nor did Logger Lake, which is located 100 km upwind of the emission sources and served as a control.

In Batchawana Lake (Delorme et al., 1986), a slightly acidic lake located northeast of Lake Superior, diatom changes predate cultural activities in the area and are interpreted as being caused by climatic changes. There is no evidence of any acidification trend over the past 150 years.

Taylor (1986) studied the diatom assemblages in the surficial sediments of 60 poorly buffered lakes in central Ontario. Using multivariate statistics, lakewater pH and humic content were identified as the two most important chemical variables affecting diatom distributions in his calibration set. The humic content of lakes has influenced the accuracy of these equations; also physical factors and some metal factors were stressed.

Dixit (1986b; Dixit and Evans, 1986; Dixit et al., 1987a, 1988a, 1988b) used diatoms and chrysophytes and a 30-lake calibration set to infer the pH history of Hannah and Clearwater Lakes. In Hannah Lake, acidification started soon after the roasting of ore began at Copper Cliff in the 1880s. Between about 1880 and 1975, the DI pH declined from about 6.0 to about 4.6. However, after liming in 1975, the Di pH increased. In Clearwater Lake, the diatom assemblages indicated that acidification occurred after the installation of tall stacks at Copper Cliff in the 1920s. The pH declined from about 6.0 in about 1930 to about 4.2 by about 1970. Acidification appears to have stopped after that time.

Chrysophyte scale remains from a variety of Ontario lakes are being used to infer pH change (K. Nicholls, Ontario Ministry of the Environment; J. P. Smol, unpublished data). Chrysophytes indicate a decrease in lakewater pH in Algonquin Park (Maggie and Pincher Lakes), in the Parry Sound region (Lady and Raven Lakes), and in Pukaskwa National Park (two unnamed lakes). The pH declines are currently being quantified.

Quirke Lake (near Elliot Lake, Ontario) apparently acidified as a result of local mining and milling of uranium, although acidic precipitation may also have been involved (McKee et al., 1987). The pH declined in the 1950s and reached a minimum in the early 1960s, when it dropped to about 4.5 to 5.0. With the collapse in the uranium market, the mine closed, and the diatoms recorded the recovery of the lake to its present pH of 6.4.

Davidson (1984) studied artificially acidified Lake 223 in the Experimental Lake Area using a variety of techniques (e.g., tape peels and development of pH inference techniques based on a calibration set of plankton samples taken from the lake during the period of artificial acidification). The diatom flora in a core from the lake showed good agreement with the known plankton history. These results further document the rapid response of diatom flora to changing water chemistry, and indicate that the changes are recorded in the sediment. Dickman and others (1988) also studied a core from this lake and found that diatom flora shifts and inferred pH trends correlated with the acidification trend.

b. Quebec

Cores from Key Lake (pH 5.2) and Lake C-22 (pH 5.8), in the Matamek watershed in northeastern Quebec, were studied by Hudon and others (1986). In Key Lake's recent sediments, there was a decrease in the number of indifferent taxa and an increase in acidobiontic diatoms; however, no statistical increase in acidity could be inferred. Similarly, in Lake C-22 there is a slight tendency to greater acidification in the last 20 years, but the trend is less distinct than that shown in Key Lake.

Dixit and others (1987b, 1988c, 1988d) have conducted paleolimnological work (diatoms, chrysophytes, geochemistry, ^{210}Pb dating) on a series of LRTAP (Long Range Transport of Air Pollutants) lakes in Quebec. Located in a strip 150 km wide north of the St. Lawrence River between the Ottawa and Saguenay Rivers, the 25 lakes chosen for this study have been extensively monitored by the Inland Waters Directorate. All the lakes are remote. In most of the study lakes, there was either a small change or no change at all in inferred pH. Four of the lakes suggested a recent minor pH reduction. Interestingly, chrysophytes appeared to change much more strikingly in recent history in several of the study lakes (e.g., Figure 8–6) but again suggesting only minor acidification.

c. Atlantic Provinces

Vaughan and others (1982) compared surface sediment diatom assemblages of four Halifax, Nova Scotia, area lakes with assemblages recorded by G. H. Yezdani (unpublished) in the same lakes in 1971. No attempt was made to reconstruct past lake water chemistry quantitatively, but Vaughan and others (1982) showed that there was a marked relative increase in acidophilic and acidobiontic diatom taxa in the 1980 samples, as compared to those recorded by Yezdani in 1971.

Delorme and others (1983) inferred recent DI pH trends for Kejimkujik Lake

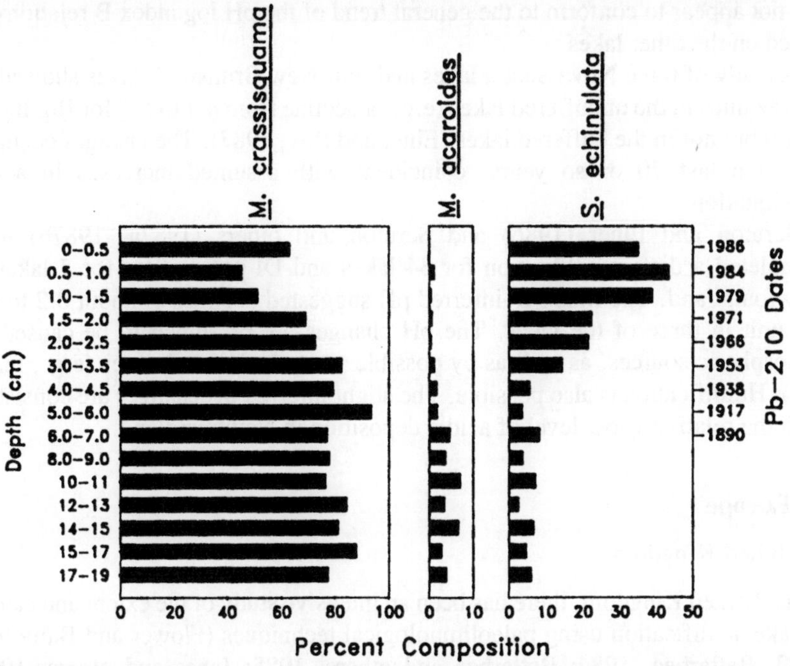

Figure 8–6. Chrysophyte taxa profiles for Lake Bonneville, Quebec. See Table 8–1 and Figure 8–3 for location and DI pH data. *Mallomonas crassisquama* generally occurs in higher pH lakes than does *Synura echinulata*. Reprinted by permission of Kluwer Academic Publishers. (From Dixit et al., 1988d.)

(central Nova Scotia), a presently acidic (pH 4.8) site in a drainage basin of organic-rich soils and bogs, overlying chemically weathering resistant igneous and metamorphic rocks. Since those results were published, the core has been reanalyzed using more recent taxonomic information. Instead of a significant increase in DI pH, a slight increase is now inferred (H. Duthie, personal communication). The lake receives wet precipitation with an average weighted mean pH of 4.6. Cultural impact in the drainage began in the mid-1800s with activities such as deforestation and the burning of lumber refuse. The present low pH of the lake cannot be attributed entirely to atmospheric loading; acidic conditions have persisted for at least the last 1,000 years, probably because of organic acids.

Walker and Paterson (1986) described the distribution of diatoms in the surficial sediments of 22 lakes and 6 peatland pools from eastern New Brunswick and adjacent Nova Scotia. This study provides data pertinent to natural acidification processes in Atlantic Canada and to paleoecological studies of recent, anthropogenic acidification. Calibrations based on Index B and multiple regressions were performed. The peatland pool data were not used in the calibration because they did not appear to conform to the general trend of the pH log index B relationship based on the other lakes.

A study of three Nova Scotia lakes and four New Brunswick lakes showed DI pH declines in the unbuffered lakes (e.g., a decline from 6.1 to 5.3 for Big Indian Lake) but not in the buffered lakes (Elner and Ray, 1987). The changes occurred over the last 70 or so years, coincident with assumed increases in acidic precipitation.

Scruton and Elner (1986) and Scruton and others (1987a, 1987b) have completed a diatom calibration for 34 lakes and DI pH profiles for 7 lakes in Newfoundland. The Index B–inferred pH suggested declines of about 0.2 to 0.3 pH unit in three of the lakes. The pH changes are presumed to be caused by atmospheric sources, as well as by possible watershed disturbances (e.g., forest fire). Humification is also possible. The slight inferred pH changes are consistent with the relatively low level of acidic deposition in Newfoundland.

C. Europe

1. United Kingdom

In the United Kingdom, there has been an intensive study of the extent and causes of lake acidification using paleolimnological techniques (Flower and Battarbee, 1983; Battarbee, 1984; Battarbee and others, 1985; Jones and others, 1986; Flower and others, 1987, 1988; Battarbee et al., 1988a, 1988b). The results as of early 1988 for the United Kingdom as a whole are described in Battarbee and others (1988, Report). Figure 8–7 shows the sites studied, and Table 8–1 summarizes the data.

a. Diatom Trends

Diatom analysis of the recent sediments of acidified lakes in the United Kingdom shows that in almost all cases the preacidification (before 1850) floras of these lakes were remarkably stable. The same species occurred in approximately the same proportions at each site for hundreds of years, despite shifting land-management patterns and climatic changes.

At many sites during the preacidification period, a significant proportion of the diatom assemblage was composed of planktonic forms, especially *Cyclotella kützingiana* Thwaites. At some sites, however, plankton was absent or rare despite apparently stable conditions and a relatively high pH. So far, the reasons for this difference between sites are not known, although auxiliary factors such as water depth and nutrient availability may be important.

In cores from all acidified sites studied, *C. kützingiana,* where originally present, is the first species to decline. In most cases (e.g., Llyn Hir, Loch Laidon, Scoat Tarn), the decline is abrupt and absolute.

The nonplanktonic floras at most sites prior to acidification are dominated by *Achnanthes minutissima* Kütz, *Anomoeoneis vitrea* (Grun.) Ross, and *Fragilaria virescens* Ralfs (e.g., Figure 8–8). These species have mainly circumneutral pH

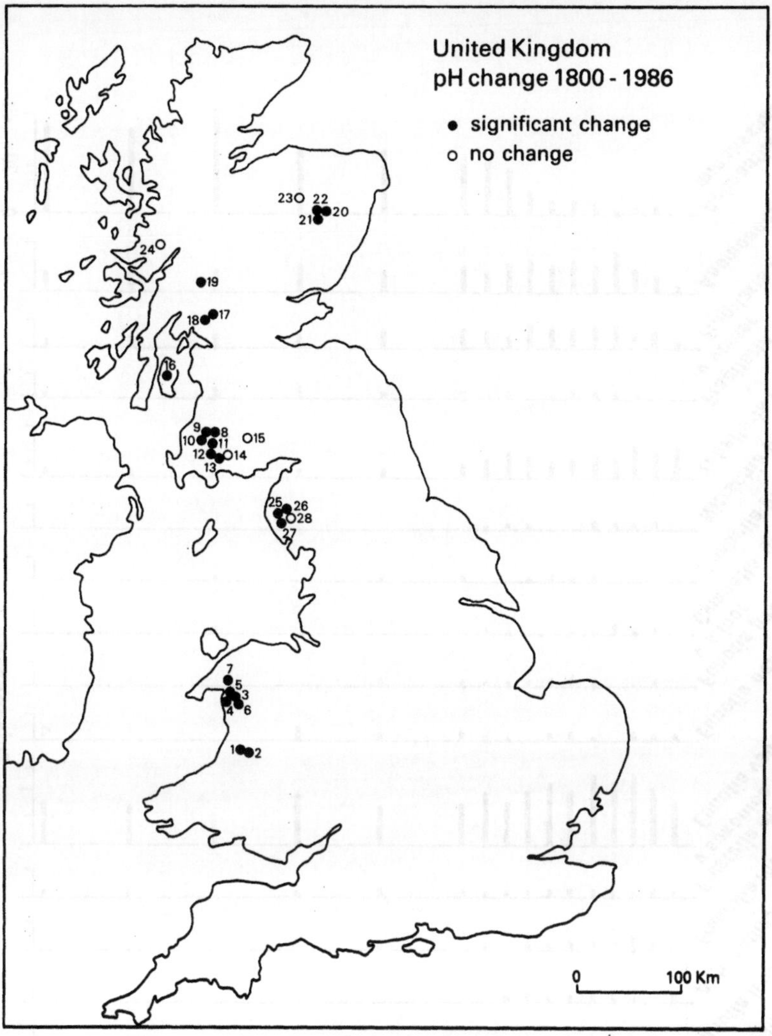

Figure 8–7. Lochs and tarns in the United Kingdom studied to infer past lake water pH by diatom analysis. Lake names corresponding to map numbers, current pH, and DI pH data are listed in Table 8–1. Lakes that have acidified recently are indicated by a black dot; those that have not changed, by an open circle.

distributions but are also found in abundance mixed with more acidophilous forms between pH 5.5 and 7.0.

As the sensitive circumneutral species decline, a proportionate increase in acidophilous species occurs. These species are invariably already present in reasonable abundance in the preacidification flora and include *Tabellaria flocculosa* (Roth) Kütz, *Frustulia rhomboides* (Ehr.) Det., and varieties, *Eunotia veneris* (Kg.) O. Müll., [= *E. incisa* (W. Smith) ex. Greg.], *Achnanthes*

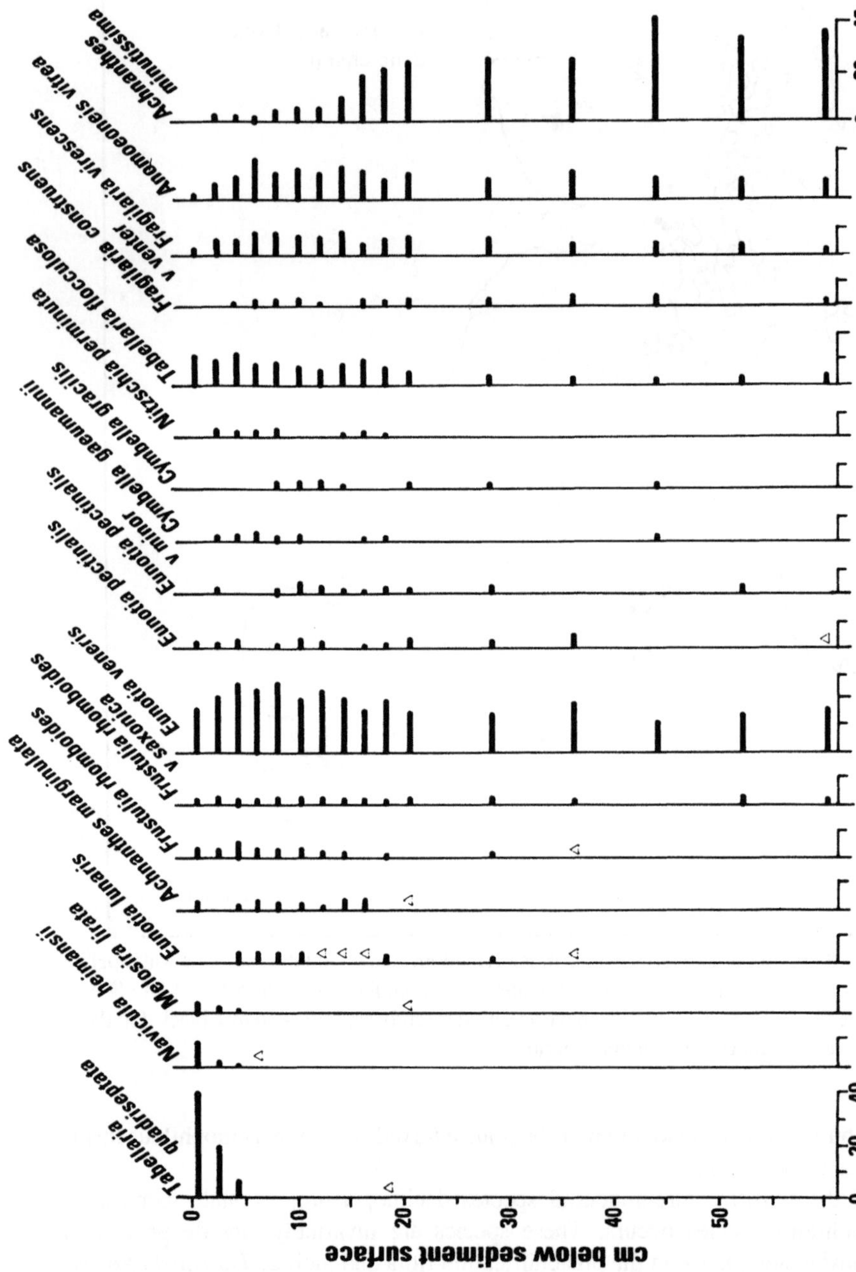

Figure 8–8. Diatom taxa profiles for Llyn Llagi, Wales, showing changes in dominant species. Synonyms for some taxa are *Navicula heimansii* (= *N. leptostriata*), *Eunotia lunaris* (= *E. curvata*), and *E. veneris* (= *E. incisa*). (From Battarbee et al., 1988c.)

marginulata Grun., and *Navicula heimansii* van Dam & Kooyman (= *N. leptostriata* Jørgensen).

Further acidification leads to the elimination of circumneutral taxa, to the continued expansion of some acidophilous species, and to the appearance and increase of acidobiontic taxa such as *Tabellaria quadriseptata* Knud., and *T. binalis* (Ehr.) Grun. However, the response of any one site is variable.

At the two higher-pH sites chosen for analysis, Loch Urr (pH 6.8) in Galloway and Burnmoor Tarn (pH 6.4) in Cumbria (Table 8–1), only slight changes in diatom assemblages have been recorded, and in both cases, the floras are dominated by planktonic taxa, including *C. kuetzingiana* and circumneutral nonplanktonic taxa, especially *Achnanthes minutissima* and *Anomoeoneis vitrea*.

b. pH Trends and Extent of Surface Water Acidification

At all sites except Loch Urr and Burnmoor Tarn, a clear post-1850 acidification is indicated by the trends in diatom assemblages. Quantification of these trends using the multiple regression approach (Flower, 1987) gives results in good agreement with measured pH for modern-day samples (Table 8–1). However, at some sites (e.g., Llyn Hir) a full reconstruction is not possible because of the numerical importance of unknown species. In this case, the approximate pH change can be inferred from the difference between the diatom-based pH for the preacidification period and the measured pH of today.

At most of the acidified sites, those with present pH levels <5.0 had pH levels about 6.0 prior to 1850. Only two sites, Loch Enoch and Loch Tanna, had pH values <5.5 at that time. These data show that sites with zero or negative alkalinity prior to 1850 or so were exceptionally rare.

The largest inferred pH declines occurred in the Welsh lakes in the Rhinog area. Here pH has probably decreased by 1.0 to 1.5 pH units. These lakes have extremely clear water (TOC <1 mg/L), and the lack of a significant organic acid input from the catchment was probably the reason for their relatively high preacidification pH. In general, pH declines since 1850 have varied from about 0.5 to 1.5 pH units, depending on the initial sensitivity of the lake to acidification and the intensity of acidic deposition.

Although sensitive areas in Scotland, northern England, and Wales have been acidified to a similar extent and over the same broad time period (since 1850), there is considerable variation among sites in the timing of the change and in the acidification trajectory (Figure 8–9).

In Wales, acidification began in the mid- to late nineteenth century at Llyn Hir, Llyn Dulyn, and Llyn Llagi, in the early twentieth century at Llyn y Bi, but not until the mid-1940s at Llyn Eiddew Bach and Llyn Gynon. In Scotland, acidification began at many sites in the mid-nineteenth century, but at other sites it was delayed until the turn of the century (e.g., Loch Dee) or later (Loch Grannoch and Loch Chon). In Cumbria, some upland tarns started to acidify in the mid-nineteenth century (Scoat Tarn), but other sites (e.g., Greendale Tarn) did not begin to change until about 1900. At some sites, although the first sign of

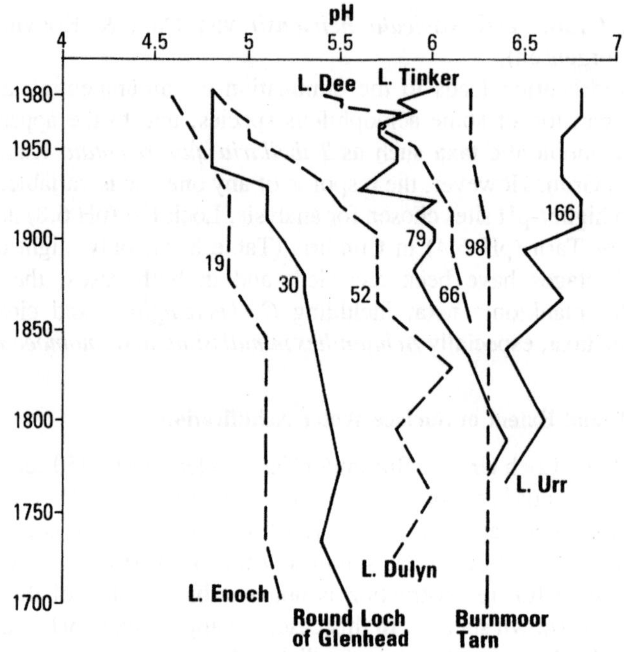

Figure 8–9. Comparison of diatom-inferred pH profiles for a series of sites in the United Kingdom with increasing lake water calcium values (Ca^{2+} as μeq/L). (From Battarbee et al., 1988c.)

acidification occurred in the mid-nineteenth century, an acceleration in the rate of acidification occurred after 1940 (e.g., Llyn Llagi).

Although we have as yet insufficient catchment information to explain in detail why this kind of variation has occurred, it is clear that the response of any site depends on the historical pattern and intensity of acidic deposition and the neutralizing capacity of the catchment soils and groundwater.

The most sensitive sites (Loch Enoch, Loch Valley, Lochnagar, all with Ca^{++} <35 μeq/L) appear to have acidified continuously from about 1850 until about the present time (Figure 8–9). At less sensitive sites (Loch Dee, Loch Tinker, with Ca^{++} 50–80 μeq/L) or at sites with lower acidic deposition (Loch Laidon, with Ca^{++} 40 μeq/L), the acidification has been slow or has been delayed; in the case of Burnmoor Tarn and Loch Urr, with Ca^{++} >150 μeq/L, acidification has not occurred despite their location in areas of high acidic deposition (Battarbee et al., 1988c).

c. Extent of Surface Water Acidification

Earlier work in Galloway, southwestern Scotland, showed that recent (post-1850) acidification had occurred at lakes lying on the two upland granitic areas in that region (Flower and Battarbee, 1983; Battarbee et al., 1985; Flower et al., 1987).

The extension of this work to other regions of the United Kingdom shows that similar acidification has occurred at lakes in central and northern Wales on Lower Paleozoic sedimentary and metamorphic rocks, in Cumbria on Borrowdale volcanic strata, and throughout Scotland southeast of Loch Ness on granites. Exceptionally acidic waters also occur in the English Pennines.

Only a few water bodies have been analyzed in each of the regions of the United Kingdom considered. However, the patterns are so consistent, both among and within regions, that this interpretation can be extrapolated confidently to other water bodies with similar modern water chemistry in these areas. Although many of the sites are quite remote, almost all are of special interest because they are situated in National Parks, National Nature Reserves, Sites of Special Scientific Interest, Man and Biosphere Reserves, and National Trust areas.

The full geographical extent of acidification is not yet known. Work in progress (Figure 8–7) is designed to reconstruct the recent history of sensitive sites in the northwest of Scotland, in Ireland, and in parts of lowland England. Preliminary data suggest that little acidification has taken place in northwestern Scotland and western Ireland, but that in eastern Ireland there may be acidified lakes in the Mourne and Wicklow Mountains.

d. Causes of Surface Water Acidification

It has already been established that the primary cause of lake acidification in Galloway is acidic deposition (Battarbee et al., 1985; Jones et al., 1986, 1988). More recent data from other parts of Scotland, Wales, and England (Battarbee et al., 1988a, 1988b) support this conclusion. Not only is the evidence consistent with this hypothesis but also the overall pattern of observations cannot be accounted for by alternative hypotheses. There are a number of cases where recent (post-1950) conifer afforestation has played a role, but its effect is variable from site to site.

The evidence for the effects of acidic deposition is as follows (Battarbee et al., 1988a):

1. At all acidified sites, the beginning of acidification is consistent with the expansion of industry in Britain from the late eighteenth to the early nineteenth century. At no sites did acidification begin before the first evidence of trace metal contamination.
2. Cores from all acidified sites have high concentrations of carbonaceous particles that are derived from fossil fuel combustion.
3. Although no cores from areas of very low acidic deposition have yet been analyzed, sensitive sites in northwestern Scotland, such as Loch Corrie nan Arr and Lochan Dubh ($Ca^{++} < 50$ μeq/L), have pH values > 5.5, and the only lochs with pH < 5.0 in this rather low deposition area are shallow, brown-water sites (TOC > 8 mg/L).
4. Although there have been changes in burning and grazing regimes in many of the catchments studied, this has led in most cases to a decline in heathland vegetation rather than to an increase. In Llyn y Bi, the only case with a

well-documented increase in catchment *Calluna*, acidification began before the no-burning policy was implemented when the lake became National Trust property. It has been clearly shown elsewhere that vegetation change of this kind cannot be used to explain intense surface water acidification (Jones et al., 1986).

5. There is no evidence that liming has ever taken place in the catchments of any of the sites studied. Cessation of that practice may therefore be excluded as a potential acidification mechanism.
6. Acidification at Llyn cwm Mynach, a site with an afforested catchment, began before afforestation. This was also the case at sensitive Galloway sites (Flower et al., 1987).

The evidence for the additional effects of afforestation is:

1. At Llyn Berwyn, documentary evidence and limited core data suggest that soil disturbance following deep drainage caused not only soil erosion and inwash but also acidification and damage to fish stocks.
2. At Loch Fleet (Anderson et al., 1986b), acidification followed afforestation. At this unusual site, acidification may be due to a combination of factors, including the scavenging effect of the forest and the erosion of peat following land drainage. The latter process caused a thick layer of acidic organic peat-derived sediment to be deposited across the bed of the loch, and this may have sealed the lake from a source of calcium-rich groundwater.
3. Afforestation can cause acidification indirectly through a scavenging mechanism. At Loch Chon, afforestation occurred in the 1950s. Because planting was carried out on the steep catchment slopes by hand, soil disturbance did not occur. However, acidification accelerated about 15 years after planting, although no further acidification occurred at this time at Loch Tinker, a control site with a moorland catchment.

e. Reversibility

Where sediment accumulation rates are sufficiently rapid, contemporary trends in acidity can be monitored by repeat coring of the lakes described here. Already the uppermost sediments of some Scottish lakes show slight signs of improvement (Battarbee et al., 1988a; Neal et al., 1988), probably in response to post-1970 reductions in acidic deposition in Scotland (Harriman and Wells, 1985). These preliminary data indicate that a large reduction in acidic deposition is likely to lead to a rapid response in lake chemistry and biota.

2. Fennoscandia

Diatom analysis of sediment cores for the assessment of recent lake acidification history was initiated in Sweden and Norway in the early 1970s, as a response to the then growing debate and concern that acidic precipitation may have been causing extensive damage to aquatic environments. Miller (1973) in Sweden and Berge

(1975) in Norway were pioneers of the technique, and both applied methods developed by Hustedt (1939), Nygaard (1956), and Meriläinen (1967). These investigations were followed by several others. Figure 8–10 shows the locations of the lakes studied; pre-1800 and current diatom-inferred pH values are in Table 8–1.

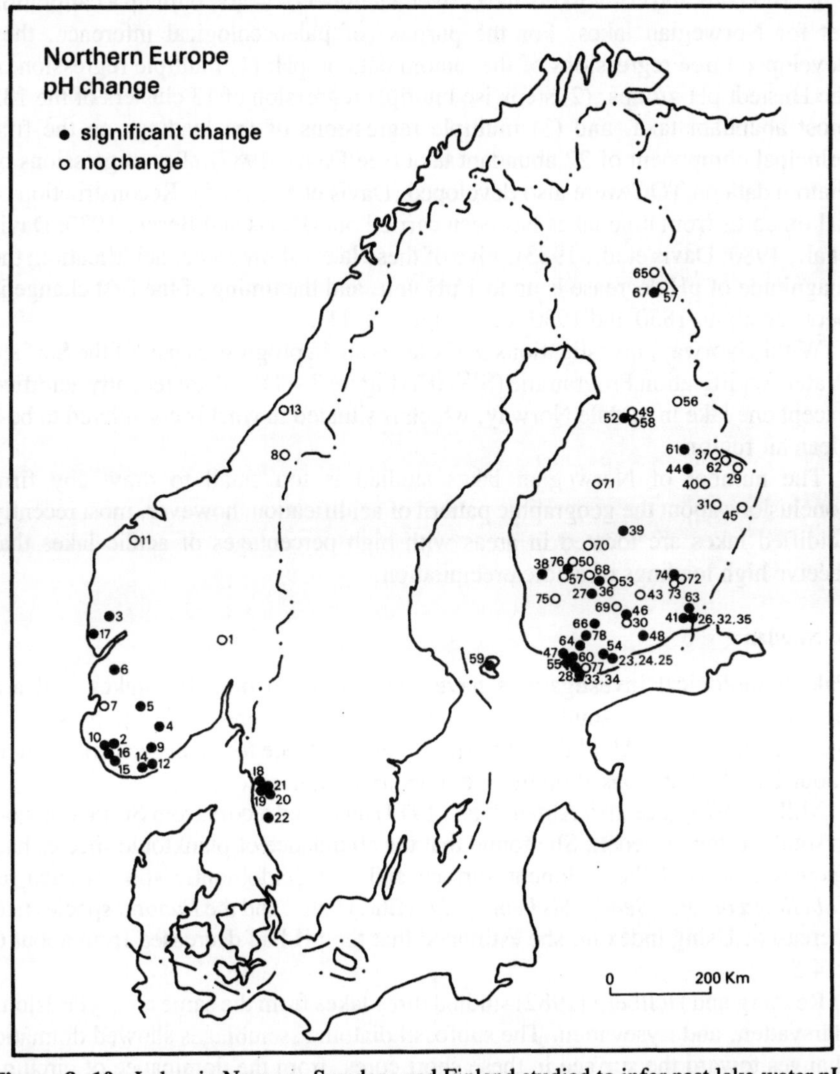

Figure 8–10. Lakes in Norway, Sweden, and Finland studied to infer past lake water pH by diatom analysis. Lake names corresponding to map numbers; current pH and inferred pH data are listed in Table 8–1. Lakes that have acidified recently are indicated by black dots; those that have not changed, by an open circle. Lake 42 is in the same location as lake 36.

a. Norway

Berge (1975) found no recent acidification of Langtjern in central Norway, but in a later study (Berge, 1979), he observed a clear recent acidification in Övre Målmesvatn, a lake in southwestern Norway, together with indications of acidification in two lakes where only core tops and bottoms were analyzed. He has also presented ecological data on important diatom taxa that support his inferences from analyses of sediment cores (Berge, 1975, 1985).

R. B. Davis and co-workers have, over a number of years, built up a calibration set for Norwegian lakes. For the purpose of paleoecological inference, they developed three regressions of the diatom data on pH: (1) multiple regression of the Hustedt pH groups, (2) stepwise multiple regression of 13 clusters of the 135 most abundant taxa, and (3) multiple regressions of the loadings on the first principal component of 22 abundant taxa (see Davis, 1987). Two regressions of diatom data on TOC were also developed (Davis et al., 1985). Reconstruction of pH on cores from nine lakes has been carried out (Davis and Berge, 1980; Davis et al., 1980; Davis et al., 1983). Five of these lakes show recent acidification; the magnitude of pH decrease is up to 1 pH unit, and the timing of the first change is between about 1850 and 1930 (e.g., Figure 8–11).

Within Norway, investigations of six lakes are in progress as part of the Surface Water Acidification Programme (SWAP) (Figure 8–11). All are recently acidified except one lake in middle Norway, which is situated in what is considered to be a clean air region.

The number of Norwegian lakes studied is too small to draw any firm conclusions about the geographic pattern of acidification; however, most recently acidified lakes are located in areas with high percentages of acidic lakes that receive high loadings of acidic precipitation.

b. Sweden

Paleolimnological investigations have been made on only five lakes. All are situated in an area of southwestern Sweden that is severely affected by surface water acidification. All of the lakes show clear evidence for recent acidification of about 1.5 pH units, based on the sedimentary diatom record.

Miller (1973) (see also Almer et al., 1974) analyzed a core from Stora Skarsjön in southwestern Sweden. She found that the abundance of planktonic species had decreased toward the sediment surface and that acidobiontic species such as *Tabellaria binalis, Semiorbis hemicyclus* (Ehr.) Patr., and the *Eunotia* species had increased. Using index ω, she estimated that the pH had decreased from about 6 to 4.5.

Renberg and Hellberg (1982) studied three lakes from the same area: Gårdsjön, Härsvatten, and Lysevatten. The subfossil diatom assemblages showed dramatic changes toward the surface in these short cores, from the dominance of circumneutral planktonic species such as *Cyclotella kuetzingiana* to acidophilous and acidobiontic periphytic species (Figure 8–12). Index B and a transfer function obtained from a calibration set of 30 lakes were used to infer past pH, which had

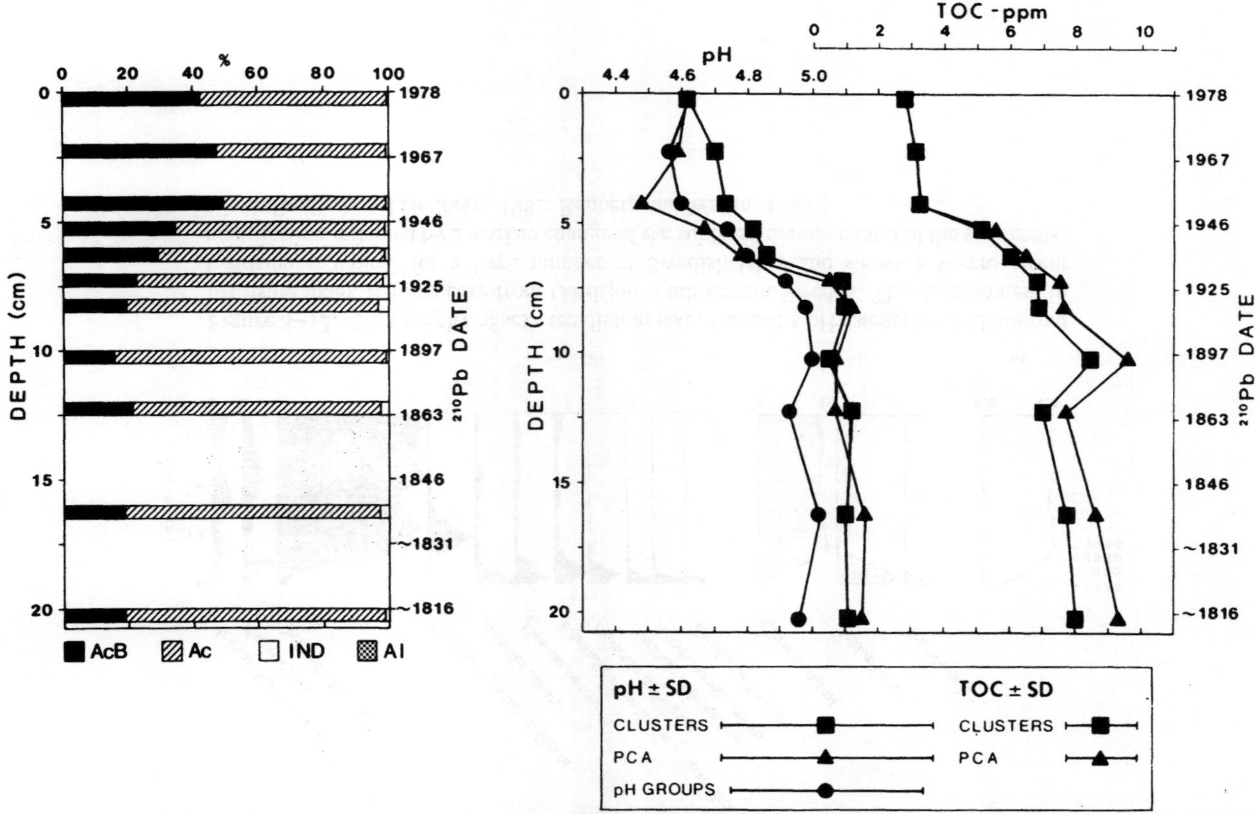

Figure 8–11. Diatom pH spectra in a sediment core from Holmvatn, southern Norway (left), and diatom-inferred pH and TOC (right). In the 1970s, the average measured pH was 4.7, and the average measured TOC was 2.2 ppm. Based on diatom studies, Holmvatn was a naturally acidic, moderately humic lake that has become more acidic and much less humic since about the 1930s. The absence of watershed changes correlated with the acidification suggests that acidic deposition was the cause. AcB = acidobiontic; Ac = acidophilic; IND = "indifferent"; and Al = alkaliphilic. (From Davis, 1987. Reprinted with permission.)

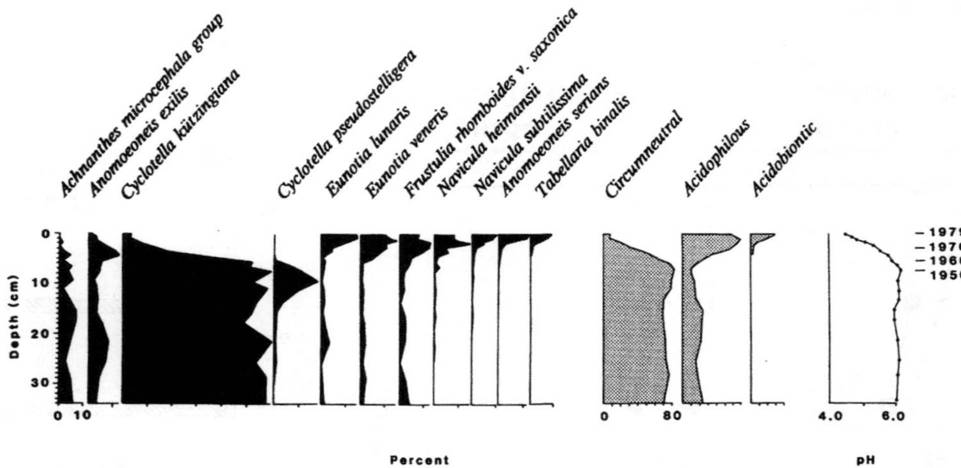

Figure 8–12. Stratigraphy of selected diatom taxa, Hustedt's pH categories, and inferred pH using index B for a core from Gårdsjön southwestern Sweden. The development in Gårdsjön is typical for a large number of Swedish lakes and shows a severe recent acidification reflected by a marked change of the subfossil diatom record of the sediments. (From Renberg and Hellberg, 1982; Renberg and Wallin, 1985.)

decreased from about 6.0 to about 4.5 during the postwar period. The dating was obtained by comparing the diatom and oil soot records of the sediments (Renberg and Wik, 1985). Lysevatten, which was limed in 1974, has had a different recent pH history, reflecting the changes caused by the lime treatment. Renberg and Wallin (1985) discussed the results from Gårdsjön further, in particular the problem of sediment record response to rapid pH changes, such as those associated with the acidification process, and also possible causes of the acidification of the lake. Acidic precipitation is believed to be the most likely cause.

Another lake, Lilla Öresjön, a little farther to the south, has now been analyzed as part of SWAP, and the recent trends there are similar to the results from the other lakes.

c. Finland

Paleolimnological research concerning the effects of lake acidification in Finland has undergone a rapid expansion recently. In addition to the core studies briefly reviewed here, the relationships between water quality and sedimentary parameters, including those between pH and diatoms, and perhaps Chrysophytes, are being investigated by Meriläinen, Huttunen, and others (e.g., Huttunen and Meriläinen, 1983, 1986a, 1986b; Christie et al., 1988; Oksanen et al., 1988).

The results of the analyses of more than 50 sediment cores from Finland show that recent acidification is widespread throughout the country. Dating seems to indicate that the acidification started later in Finland than in Norway.

Tolonen and Jaakkola (1983) found that three lakes near Helsinki in southern Finland had all acidified by about 1 pH unit. Dating by ^{210}Pb indicated that the acidification started during the 1960s (Hauklampi) or later.

Simola and others (1985) investigated the diatom stratigraphy of cores from ten lakes dated using ^{210}Pb; pH was inferred using Hustedt's pH categories and index B. On the basis of their pH development, the lakes fell into three groups. The first group contained three lakes in which the pH had declined by 0.3 to 0.8 pH units from a reasonably steady background level. According to ^{210}Pb dating, the decline in two of these lakes had occurred since the 1960s. The second group, also composed of three lakes, showed a similar recent decline, but in these lakes the oldest DI pH obtained was similar to that near the surface. These trends were interpreted as a reacidification after a period of cultural influence that had caused pH to increase. The third group consisted of four lakes. One was limed, two showed no significant inferred pH change, and one had badly corroded diatoms.

Tolonen and others (1986) analyzed the diatom records in sediments from ten forest lakes. A striking decline in DI pH (Index B, Index α) occurred in six of those during about the last 30 years. Analysis of chrysophycean scales in one of the lakes (Munajärvi) supported the results of the diatom analysis. In some cases, atmospheric deposition was considered the most likely cause of acidification, but in most cases effects of land use changes could not be ruled out as important factors.

Work is in progress on a large number of lakes in Finland (Table 8–1, Figure 8–10), and preliminary results have kindly been made available to us. The investigations include a study by Tolonen and Suksi (personal communication) of the three lakes mentioned above that have all recently acidified, a study of 24 lakes by M. Liukkonen (personal communication), and a study of six lakes by J. Turkia (personal communication). Analysis of cores from about 50% of the lakes in the two latter studies indicate recent acidification.

3. Western and Central Europe

The small-scaled landscape of central Europe has been cultivated for a very long time. Consequently, the region no longer contains any pristine water bodies, although there are several relatively undisturbed small lakes and pools in nature reserves. Because past human activities (e.g., agricultural drainage water, foddering of ducks and fishes) in and near pools and lakes are not always well documented, it is often difficult to determine the relative importance of land use changes versus acidic deposition in causing DI pH trends.

The majority of the acid-sensitive water bodies in the North Sea lowlands are moorland pools. The open water bodies in many of these pools were formed by excavation of peat. Most moorland pools are seepage pools with a perched water table. Consequently, the catchment area is only slightly larger than the area of the water surface. Nevertheless, land use around these pools can have a profound influence on their water quality (e.g., the litter of trees falling in the pools or the washing of sheep in the pools). The problems associated with paleoecological investigations of these shallow waters, such as wind mixing, bioturbation, desiccation, and presence of a dense aquatic vegetation, have been discussed by Dickman and others (1987). To avoid these problems, and also the problem of dating, researchers have assessed acidity changes in many of these pools by comparing preserved diatom samples collected in the early twentieth century and diatoms removed from the surface of old aquatic macrophyte herbarium material with recent samples that were collected in the same ways.

Many of the central European water bodies investigated (Figure 8–13; Tables 8–1 and 8–2) have a pH below 4.5 and contain only acidobiontic and acidophilous diatoms. Index B (Renberg and Hellberg, 1982), commonly used to infer past pH, does not work well in these situations. Therefore, a new index was developed based on weighted averaging of the pH optima of diatom taxa and calibrated using a set of 97 pools and lakes of this region (Ter Braak and Van Dam, 1988).

a. Denmark

Old and modern diatom assemblages from acidic Danish lakes have been analyzed and compared to determine historical pH changes (Table 8–2). The lakes and pools where no decrease or even an increase of DI pH is observed are more or less humic. Tormål, which is included here as a reference (or control), has the greatest alkalinity of the investigated sites (Ca^{2+}, 591 μeq/L; HCO_3, 414 μeq/L), and the increase of pH may be caused by hardening of the water, as has been observed in dune pools in the Netherlands by J. G. M. Roelofs (Catholic University, Nijmegen, personal communication). In general, DI pH decreases are less than 1

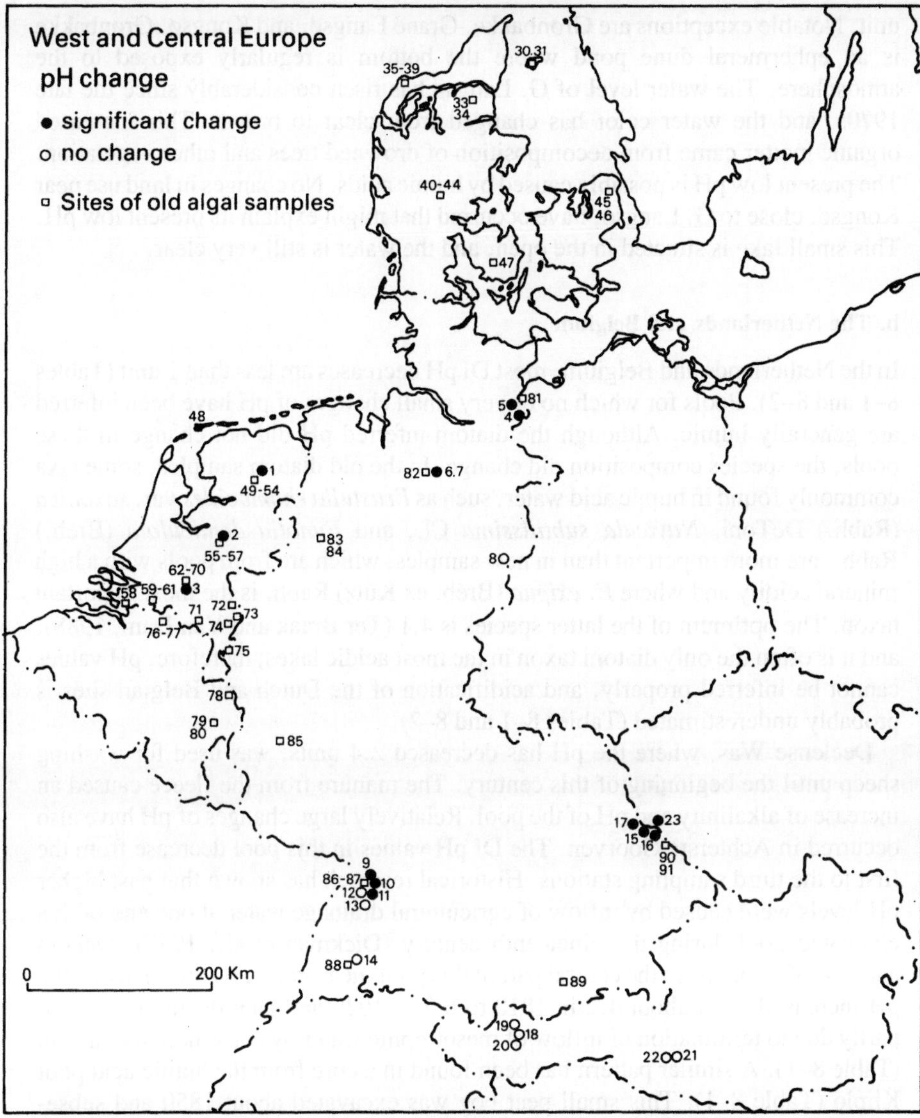

Figure 8–13. Lakes and pools in western Europe studied to infer past lake water pH by diatom analysis. Lake names corresponding to map numbers; current pH and inferred pH data are listed in Tables 8–1 (sediment cores, open and solid circles on map; lakes that have acidified recently are indicated by a black dot, and those that have not changed, by an open circle) and 8–2 (comparison of old and new algal samples, open squares on map). Numbers 24–29 are not used on map.

unit. Notable exceptions are Grønbakke, Grane Langsø, and Kongsø. Grønbakke is an ephermeral dune pond where the bottom is regularly exposed to the atmosphere. The water level of G. Langsø has risen considerably since the late 1970s, and the water color has changed from clear to brown. This increased organic matter came from decomposition of drowned trees and other vegetation. The present low pH is possibly caused by humic acids. No changes in land use near Kongsø, close to G. Langsø, have occurred that might explain its present low pH. This small lake is situated in the open, and the water is still very clear.

b. The Netherlands and Belgium

In the Netherlands and Belgium, most DI pH decreases are less than 1 unit (Tables 8–1 and 8–2). Pools for which no or very small changes of pH have been inferred are generally humic. Although the diatom-inferred pH did not change in these pools, the species composition did change. In the old diatom samples, some taxa commonly found in humic acid water, such as *Frustulia rhomboides* var. *saxonica* (Rabh.) DeToni, *Navicula subtilissima* Cl., and *Eunotia denticulata* (Bréb.) Rabh., are more important than in new samples, which are from pools with a high mineral acidity and where *E. exigua* (Bréb. ex Kütz) Rabh. is the most important taxon. The optimum of the latter species is 4.1 (Ter Braak and Van Dam, 1988), and it is often the only diatom taxon in the most acidic lakes; therefore, pH values cannot be inferred properly, and acidification of the Dutch and Belgian sites is probably underestimated (Tables 8–1 and 8–2).

Deelense Was, where the pH has decreased 2.4 units, was used for washing sheep until the beginning of this century. The manure from the fleece caused an increase of alkalinity and pH of the pool. Relatively large changes of pH have also occurred in Achterste Goorven. The DI pH values in this pool decrease from the first to the third sampling stations. Historical research has shown that past higher pH levels were caused by inflow of agricultural drainage water at one end of this elongated pool during the nineteenth century (Dickman et al., 1987). Diatom analysis of a core from the central part of this pool (near station B) revealed that the pH increased from about 4.2 in 1810 to 6.2 in 1920 and then decreased to 4.5, partly due to termination of inflow of mesotrophic water by construction of a dam (Table 8–1). A similar pattern has been found in a core from the humic acid pool Kliplo (Table 8–1). This small peat bog was excavated about 1850 and subsequently eutrophied, presumably due to washing of sheep and/or foddering and catching of ducks. The pH increased to a peak of 6.2 in about 1910 and decreased to 5.0 in 1985. High pH values in the pool Gerritsfles in the early twentieth century (Tables 8–1 and 8–2) are due to sheep washing in the late nineteenth century (Van Dam et al., 1988).

Significant human activities have also occurred in and near some of the other pools listed in Table 8–1. These include soaking of timber (Kempesfles), bathing (Groot Huisven, Diepmeerven), and retting of flax (Onderste Rolven-E). Relatively large changes in DI pH ($\geq$0.7 units) have been observed for these pools. A part of the pH decline inferred for these other pools may have been caused by termination of traditional agricultural activities, but this cannot be determined with certainty because past human disturbance has not been documented.

c. West Germany, Austria, and Czechoslovakia

In West Germany, large DI pH decreases have been observed in Erdfallsee and Heideweiher (Table 8–2). Both water bodies received inflow of agricultural drainage water until a few decades ago. Comparison of the old and new samples from Plötschersee suggests an increase of 0.8 pH units, which could have been caused by foddering of fishes. However, the coring results suggest a decrease of 0.5 pH units. Such differences are occasionally found, but normally DI pH changes calculated from old and new samples and from cores are in good agreement (Arzet and Van Dam, 1986; Dickman et al., 1987; Van Dam, 1988).

Some of the small lakes investigated in the Black Forest and the Bavarian Forest of southern Germany (e.g., Grosser and Kleiner Arbersee) have been used as reservoirs to increase the discharge of the outlet for wood-rafting purposes. This activity caused major fluctuations in water level and lake surface area until about the beginning of the present century. Forestry in the catchment areas has caused a shift from hardwood to softwood trees since the Middle Ages. The formerly mixed forest around Pinnsee was partially logged after World War II and reforested with conifers. Also, mining and smelting of heavy metals (Arbersees) and presence of local glass industry (Arbersees, Herrenwiesersee) may have changed the acidity status of some lakes in the past, including Teufelssee in Czechoslovakia (Arzet, 1987). Although some of these activities have left traces in the sediment profiles as small declines of DI pH, the acidification by acidic atmospheric deposition is clearly discernible as a rapid decrease of DI pH values since about 1950 (e.g., Grosser Arbersee; Figure 8–14) and shifts in chrysophyte species composition (Kleiner Arbersee; Hartmann and Steinberg, 1987). The recent acidification of some lakes in the northern Black Forest (Herrenwiesersee, Huzenbachersee, and Schurmsee) started between 1800 and 1850. *Asterionella ralfsii* decreased and *Eunotia exigua* increased in abundance during the period of acidification (Steinberg et al., 1984; Arzet et al., 1986b; Arzet, 1987).

There is no evidence of a decrease in pH in the high Alpine lakes of Austria that have been investigated (Table 8–2; Arzet et al., 1986a).

IV. Discussion

Considerable high-quality data now exist on which to make assessments concerning acidification trends in North America and Europe. As of 1988, sediment core DI pH data exist for at least 100 lakes on both continents (Tables 8–1 and 8–2). This number will probably double within the next two years. The pH inference equations on which the DI pH data are calculated are based on 15 calibration data sets for North America and 10 for Europe, involving at least 500 and 300 lakes, respectively. The number of calibration lakes will also increase substantially in the next few years.

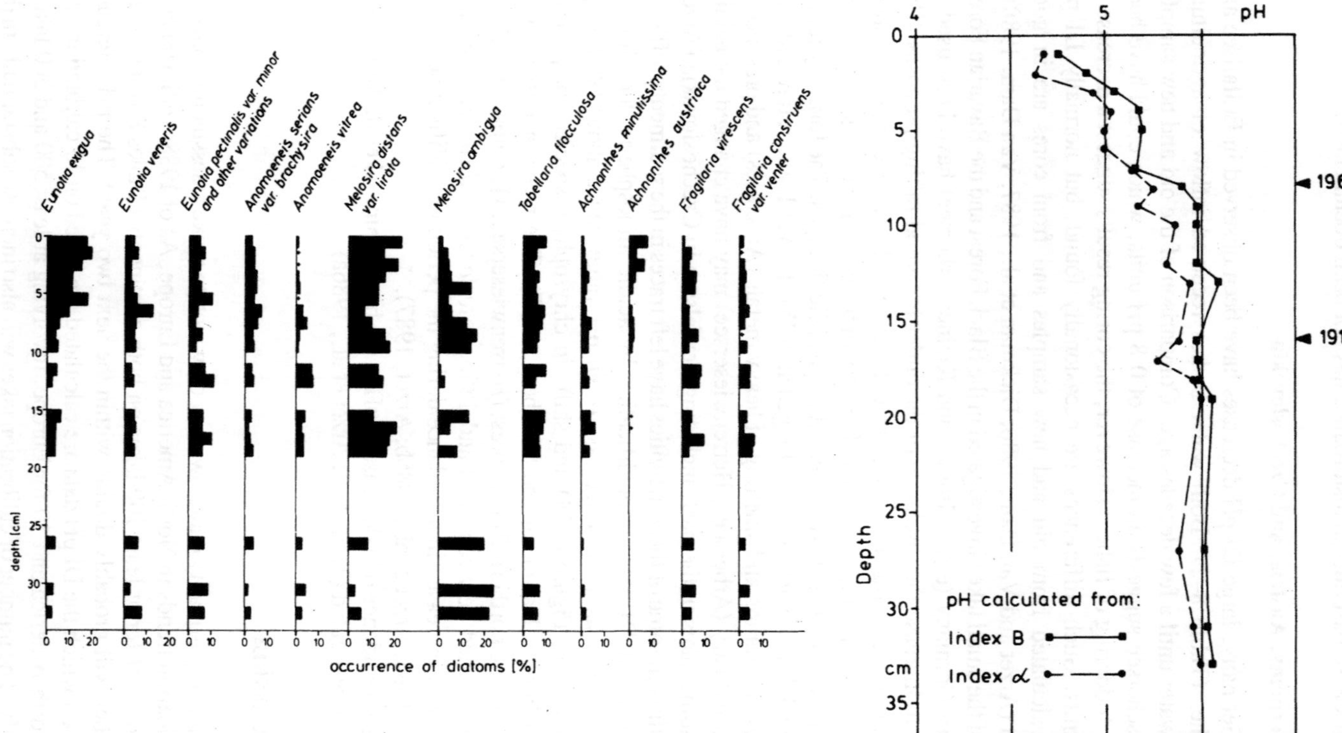

Figure 8–14. Relative frequency of selected diatom taxa and inferred pH based on Index α and Index B for Grosser Arbersee. (From Arzet and others, 1986b.) *Eunotia veneris* is a synonym of *E. incisa*.

The ability of the existing predictive equations to infer pH is demonstrated in Table 8–1. With few exceptions, the surface sediment DI pH values agree within about 0.1 to 0.4 units of current measured pH.

A. Trends in pH and Extent of Acidification

The quantity and quality of diatom, chrysophyte, and other paleoecological data now available are sufficient to conclude that extensive regional lake acidification has occurred in low-alkalinity lakes of North America and Europe (Figure 8–15). However, it appears that sensitive water bodies in Europe are affected more by acidification than those in North America. At least 30% of the lakes studied in Europe had a DI pH decrease of >0.6 pH units, whereas only one region in North America had more than 20% of its lakes in this category.

The lakes listed in Table 8–1 are not representative of all the lakes in their regions, but based on the studies done to date as summarized in this chapter, there is clear evidence of widespread acidification of low-alkalinity lakes in the Adirondack Mountains of New York, southeastern Canada, several regions in the United Kingdom, southern Sweden, Norway, and Finland, the Netherlands, and northern Germany.

B. Causes of Lake Acidification

The primary cause of most recent lake acidification on both continents is atmospheric deposition of strong acids derived from combustion of fossil fuels, although watershed changes and probably natural long-term processes also play a role. The relative importance of these potential causal mechanisms can be evaluated using the hypothesis-testing criteria described in section II, C, 4. First, declines in pH in areas mentioned in section IV, A cannot be accounted for by natural long-term processes (the rates of change are too fast) or watershed change, although the latter contributes, at least in some cases. Second, in the majority of cases, declines occurred after the onset of acidic deposition, and in many cases there is paleoecological evidence (e.g., Pb, V, polycyclic aromatic hydrocarbons, spherical carbonaceous particles) demonstrating that the fossil-fuel-derived inputs began before the pH declines. Third, all regions where pH declines are substantial (see above) receive relatively high atmospheric loading of strong acids.

The importance of watershed disturbance and change in affecting lake chemistry varies in kind and magnitude among the regions. In the northern United States and Canada, the most common disturbances are due to deforestation, forest fires, and blowdown. In Ontario, smelters have caused acidification of nearby lakes. In the western United States, factors such as increased deposition of windblown dust particles due to land use changes, ash falls from volcanoes (long term to the present), and localized concentrations of automobile emissions may be important. Recent acidification of some Florida lakes may be due partly or wholly to natural or anthropogenic lowering of the water table, leading to reduced inflow of higher pH groundwater. In Europe, different types of recent anthropogenic land use

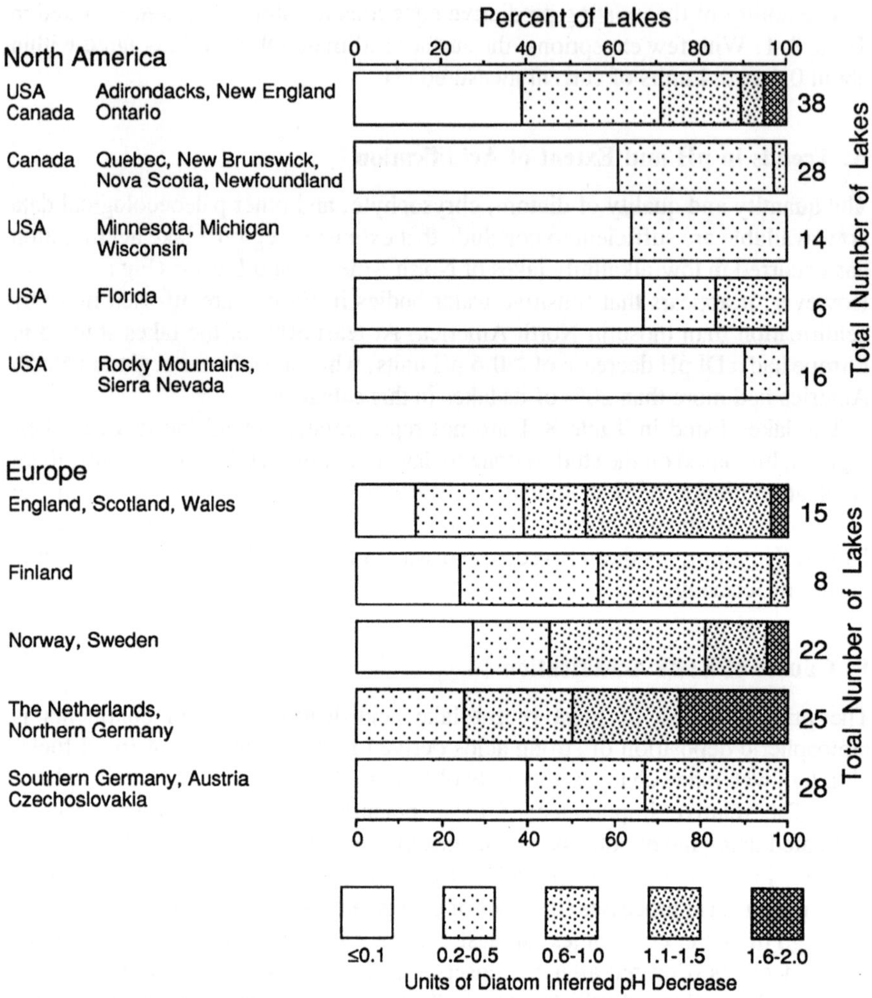

Figure 8–15. Decrease of DI pH in ten different regions. The length of each graph is 100%. The number of localities in each region is shown at right. All lakes listed in Table 8–1 are included, except those artificially acidified. Five classes of decrease of pH from preindustrial times until about 1980 to 1985 are distinguished. The lakes were not selected randomly and do not represent all lakes in a region. They show results only for a particular group of lakes studied to date.

changes are important and have been occurring over a longer period of time. In the United Kingdom, afforestation rather than deforestation is of some importance. Decline in traditional agricultural practices, leading to regrowth of forest in former fields, has occurred in much of northern Europe and the United Kingdom; washing of sheep and foddering of ducks in the Netherlands have also contributed to change in pH. However, in none of the areas where considerable acidification has occurred can watershed change account for the inferred regionwide acidification.

C. Patterns of pH Trends

Although DI pH profiles exhibit a continuum of acidification trends (Figures 8–4 and 8–9), they can be classified arbitrarily into four basic patterns (Charles and Norton, 1986). Fluctuations in these patterns are caused by natural variability in diatom composition, watershed events, or inconsistencies introduced by pH inference methodologies. Two represent trends, and two represent no trend. The patterns are (1) background pH greater than about 6.0 and no overall trend in pH, (2) relatively rapid decline in pH from the range of 5.7 to 6.0 to less than 5.0 (some pH declines start in pH ranges above 6.0), (3) pH about 4.8 to 5.2 throughout the entire profile and no overall trend, and (4) decline from the range of pH 4.8 to 5.3 to as low as about 4.3.

The greatest changes in pH and sediment core diatom composition occur when DI pH decreases from a pH above approximately 5.5 to a pH below 5.5 (pattern 2). A smaller pH drop is observed for lakes with a pre-1850 DI pH of less than 5.5 (pattern 4). These patterns are consistent with the logarithmic nature of the pH scale and the response of sample lake water pH to titration with strong acid in the laboratory. The pre-1850 water chemistry of lakes with patterns 1 and 2 and patterns 3 and 4 represents the dominant pre-1850 buffering systems in the studied regions: bicarbonate in the higher pH lakes, and organic acids and Al species in the lower pH lakes. Declines in pH in the latter group (pattern 4) should be interpreted carefully. Though pH may have decreased only slightly, alkalinity may have decreased substantially. This is because the change in alkalinity per pH unit is greater in the pH range 4.5 to 5.0 than in the range 5.5 to 6.0. Also, an increase in Al concentration, which would be greater in the pH range 4.5 to 5.0, could have major impact on biota, much greater than the decline in pH alone. The effect on biota could be especially large if accompanied by a decrease in organic compounds that could complex with the Al.

D. Naturally Acidic Lakes

Lakes with pre-1850 DI pH less than 5.5 deserve special attention. First, they represent what most would consider "naturally" acidic lakes; second, it is more difficult to explain their past chemistry and how it changed during the process of acidification. Many lakes of this type have been studied. For several regions (Tables 8–1 and 8–2), at least 40% of the lakes have pre-1850 DI pH <5.5: Adirondacks, northern New England, northern Florida, Norway, Denmark, the Netherlands, Belgium, and Germany. There are two probable reasons for this high percentage. First, a region could have a high proportion of naturally acidic lakes due to its geologic setting (e.g., Florida and the Netherlands); second, because of conscious site-selection criteria, only the more acidic lakes in a region were chosen for study, many of which turned out to be naturally acidic (e.g., New England).

Three important questions arise with respect to these naturally acidic lakes: (1) What was the chemistry of these lakes? (2) In particular, what were the dominant anions associated with the low pH? (3) What specific chemical changes could have occurred in these lakes in response to increased deposition of strong acids, and how would those affect hydrogen ion concentration and alkalinity?

In regions with low-alkalinity waters that receive precipitation with pH greater than about 5.0, such as northeastern Minnesota and Wisconsin, western Ontario, Labrador (Barrie, 1984), and northwestern Scotland (Barrett et al., 1983, 1987), clearwater lakes with pH less than 5.5 are rare (Wright, 1983). However, some lakes in these regions have pH less than 5.5. These lakes are typically moderately to highly colored (platinum-cobalt units >25) (Lillie and Mason, 1983; Gorham et al., 1985), associated with peatland vegetation, and only slightly influenced by groundwater input. Plants such as *Sphagnum* can remove cations from surrounding water (Clymo, 1963), and decay products of plants yield organic acids, which may or may not be colored (Vitt and Bayley, 1984). These organic acids can account for a significant portion of total anions in lake water and cause pH values to be less than 5.0 (Oliver et al., 1983). It seems probable, therefore, that organic acids were an important cause of acidic conditions in many lakes in acid-sensitive areas before 1850 (Patrick et al., 1981; Arzet, 1987; Davis et al., 1985). Also, if the pH was about 5.0, Al concentrations may have been moderate at that time, although much of the Al would have been complexed with organic compounds (Driscoll et al., 1984).

As low-pH (5.0 to 5.5) lakes have become more acidic, changes in Al and organic acid concentrations have probably been the most important factors affecting pH and alkalinity (e.g., Sullivan et al., 1989). The output of dissolved Al from these lakes' watersheds should have increased with increases in inputs of strong acids (Almer et al., 1978; Johnson et al., 1981; Driscoll et al., 1984), except perhaps for seepage lakes in outwash deposits. This is consistent with the fact that most of the lakes examined for this chapter that have current pH less than 5.5 have concentrations of total Al greater than 100 to 200 μg/L; the value in some lakes is as high as 600 to 700 μg/L (e.g., Appendix Table E.3; National Research Council, 1986; Van Dam et al., 1981; Leuven, 1988; Van Dam, 1988).

Increased Al concentration, at least in some Adirondack lakes, is suggested by the increase in percentages of diatom taxa such as *Navicula tenuicephala* (= *?Stauroneis gracillima*) and *Fragilaria acidobiontica* toward the top of lake sediment cores (Charles et al., 1989). In the Adirondacks, these taxa are abundant only in lakes with pH less than 5.5 and total Al concentrations greater than 200 μeq/L (Charles, 1985, 1986). Also, computer modeling efforts based on current lake water chemistry and DI pH suggest that the Al concentration in Big Moose Lake increased as pH decreased (Charles, 1984).

Increased Al loading is important because it could lead to changes in alkalinity and reduced organic acids and water color. A relatively minor decline of pH in the range 4.5 to 5.5 can cause a major change in Al speciation and a substantial increase in total Al charge (Driscoll et al., 1984). Because Al acts as a buffer

(Johannessen, 1980), increased concentrations can act further to minimize pH change as a result of increased acidic loading.

Organic acid concentrations and water color can decrease as a result of decreasing pH and increasing Al concentrations. Support for these processes includes theoretical analyses, laboratory studies, and observations that water color has decreased over time in lakes that have apparently acidified recently (Schofield, 1976; Almer et al., 1978; Dickson, 1980). Also, studies of limed lakes indicate that, as alkalinity increases, color increases and concentration of Al decreases (Hultberg and Andersson, 1982; Yan, 1983; Wright, 1984). Van Dam (1987) studied the recovery of some moorland pools after the severe drought of 1976. The concentration of organic acids and water color increased, along with decreasing concentration of Al. At the same time, the proportional abundance of *Eunotia exigua,* characteristic of high mineral acids, decreased in favor of *Frustulia rhomboides* var. *saxonica,* which is commonly found in humic, acidic waters.

E. Taxonomic Composition

Many diatom taxa occur in all the regions discussed in this chapter, and in general they all have similar ecological characteristics. The large and growing body of ecological data on diatoms in these regions provides additional support for the interpretations of a changing water chemistry inferred for any one region. The abundances of several taxa change consistently in sediment core profiles both within and among regions. Acidobiontic and acidophilic taxa that generally increase with increasing acidification in the Adirondacks, New England, eastern Canada, United Kingdom, Fennoscandia, and central Europe include *Tabellaria binalis* (except central Europe), *Tabellaria quadriseptata, Frustulia rhomboides* var. *saxonica, Eunotia curvata* (= *E. lunaris*), *Navicula heimansii* (= *N. leptostriata*), *Navicula subtilissima* types, and *Eunotia exigua*. In some regions, one or more taxa may be more important than others; for example, *Fragilaria acidobiontica* and *Navicula tenuicephala* are important in currently acidic lakes with high Al in the Adirondacks, but the former has not been found in any European lakes. *Eunotia exigua* and other small *Eunotia* seem to be of much more importance in pools and lakes in the Netherlands and Germany than elsewhere.

Taxa that consistently decline with acidification in the above regions include *Cyclotella kuetzingiana* (primarily Europe), *Cyclotella stelligera* and *Cyclotella comta* (primarily North America and Fennoscandia), *Achnanthes microcephala* group, *Anomoeoneis vitrea,* and *Fragilaria virescens*. Many more taxa exhibit trends similar to those discussed above.

One of the most striking characteristics of diatom distributions both in modern assemblages and in sediment cores is the nearly universal absence of euplanktonic taxa where measured or DI pH is <5.5 to 6.0. However, there are exceptions to this pattern. *Asterionella ralfsii* var. *americana* is apparently one of the few euplanktonic taxa found in lakes with pH <5.5, at least in North America (e.g., Sweets, 1983). Also, recent studies of surface sediment diatom assemblages of lakes in the Killarney region of Ontario show that euplanktonic *Cyclotella* species are common in lakes with pH <5.5. Lake morphometry may be an important reason for the difference in occurrence of euplanktonic diatoms in these larger, deeper lakes, as compared with others (Taylor et al., 1987).

Chrysophyte assemblages also appear to change in a consistent manner with increasing acidification. In many lake regions, the preacidification chrysophyte floras are dominated by taxa such as *Mallomonas crassisquama*. However, as lakewater pH decreases, taxa such as *Synura sphagnicola* and *S. echninulata* commonly begin to increase. If acifification continues (e.g., to below pH of 5.0), even these species may be replaced by more acidobiontic forms such as *M. hindonii, M. canina,* and *M. hamata*. The often precipitous decline in *M. crassisquama* may be a useful marker of the onset of acidification, as is the decline in planktonic diatoms in many regions. Interestingly, changes in chrysophytes often predate shifts in diatoms by about 10 years. This may reflect a greater sensitivity of chrysophytes to subtle pH changes, perhaps because they are often vernal blooming and planktonic.

F. Future Research

Despite the power of the diatom- and chrysophyte-based paleoecological approach to inferring past acidification trends, improvements in methodology are possible. More research is needed on diatom and chrysophyte taxonomy, physiology, and ecology; in-lake transport and sedimentation processes; quantitative techniques for inferring water chemistry characteristics, including pH, alkalinity, dissolved organic carbon, and Al; statistical analysis of sediment core data; and lake selection criteria. For specific recommendations see Battarbee (1986), Charles and Norton (1986), and Davis and Smol (1986).

V. Conclusions

1. Analysis of sediment diatom and chrysophyte assemblages is the best technique currently available for inferring past lake water pH trends, and use of this approach is increasing rapidly.
2. Sediment-core-inferred pH data exist for at least 100 lakes in both North America and Europe. This number will approximately double in the next 2 years. The pH inference equations are based on at least 15 calibration data sets for North America and 10 for Europe, involving totals of at least 500 and 300 lakes, respectively.
3. Paleoecological studies indicate that recent acidification has been caused by acidic deposition in the Adirondack Mountains (New York), northern New England, Ontario, and the Atlantic Provinces in North America; England, Scotland, Wales, Norway, Sweden, Finland, the Netherlands, and West Germany in Europe. Inferred pH decreases are commonly 0.5 to 1.5 pH units. Paleoecological and other data for several of the regions indicate combustion of

fossil fuels as the source of acidic deposition. Watershed land use changes have caused some pH changes but cannot account for the pH changes on a regional scale.

4. No acidification trends were observed in regions currently receiving low deposition of strong acids (e.g., Rocky Mountains and Sierra Nevada in the western United States, northwestern Norway, and northwestern Scotland), and slight trends toward decreasing pH were observed in regions receiving moderately acidic deposition (upper Midwest and northern Florida, United States, and Atlantic Provinces of Canada).
5. The magnitude and extent of acidification of lakes appears to be greater in low-alkalinity regions of Europe than in North America. This is probably due to greater acidic loading in Europe, regional differences in the ability of watersheds to neutralize acids, and representation of lakes chosen for study.
6. Naturally acidic lakes (pre-1800 pH < 5.5) were apparently common in some regions but rare in others.
7. Acidic deposition has decreased in some regions, but there is evidence of continuing acidification for some lakes.
8. Aluminum and organic acids are important in buffering pH changes, particularly in acidic waters.

Acknowledgments

We thank Dennis Anderson, Klaus Arzet, David Beeson, Frode Berge, Brian Cumming, Ronald Davis, L. Denis Delorme, Anthony Del Prete, Sushil Dixit, Hamish Duthie, Jesse Ford, John Kingston, Mikko Liukkonen, Roger Sweets, Kimmo Tolonen, Jaana Turkia, and Mark Whiting for providing access to their unpublished data. They also provided interpretation of their results and many useful comments on the manuscript. Jill Baron, Charles Driscoll, Jesse Ford, and Stephen Norton contributed water chemistry and other information on many of the lakes mentioned in this section.

Preparation of this chapter has been funded partially by the U.S. Environmental Protection Agency through Contract #68-C8-0006 with NSI Technology Services at the Environmental Research Laboratory in Corvallis, Oregon, and through a cooperative agreement CR-813933-01-1 with Indiana University. The chapter has been subjected to the agency's peer and administrative review and approved for publication.

Some of the unpublished Adirondack region data were obtained as part of research efforts supported by the Electric Power Research Institute, Palo Alto, California.

Rick Battarbee is grateful for financial support from the Central Electricity Generating Board, Natural Environment Research Council, Department of the Environment, Nature Conservancy Council, and the Royal Society (through the Surface Water Acidification Programme, SWAP). He also thanks N. J. Anderson, P. G. Appleby, R. J. Flower, S. C. Fritz, E. Y. Haworth, S. Higgitt, V. J. Jones,

A. Kreiser, M. A. R. Munro, J. Natkanski, F. Oldfield, S. T. Patrick, N. G. Richardson, B. Rippey, and A. C. Stevenson for help with the production and collection of the United Kingdom data.

Ingemar Renberg thanks the National Swedish Environment Protection Board, the Swedish Natural Science Research Council, and the Royal Society (through SWAP) for financial support.

Herman van Dam is greatly indebted to the late Dr. N. Foged (Odense), Prof. Dr. E. G. Jørgensen (Lyngby), Prof. Dr. J. J. Symoens (Brussels), Prof. Dr. R. Düll (Duisburg), and the curators of the collections of algae and macrophytes at the Hugo de Vries-laboratorium (Amsterdam), the Rijksherbarium (Leiden), the Laboratorium voor Dierkunde (Gent), the Institut Royal des Sciences Naturelles de Belgique (Brussels), theWestfälisches Museum für Naturkunde (Münster), the Institut für Meeresforschung (Bremerhaven), the Landessammlungen für Naturkunde (Karlsruhe), the Institut für Biologie 1 (Tübingen), and the Botanische Staatssammlung (Munich) for making available old material. Messrs. C. N. Beljaars and J. C. Buys provided technical assistance. The research was supported in part by the European Community (Contract ENV-650-N(N)).

John Smol is grateful to the Natural Sciences and Engineering Research Council of Canada for operating and strategic grants and also wishes to acknowledge the assistance of D. Scruton, H. Duthie, K. Nicholls, S. Dixit, I. Walker, and W. Keller for help in compiling portions of this review.

We thank Susan Christie for editing and Roze Royce for typing the manuscript.

References

Alhonen, P. 1967. Acta Bot Fenn 76:1–59.

Alhonen, P. 1968. Memo Soc Fauna Flora Fenn 44:13–20.

Almer, B., W. Dickson, C. Ekström, and E. Hörnström. 1978. *In* J. O. Nriagu, ed. *Sulfur in the environment,* Part II. *Ecological impacts,* 271–311. John Wiley and Sons. New York.

Almer, B., W. Dickson, C. Ekström, E. Hörnström, and U. Miller. 1974. Ambio 3:30–36.

Altshuller, A. P., and R. A. Linthurst. 1984. *The acidic deposition phenomenon and its effects: Critical assessment review papers.* EPA/600/8-3/016BF. U.S. Environmental Protection Agency, Washington, DC.

Andersen, S. T. 1966. Paleobotanist 15:117–127.

Andersen, S. T. 1967. *In* R. Tüxen, ed. *Pflanzensoziologie und palynologie,* 106–118. Junk, Dordrecht, Netherlands.

Anderson, D. S., R. B. Davis, and F. Berge. 1986a. *In* J. P. Smol, R. W. Battarbee, R. B. Davis, and J. Meriläinen, eds. *Diatoms and lake acidity,* 97–113. Junk Dordrecht, Netherlands.

Anderson, N. J., R. W. Battarbee, P. G. Appleby, A. C. Stevenson, F. Oldfield, J. Darley, and G. Glover. 1986b. *Palaeolomnological evidence for the acidification of Loch Fleet.* Research Paper No. 17, Paleoecology Research Unit, University College London, London.

Arzet, K. 1987. Diatomeen als pH-Indikatoren in subrezenten Sedimenten von Weichwasserseen. Dissertation aus der Abteilung für Limnologie des Institut für Zoologie der Universität Innsbruck. 266 p.

Arzet, K., Krause-Dellin, D., and. C. Steinberg. 1986b. *In* J. P. Smol, R. W. Battarbee, R. B. Davis, and J. Meriläinen, eds. *Diatoms and lake acidity,* 227–250. Junk, Dordrecht, Netherlands.

Arzet, K., C. Steinberg, P. Psenner, and N. Schulz. 1986a. Hydrobiologia 143:247–254.

Arzet, K., and H. van Dam. 1986. *In* M. Ricard, ed. *Proceedings of the Eighth International Diatom Symposium,* 748–749. Koeltz, Koenigstein.

Asman, W. A. M., B. Drukker, and A. J. Janssen. 1988. Atmos Environ 22:725–735.

Baron, J., S. A. Norton, D. R. Beeson, and R. Herrmann. 1986. Can J Fish Aq Sci 43:1350–1362.

Barrett, C. F., D. H. F. Atkins, J. N. Cape, J. Crabtree, T. D. Davies, R. G. Derwent, B. E. A. Fisher, D. Fowler, A. S. Kallend, A. Martin, R. A. Scriven, and J. G. Irwin. 1987. *Acid deposition in the United Kingdon 1981–1985: A second report of the United Kingdom Review Group on Acid Rain.* Warren Spring Laboratory.

Barrett, C. F., D. H. F. Atkins, J. N. Cape, D. Fowler, J. G. Irwin, A. S. Kallend, A. Martin, J. I. Pitman, R. A. Scriven, and A. F. Tuck. 1983. *Acid deposition in the United Kingdom: Report of the United Kingdom Review Group on Acid Rain.* Warren Spring Laboratory.

Barrie, L. A. 1984. Tellus 36B:333–355.

Battarbee, R. W. 1984. Phil Trans Royal Soc London, Series B 305:451–477.

Battarbee, R. W. 1986. *In* B. Berglund, ed. *Handbook of Holocene palaeoecology and palaeohydrology.* John Wiley and Sons, New York. p. 527–570.

Battarbee, R. W., N. J. Anderson, P. G. Appleby, R. G. Flower, S. C. Fritz, E. Y. Haworth, V. J. Jones, A. Kreiser, J. Natkanski, F. Oldfield, S. T. Patrick, N. Richardson, B. Rippey, and A. C. Stevenson. 1988c. *Lake acidification in the United Kingdom: Evidence from the analysis of lake sediments.* Ensis Press, London. 68 p.

Battarbee, R. W., and D. F. Charles. 1986. Water Air Soil Pollut 30:347–354.

Battarbee, R. W., and D. F. Charles. 1987. Prog Phys Geog 11:552–580.

Battarbee, R. W., R. J. Flower, A. C. Stevenson, V. J. Jones, R. Harriman, and P. G. Appleby. 1988a. Nature 332:530–532.

Battarbee, R. W., R. J. Flower, A. C. Stevenson, and B. Rippey. 1985. Nature 315: 350–352.

Battarbee, R. W., and I. Renberg. 1985. *Royal Society Surface Water Acidification Project (SWAP): Palaeolimnology programme.* Working Paper No. 12. Paleoecology Research Unit, University College London, London.

Battarbee, R. W., J. P. Smol, and J. Meriläinen. 1986. *In* J. P. Smol, R. W. Battarbee, R. B. Davis, and J. Meriläinen, eds. *Diatoms and lake acidity,* 5–14. Junk, Dordrecht, Netherlands.

Battarbee, R. W., A. C. Stevenson, B. Rippey, C. Fletcher, J. Natkanski, M. Wik, and R. J. Flower. 1988b. J. Ecol O. (In press.)

Beeson, D. R. 1984. Bull Ecol Soc Amer 65:135.

Beeson, D. R. 1988. *Recent history of three lakes in the Wind River Mountains, Wyoming, based on sediment diatom remains.* Report submitted to the U.S. Forest Service, Bridger/Teton National Forest, Jackson, WY (unpubl.).

Berge, F. 1975. *pH-förändringer og sedimentasjon av diatomeer i Langtjern.* SNSF prosjektet IR 11/75:1–18. Aas-NLH, Norway.

Berge, F. 1976. *Kiselalger og pH i noen elver og innsjöer i Agder og Telemark. En sammenlikning mellom ärene 1949 og 1975 [Diatoms and pH in some rivers and lakes in Agder and Telemark (Norway): A comparison between the years 1949 and 1975].* SNSF prosjektet, IR18/76:1–36, Aas-NLH, Norway.

Berge, F. 1979. *Kiselalger og pH i noen innsjöer i Agder og Hordaland [Diatoms and pH in some lakes in the Agder and Hordaland Counties, Norway]*. SNSF prosjektet IR42/79:1–64, Aas-NLH, Norway.
Berge, F. 1985. Relationships of diatom taxa to pH and other environmental factors in Norwegian soft-water lakes. Doctoral dissertation, University of Maine, Orono, ME.
Berglund, B., and N. Malmer. 1971. Geol Fören Stockh Förh 93:575–586.
Binford, M. W. 1988. *In* D. F. Charles and D. R. Whitehead, eds. *Paleoecological investigation of recent lake acidification*. Developments in hydrobiology series. Junk, Dordrecht, Netherlands. (In press.)
Birks, H. J. B. 1987. *Methods for pH-calibration and reconstruction from palaeolimnological data: Procedures, problems, potential techniques*. Proceedings of the Surface Water Acidification Programme (SWAP) Mid-Term Review Conference, Bergen, June 22–26, 1987. Norwegian Academy of Science and Letters, Royal Society, Royal Swedish Academy of Sciences. 370–380.
Bormann, F. H., and G. E. Likens. 1979. *Pattern and process in a forested ecosystem*. Springer-Verlag, New York.
Brimblecombe, P., and D. H. Stedman. 1982. Nature 298:460–462.
Brugam, R. B. 1980. Quaternary Research 13:133–146.
Brugam, R. B. 1983. Hydrobiologia 98:223–235.
Charles, D. F. 1984. Mitt Int Ver Theor Angew Limnol 22:559–566.
Charles, D. F. 1985. Ecology 66:994–1011.
Charles, D. F. 1986. *In* J. P. Smol, R. W. Battarbee, R. B. Davis, and J. Meriläinen, eds. *Diatoms and lake acidity,* 35–44. Junk, Dordrecht, Netherlands.
Charles, D. F., M. W. Binford, B. D. Fry, E. Furlong, R. A. Hites, M. J. Mitchell, S. A. Norton, M. J. Patterson, J. P. Smol, A. J. Uutala, J. R. White, D. R. Whitehead, and R. J. Wise. 1989. *In* D. F. Charles and D. R. Whitehead, eds. *Paleoecological investigation of recent lake acidification*. Developments in hydrobiology. Junk, Dordrecht, Netherlands. (In press.)
Charles, D. F. and S. A. Norton. 1986. *Paleolimnological evidence for trends in atmospheric deposition of acids and metals. Acid deposition: Long-term trends,* 335–435. National Academy Press, Washington, DC.
Charles, D. F., and J. P. Smol. 1988. Limnol Oceanog 33:1451–1462.
Charles, D. F., and D. R. Whitehead. 1986a. Hydrobiologia 143:13–20.
Charles, D. F., and D. R. Whitehead. 1986b. *Paleoecological investigation of recent lake acidification: Methods and project description*. EPRI EA-4906. Electric Power Research Institute, Palo Alto, CA. 228 p.
Charles, D. F., D. R. Whitehead, D. R. Engstrom, B. D. Fry, R. A. Hites, S. A. Norton, J. S. Owens, L. A. Roll, S. C. Schindler, J. P. Smol, A. J. Uutala, J. R. White, and R. J. Wise. 1987. Biogeochemistry 3:267–296.
Cholnoky, B. J. 1968. *Die ökologie der diatomeen in binnengewässern*. J. Cramer, Lehre, Germany.
Christie, C. E., and J. P. Smol. 1986. Hydrobiologia 143:355–360.
Christie, C. E., J. P. Smol, P. Huttunen, and J. Meriläinen. 1988. Hydrobiologia 161:237–243.
Cleve, P. T. 1891. Acta Soc Fauna Flora Fenn 8:1–68.
Clymo, R. S. 1963. Ann Bot 27:309–324.
Colingsworth, R. F., M. H. Hohn, and G. B. Collins. 1967. Papers of the Michigan Academy of Science, Arts, and Letters 52:19–30.
Conger, P. S. 1939. Am J Sci 237:324–340.

Crabtree, K. 1969. Mitteilungen, Int Ver Theor Angew Limnol 17:165–171.
Crocker, R. L., and J. Major. 1955. J Ecol 43:427–448.
Davidson, G. A. 1984. Paleolimnological reconstruction of the acidification history of an experimentally acidified lake. M.Sc. Thesis, The University of Manitoba. 186 p.
Davis, R. B. 1987. Quaternary Sci Reviews 6:147–163.
Davis, R. B., and D. S. Anderson. 1985. Hydrobiologia 120:69–87.
Davis, R. B., D. S. Anderson, and F. Berge. 1985. Nature 316:436–438.
Davis, R. B., D. S. Anderson, D. F. Charles, and J. N. Galloway. 1988. *In* W. J. Adams, G. A. Chapman, and W. G. Landis, eds. *Aquatic toxicology and hazard assessment: 10th vol.*, ASTM STP 971, 89–111. American Society for Testing and Materials, Philadelphia.
Davis, R. B., and F. Berge. 1980. *In* D. Drabløs, and A. Tollan, eds. *Ecological impact of acid precipitation,* 270–271. SNSF Project, Oslo, Norway.
Davis, R. B., S. A. Norton, D. F. Brakke, F. Berge, and C. T. Hess. 1980. *In* D. Drabløs and A. Tollan, eds. *Ecological impact of acid precipitation,* 274–275. SNSF Project, Oslo, Norway.
Davis, R. B., S. A. Norton, C. T. Hess, and D. Brakke. 1983. Hydrobiologia 103:113–123.
Davis, R. B., S. A. Norton, J. S. Kahl, M. C. Whiting, J. Ford, and J. P. Smol. 1989. *In* D. F. Charles and D. R. Whitehead, eds. *Paleoecological invvestigation of recent lake acidification*. Junk, Dordrecht, Netherlands. (In press.)
Davis, R. B., and J. P. Smol. 1986. *In* J. P. Smol, R. W. Battarbee, R. B. Davis, and J. Meriläinen, eds. *Diatoms and lake acidity,* 291–300. Junk, Dordrecht, Netherlands.
Davis, R. B., and P. M. Stokes. 1986. Water Air Soil Pollut 30:311–318.
Delorme, L. D., H. C. Duthie, S. R. Esterby, S. M. Smith, and N. S. Harper. 1986. Archiv für Hydrobiol 108:1–22.
Delorme, L. D., S. R. Esterby, and H. Duthie. 1983. Int Rev gesamten Hydrobiol 69:41–55.
Del Prete, A., and J. N. Galloway. 1983. *Temporal trends in the pH of Woods, Sagamore and Panther lakes as determined by an analysis of diatom populations. The Integrated Lake-Watershed Acidification Study: Proceedings of the ILWAS Annual Review Conference*. EPRI EA-2827, Electric Power Research Institute, Palo Alto, CA.
Del Prete, A., and C. Schofield. 1981. Archiv Hydrobiol 91:332–340.
Dickman, M., and J. Fortescue. 1984. Verh Int Verein Limnol 22:1345–1356.
Dickman, M., H. G. Thode, S. Rao, and R. Anderson. 1988. Environ Pollut 49:265–288.
Dickman, M. D., S. Dixit, J. Fortescue, B. Barlow, and J. Terasmae. 1984. Water Air Soil Pollut 21:375–386.
Dickman, M. D., and H. G. Thode. 1985. Water Air Soil Pollut 26:233–253.
Dickman, M. D., H. van Dam, B. van Geel, A. G. Klink, and A. van der Wijk. 1987. Archiv Hydrobiol 109:377–408.
Dickson, W. 1980. *In* D. Drabløs and A. Tollan, eds. *Ecological impact of acid precipitation,* 75–83. SNSF Project, Oslo, Norway.
Digerfeldt, G. 1972. Folia Limnol Scand 16:1–104.
Digerfeldt, G. 1975. Feol Fören Stockh Förh 97:13–28.
Digerfeldt, G. 1977. *Regional vegetation history and paleolimnology. The Flandrian development of Lake Flarken,* Report 13. Department of Quaternary Geology, University of Lund, Sweden.
Dillon, P. J., and R. D. Evans. 1982. Hydrobiologia 91:121–130.
Dixit, A. S., S. S. Dixit, and R. D. Evans. 1988c. J Paleolimnol 1:23–38.

Dixit, S. S. 1983. *The utility of sedimentary diatoms as a measure of historical lake pH*. M.Sc. Thesis, Brock University, Ontario.
Dixit, S. S. 1986a. Can J Bot 64:1129–1133.
Dixit, S. S. 1986b. Algal microfossils and geochemical reconstructions of Sudbury lakes: A test of the paleo-indicator potential of diatoms and chrysophytes. Ph.D. Thesis, Queen's University, Kingston, Ontario. 190 p.
Dixit, S. S., and M. D. Dickman. 1986. Hydrobiologia 131:133–143.
Dixit, S. S., A. S. Dixit, and R. D. Evans. 1987a. Sci Total Environ 67:53–67.
Dixit, S. S., A. S. Dixit, and R. D. Evans. 1987b. *Paleolimnological study of LRTAP network lakes, Quebec region*. Report prepared for Department of Supply and Services, Canada. Trent Aquatic Research Centre, Trent University, Peterborough, Ontario. 121 pages (unpubl.).
Dixit, S. S., A. S. Dixit, and R. D. Evans. 1988a. Can J Fish Aq Sci 45:1411–1421.
Dixit, S. S., A. S. Dixit, and R. D. Evans. 1988b. Hydrobiologia. 169:135–148.
Dixit, S. S., A. S. Dixit, and R. D. Evans. 1988c. Water Air Soil Pollut 38:97–104.
Dixit, S. S., and R. D. Evans. 1986. Can J Fish Aq Sci 43:1836–1845.
Dixit, S. S., J. P. Smol, R. B. Davis, and D. S. Anderson. 1989. *In* D. F. Charles and D. R. Whitehead, eds. *Paleoecological investigation of recent lake acidification*. Developments in hydrobiology. Junk, Dordrecht, Netherlands. (In press.)
Driscoll, C. T., J. P. Baker, J. J. Bisogni, C. L. Schofield. 1984. *In* O. P. Bricker, ed. *Geological aspects of acid deposition,* 55–75. Butterworth, Boston.
Eilers, J. M., D. F. Brakke, D. H. Landers, and P. E. Kellar 1988. Verh Internat Verein Limnol 23:144–151.
Elner, J. K., and S. Ray. 1987. Water Air Soil Pollut 32:17–29.
EMEP/CCC. 1984. *Summary report from the chemical coordinating centre for the second phase of the co-operative program for monitoring and evaluation of long-range transmission of air-pollutants in Europe (EMEP)*. Norwegian Institute for Air Research (NILU), Lillestrøm.
Esterby, S. R., and A. H. El-Shaarawi. 1981a. J Appl Stat 30:277–285.
Esterby, S. R., and A. H. El-Shaarawi. 1981b. J Hydrobiol 53:17–30.
Evans, G. H. 1970. New Phytol 69:821–874.
Evans, G. H., and R. Walker. 1979. New Phytol 78:221–236.
Florin, M.-B. 1977. Striae 5:1–60.
Flower, R. J. 1986. *In* J. P. Smol, R. W. Battarbee, R. B. Davis, and J. Meriläinen, eds. *Diatoms and lake acidity,* 45–54. Junk, Dordrecht, Netherlands.
Flower, R. J. 1987. Hydrobiologia 143:93–104.
Flower, R. J., and R. W. Battarbee. 1983. Nature 305:130–133.
Flower, R. J., R. W. Battarbee, and P. G. Appleby. 1987. J Ecol 75:797–824.
Flower, R. J., R. W. Battarbee, J. Natkanski, B. Rippey, and P. G. Appleby. 1988. J Appl Ecol 25:715–724.
Foged, N. 1969. Dan Geol Foren 19:237–256.
Foged, N. 1972. Medd Grønl 194:1–67.
Ford, J. 1986. *In* J. P. Smol, R. W. Battarbee, R. B. Davis, and J. Meriläinen, eds. *Diatoms and lake acidity,* 131–148. Junk, Dordrecht, Netherlands.
Ford, M. S. 1984. The influence of lithology on ecosystem development in New England: A comparative paleoecological study. Doctoral dissertation, University of Minnesota, Minneapolis.
Ford, M. S. 1989. *A 10,000-year history of natural ecosystem acidification*. Ecological Monographs. (In press.)

Fortescue, J. A. C. 1984. *Interdisciplinary research for an environmental component (acid rain) in regional geochemical surveys (Wawa area)*. Algoma District, Ontario Geological Survey Map 80 713, Geochemical Series, compiled 1983.
Frey, D. G. 1969. Mitt Int Ver Theor Agnew Limnol 17:7–18.
Fritz, S. C., and R. E. Carlson. 1982. Water Air Soil Pollut 17:151–163.
Galloway, J. N., C. L. Schofield, N. E. Peters, G. R. Hendrey, and E. R. Altwicker. 1983. Can J Fish Aq Sci 40:799–806.
Gasse, F. 1986. *In* J. P. Smol, ed. *Diatoms and lake acidity,* 149–168. Junk, Dordrecht, Netherlands.
Gasse, F., and F. Tekaia. 1983. Hydrobiologia 103:85–90.
Gibson, J. H., J. N. Galloway, C. Schofield, W. McFee, R. Johnson, S. McCorley, N. Dise, and D. Herzog. (1984). *Rocky Mountain acidification study*. FWS/OBS-80/40. U.S. Fish and Wildlife Service, Division of Biological Services Eastern Energy and Land Use Team. 137 p.
Gibson, K. N., J. P. Smol, and J. Ford. 1987. Can J Fish Aq Sci 44:1584–1588.
Gorham, E., S. J. Eisenreich, J. Ford, and M. V. Santelmann. 1985. *In* W. Stumm, ed. *Chemical processes in lakes,* 339–363. Wiley-Interscience, New York.
Gorham, E., P. M. Vitousek, and W. A. Reiners. 1979. Annu Rev Ecol Syst 10:53–84.
Harriman, R., and D. E. Wells. 1985. J Water Pollut Control 84:215–222.
Hartmann, H., and C. Steinberg. 1987. Hydrobiologia 143:87–91.
Haworth, E. Y. 1969. J Ecol 57:429–439.
Haworth, E. Y. 1972. Geol Soc Am Bull 83:157–172.
Holdren, G. R., Jr., T. M. Brunelle, G. Matisoff, and M. Wahlen. 1984. Nature 311:245–248.
Holmes, R. W. 1986. *Calibration of diatom-pH-alkalinity methodology for the interpretation of the sedimentory record in Emerald Lake Integrated Watershed Study*. Final Report, California Air Resources Board, Contract A4-118-32 (pp. 1–80). Marine Science Institute, University of California, Santa Barbara.
Hudon, C., H. C. Duthie, S. M. Smith, and S. A. Ditner. 1986. Hydrobiologia 140:49–65.
Hultberg, H., and I. Andersson. 1982. Water Air Soil Pollut 18:311–331.
Hustedt, F. 1927. *Die kieselalgen. Rabenhorsts' kryptogamen-flora von Deutschland, Österreich und der Schweiz. Band 7*. Geest und Portig, Leipzig, Germany.
Hustedt, F. 1939. Arch Hydrobiol Supplement 16:274–394.
Huttunen, P., and J. Meriläinen. 1983. Hydrobiologia 103:91–97.
Huttunen, P., and J. Meriläinen. 1986a. *In* J. P. Smol, R. W. Battarbee, R. B. Davis, and J. Meriläinen, eds. *Diatoms and lake acidity,* 201–211. Junk, Dordrecht, Netherlands.
Huttunen, P., and J. Meriläinen. 1986b. *In* H. Simola, ed. *Proceedings of the Finnish-Soviet Symposium on Methods in Paleoecology and the Nordic Meeting of Diatomologists,* 47–54. Karelian Institute, Joensuu, Finland.
Huttunen, P., and J. Meriliänen. 1986c. *In* H. Simola, ed. *Proceedings of the Finnish-Soviet Symposium on Methods in Paleoecology and the Nordic Meeting of Diatomologists,* 41–46. Karelian Institute, Joensuu, Finland.
Iversen, J. 1958. Uppsala Universitets Årsskrift 6:210–215.
Johannessen, M. 1980. *In* D. Drabløs and A. Tollan, eds. *Ecological impacts of acid precipitation,* 222–223. SNSF project, Oslo.
Johnson, A. H., D. F. Charles, and S. B. Anderson. 1987. Environment 29:4–5.
Johnson, N. M. C. T. Driscoll, J. S. Eaton, G. E. Likens, and W. H. McDowell. 1981. Geochim Cosmochim Acta 45:1421–1437.
Jones, V. J., A. C. Stevenson, and R. W. Battarbee. 1986. Nature 322:157–158.

Jones, V. J., A. C. Stevenson, and R. W. Battarbee. 1988. *Acidification of lakes in Galloway southwest Scotland: A diatom and pollen study of the post-glacial history of the Round Loch of Glenhead*. Paleoecology Research Unit, University College London.

Jørgensen, E. K. 1948. K Dan Vidensk Selsk Biol Skr 5(2):1–140.

Kingston, J. C., and R. Kreis, Jr. 1987. *Wisconsin DNR lake acidification study: Final report to the Wisconsin Department of Natural Resources*. (unpubl.).

Kingston, J. C., R. G. Kreis, Jr., R. B. Cook, K. E. Camburn, S. A. Norton, R. A. Hites, L. A. Roll, E. Furlong, M. J. Mitchell, S. C. Schindler, B. D. Fry, L. Shane, and G. King. 1989. *In* D. F. Charles and D. R. Whitehead, eds. *Paleoecological investigation of recent lake acidification*. Developments in hydrobiology. Junk, Dordrecht, Netherlands. (In press.)

Kolbe, R. W. 1932. Ergeb Biol 8:221–348.

Kristiansen, J. 1986. Br Phycol J 21:425–436.

Landers, D. H., W. S. Overton, R. A. Linthurst, and D. F. Brakke. 1988. Environ Sci Technol 22:128–135.

Leuven, R. S. E. W. 1988. Impact of acidification on aquatic ecosystems in the Netherlands. Ph.D. thesis. Catholic University, Nijmegen, Netherlands.

Lillie, R. A., and J. W. Mason. 1983. *Limnological characteristics of Wisconsin lakes*. Technical Bulletin No. 138. Wisconsin Department of Natural Resources, Madison, WI.

Livingstone, D. A., K. Bryan, Jr., and R. G. Leaky. 1958. Limnol Oceanog 3:192–214.

Lowe, R. L. 1974. *Environmental requirements and pollution tolerance of freshwater diatoms*. EPA/670/4-74/005. U.S. Environment Protection Agency, Washington, DC.

Lundqvist, G. 1925. Sver Geol Unders Årsb 330:1–129.

Martin, C. W., D. S. Noel, and C. A. Federer. 1984. J Environ Qual 13(2):204–210.

McKee, P. M., W. J. Snodgrass, D. R. Hart, H. C. Duthie, J. H. McAndrews, and W. Keller. 1987. Can J Fish Aq Sci 44:390–398.

Melack, J. M., S. D. Cooper, R. W. Holmes, J. O. Sickman, K. Kratz, P. Hopkins, H. Hardenbergh, M. Thieme, and L. Meeker. 1987. *Chemical and biological survey of lakes and streams located in the Emerald Lake watershed, Sequoia National Park*. Final Report to the California Air Resources Board, Contract A3-096-32.

Meriläinen, J. 1967. Ann Bot Fenn 4:51–58.

Miller, U. 1973. Statens Naturvårdsverk Publikationer 7:43–60.

Miller, U. 1986. Striae (Uppsala) 24:9–14.

Ministry of Agriculture. 1982. A Swedish study prepared for the 1982 Stockholm conference on the acidification of the environment. Ministry of Agriculture, Stockholm, Sweden.

National Research Council. 1986 *Acid deposition: Long-term trends*. National Academy Press, Washington, DC.

Neal, C., P. Whitehead, and A. Jenkins. 1988. Nature 334:109–110.

Nilsson, S. I., H. G. Miller, and J. D. Miller. 1982. Oikos 39:40–49.

Norton, S. A., R. B. Davis, and D. S. Anderson. 1985. *The distribution and extent of acid and metal precipitation in Northern New England*. Final report. U.S. Fish and Wildlife Service Grant No. 14-16-0009-75-040. University of Maine at Orono.

Norton, S. A., R. B. Davis, and D. F. Brakke. 1981. *Responses of northern New England lakes to atmospheric inputs of acids and heavy metals*. Completion report project A-048-ME. Land and Water Resources Center, University of Maine at Orono.

Norton, S. A., and J. S. Kahl. 1987. *Geochemical analysis of sediment cores, Wind River Mountains, Wyoming*. Report submitted to U.S. Forest Service, Bridger/Teton National Forest, Jackson, WY.

Norton, V. C., and S. J. Herrmann. 1980. Trans Am Micros Soc 99:416–425.
Nygaard, G. 1956. *In* K. Berg and I. C. Petersen, eds. *Studies on the humic acid, Lake Gribsø*. Folia Limnol Scand 8:32–94.
Odén S. 1968. *The acidification of air precipitation and its consequences in the natural environment*. Energy Committee Bulletin, 1, Swedish Natural Sciences Research Council, Stockholm.
Oehlert, G. W. 1984. *Error bars for paleolimnologically inferred pH histories: A first pass*. Technical Report Number 267, Series 2. Department of Statistics, Princeton University, Princeton, NJ.
Oehlert, G. W. 1988. Can J Stat 16:51–60.
Oksanen, J., E. Läara, P. Huttunen, and J. Meriläinen. 1988. J Paleolimnol 1:39–49.
Oliver, B. G., E. M. Thurman, and R. L. Malcolm. 1983. Geochim Cosmochim Acta 47:2031–2035.
Patrick, R. 1943. Proc Acad Nat Sci Phila 95:53–110.
Patrick, R. 1954. J Protozool 1:34–37.
Patrick, R. 1977. *In* D. Werner, ed. *The biology of diatoms,* 284–332. Botanical Monographs, vol. 13. Blackwell Scientific Publications, Oxford, England.
Patrick, R., V. P. Binetti, and S. G. Halterman. 1981. Science 211:446–448.
Patrick, R., and C. W. Reimer. 1966. *The diatoms of the United States,* vol. 1. Monograph Number 13. Academy of Natural Sciences, Philadelphia.
Patrick, R., and C. W. Reimer. 1975. *The diatoms of the United States,* vol. 2, part 1. Monograph Number 13. Academy of Natural Sciences, Philadelphia.
Pennington, W. 1981. Hydrobiologia 79:197–219.
Pennington, W. 1984. Freshwater Biol Assoc, Ann Rep 52:28–46.
Pennington, W., E. Y. Haworth, A. P. Bonny, and J. P. Lishman. 1972. Philos Trans R Soc London Ser B Biol Sci 264:191–294.
Pollman, C. D., and D. E. Canfield, Jr. 1989. *In* D. F. Charles, ed. *Acidic deposition and aquatic ecosystems: Regional case studies*. Springer-Verlag, New York. (In press.)
Rawlence, D. J. 1988. J Paleolimnol 1:51–60.
Reed, S. 1982. The late-glacial and postglacial diatoms of Heart Lake and Upper Wallface Pond in the Adirondack Mountains, New York. Masters Thesis, Indiana University, Bloomington.
Rehakova, Z. 1983. Hydrobiologia 103:24–245.
Renberg, I. 1976. Early Norrland 9:113–159.
Renberg, I. 1978. Early Norrland 11:63–92.
Renberg, I., R. Hellberg, and M. Nilsson. 1985. Ecol Bull 37:219–223.
Renberg, I., and T. Hellberg. 1982. Ambio 11:30–33.
Renberg, I., and J. E. Wallin. 1985. *In* F. Andersson and B. Olsson, eds. *Lake Gårdsjon: An acid forest lake and its catchment*. Ecol Bull 37:47–52.
Renberg, I., and M. Wik. 1985. Ambio 14:161–163.
Rhodes, T. E. 1985. The diatom stratigraphy of an acidic lake. Master's thesis, University of Massachusetts, Amhurst.
Rochester, H. 1978. Late-glacial and postglacial diatom assemblages of Berry Pond, Massachusetts, in relation to watershed ecosystem development. Ph.D. dissertation, Indiana University, Bloomington.
Rosenqvist, I. T. 1978. Sci Tot Environ 10:39–49.
Round, F. E. 1957. New Phytol 56:98–126.
Round, F. E. 1961. New Phytol 60:43–59.
Ryan, D. F., and D. M. Kahler. 1987. Limnol Oceanog 32:751–757.

Salomaa, R., and P. Alhonen. 1983. Hydrobiologia 103:295–301.

Schofield, C. L. 1976. *Acidification of Adirondack lakes by atmospheric precipitation: Extent and magnitude of the problem*. Final report. Project F-28-R. New York State Department of Environmental Conservation, Albany.

Schwela, D. 1983. Staub-Reinhaltung der Luft 43:135–139.

Scruton, D. A., and J. K. Elner. 1986. *Paleolimnological investigation of freshwater lake sediments in insular Newfoundland. Part 1: Relationships between diatom sub-fossils in surface sediments and contemporary pH from thirty-four lakes*. Canadian Technical Report of Fisheries and Aquatic Sciences No. 1521. Science Branch, Department of Fisheries and Oceans, St. John's, Newfoundland. 50 p.

Scruton, D. A., J. K. Elner, and G. D. Howell. 1987a. *Paleolimnological investigation of freshwater lake sediments in insular Newfoundland. Part 2: Downcore diatom stratigraphies and historical pH profiles for seven lakes*. Canadian Technical Report of Fisheries and Aquatic Sciences No. 1521. Science Branch, Department of Fisheries and Oceans, St. Johns, Newfoundland. 67 p.

Scruton, D. A., J. K. Elner, and M. Rybak. 1987b. *In* R. Perry, R. M. Harrison, J. N. B. Bell, and J. N. Lester, eds. *Acid rain: Scientific and technical advances*. Selper Ltd., London.

Sherman, J. W. 1976. Post-pleistocene diatom assemblages in New England lake sediments. Ph.D. dissertation, University of Delaware, Newark.

Sherman, J. W. 1985. *An ecosystem approach to aquatic ecology: Mirror Lake and its environment*. 366–382. Springer-Verlag, New York.

Simola, H., K. Kenttämies, and O. Sandman. 1985. Aq Fenn 15(2):245–255.

Smol, J. P. 1980. Can J Bot 58:458–465.

Smol, J. P. 1983. Can J Bot 61:2195–2204.

Smol, J. P. 1986. *In* J. P. Smol, R. W. Battarbee, R. B. Davis, and J. Meriläinen, eds. *Diatoms and lake acidity*, 275–287. Junk, Dordrecht, Netherlands.

Smol, J. P. 1988. Palaeogeog Paleoclimatol Palaeocol 62:287–297.

Smol, J. P., R. W. Battarbee, R. B. Davis, and J. Meriläinen. 1986. *Diatoms and lake acidity*. Junk, Dordrecht, Netherlands.

Smol, J. P., and M. M. Boucherle. 1985. Arch Hydrobiol 103:25–49.

Smol, J. P., D. F. Charles, and D. R. Whitehead. 1984a. Can J Bot 62(5):911–923.

Smol, J. P., D. F. Charles, and D. R. Whitehead. 1984b. Nature 307:628–630.

Smol, J. P., and S. S. Dixit. 1989. *In* D. F. Charles and D. F. Whitehead, eds. *Paleoecological investigation of recent lake acidification*. Developments in Hydrobiology. Junk, Dordrecht, Netherlands. (In press.)

Somers, K. M., and H. H. Harvey. 1984. Can J Fish Aq Sci 41:20–29.

Sreenivasa, M. R., and H. C. Duthie. 1973. Can J Bot 51:1599–1609.

Steinberg, C., K. Arzet, and D. Krause-Dellin. 1984. Naturwissenschaften 71:631–634.

Steinberg, C., and H. Hartmann. 1986. Naturwissenschaften 73:37–39.

Stevenson, A. C., H. J. B. Birks, R. J. Flower, and R. W. Battarbee. 1988. Diatom-based pH reconstruction of lake acidification using canonical correspondence analysis. (Submitted.)

Stevenson, A. C., S. C. Fritz, S. T. Patrick, B. Rippey, P. G. Appleby, F. Oldfield, J. Darley, S. R. Higgitt, R. W. Battarbee, and R. J. Raven. 1987. *Paleoecological evaluation of the recent acidification of Welsh lakes. 6. Llyn Dulyn, Gwynedd*. Research Paper No. 22. Paleoecology Research Unit, Department of Geography, University College London.

Stuart, S. 1984. *Hydrology and aquatic chemistry of monitor lake watersheds in the Wind*

River Mountains, Wyoming. Report submitted to the U.S. Forest Service, Bridger/Teton National Forest, Jackson, WY.

Sullivan, T. J., C. T. Driscoll, S. A. Gherini, R. K. Munson, R. B. Cook, D. F. Charles, and C. P. Yatsko. 1989. Nature. 338:408–410.

Sweets, P. R. 1983. Differential deposition of diatom frustules in Jellison Hill Pond, Maine. Master's thesis, University of Maine at Orono. 156 p.

Sweets, P. R., R. W. Bienert, T. L. Crisman, and M. W. Binford. 1989. *In* D. F. Charles and D. R. Whitehead, eds. *Paleoecological investigation of recent lake acidification*. Developments in Hydrobiology. Junk, Dordrecht, Netherlands. (In press.)

Taylor, M. C. 1986. Surficial sedimentary diatoms–limnological relationships in Canadian Shield Lakes. Master's thesis. Department of Biology, University of Waterloo, Ontario. 159 p.

Taylor, M. C., H. C. Duthie, and S. M. Smith. 1986. *Errors associated with diatom-inferred indices for predicting pH in Canadian Shield lakes*. 9th Diatom Symposium. Biopress, Bristol, UK.

Taylor, M. C., H. C. Duthie, and S. M. Smith. 1987. J Phycol 23:673–676.

Ter Braak, C. J. F. 1986. Ecology 67:1167–1179.

Ter Braak, C. J. F., and H. van Dam. 1988. *Inferring pH from diatoms: A comparison of old and new calibration methods*. Hydrobiologia. (In press.)

Tolonen, K. 1967. Ann Bot Fenn 4:219–416.

Tolonen, K. 1972. Early Norrland 1:53–77.

Tolonen, K. 1980. Ann Bot Fenn 17:394–405.

Tolonen, K., and T. Jaakkola. 1983. Ann Bot Fenn 20:57–78.

Tolonen, K., M. Liukkonen, R. Harjula, and A. Pätilä. 1986. *In* J. P. Smol, R. W. Battarbee, R. B. Davis, and S. Meriläinen, eds. *Diatoms and lake acidity*, 169–199. Junk, Dordrecht, Netherlands.

Tynni, R. 1972. Aq Fenn 1972:70–82.

Van Dam, H. 1987. Acidification of moorland pools: A process in time. Ph.D. thesis. Agricultural University, Wageningen, Netherlands, 175 p.

Van Dam, H. 1988. Freshwater Biol. (In press.)

Van Dam, H. 1989a. *In Proceedings of the COST 612 workshop on effects of land use in catchments on the acidity and ecology of natural surface waters*. University of Wales, Institute of Science and Technology (UWIST), Cardiff, Wales, April 11–13, 1988. (In press.)

Van Dam, H. 1989b. *Impact of acidification on diatoms, macrophytes and chemistry of soft-water lakes and pools in Denmark, the Netherlands, Belgium, and West Germany*. Research Institute for Nature Management, Leersum, the Netherlands. (In press.)

Van Dam, H., G. Suurmond, and C. J. F. Ter Braak. 1981. Hydrobiologia 83:425–459.

Van Dam, H., B. van Geel, A. van der Wijk, J. F. M. Geelen, R. van der Heijden, and M. D. Dickman. 1988. Rev Paleobot Palynol 55:273–316.

Vaughan, H. H., J. K. Underwood, J. G. Ogden, III. 1982. Water Air Soil Pollut 18:353–361.

Vitt, D. H., and S. Bayley. 1984. Can J Bot 62:1485–1500.

Walker, R. 1978. New Phytol 81:791–804.

Walker, I. R., and C. G. Paterson. 1986. Hydrobiologia 134:265–272.

Watanabe, T., and I. Yasuda. 1982. Jap J Limnol 43:237–245.

Whitehead, D. R., D. F. Charles, S. E. Reed, S. T. Jackson, and M. C. Sheehan. 1986. *In* J. P. Smol, R. W. Battarbee, R. B. Davis, and J. Meriläinen, eds. *Diatoms and lake acidity*, 251–274. Junk, Dordrecht, Netherlands.

Winkler, M. G. 1985. *Diatom evidence of environmental changes in wetlands: Cape Cod National Seashore. Part 1. Prehistoric and historic trends in acidity of the outer cape ponds*. Technical report to the North Atlantic Regional Office, National Park Service, Boston. 97 p.

Winkler, M. G. 1988. Ecol Monogr 58:197–214.

Wright, H. E., Jr. 1981. *Fire regimes and ecosystem properties*. U.S. Forest Service, General Technical Report WO-26, pp. 421–444. U.S. Forest Service, Washington, D.C.

Wright, R. F. 1983. *Predicting acidification of North American lakes*. Report 4/1983. Norwegian Institute for Water Research, Oslo. 165 p.

Wright, R. F. 1984. *Changes in the chemistry of Lake Hovvatn, Norway, following liming and reacidification*. Report 6/1984. Norwegian Institute for Water Research, Oslo. 68 p.

Yan, N. 1983. Can J Fish Aq Sci 40:621–626.

Index

MIX
Papier aus verantwortungsvollen Quellen
Paper from responsible sources
FSC® C105338

If you have any concerns about our products,
you can contact us on
ProductSafety@springernature.com

In case Publisher is established outside the EU,
the EU authorized representative is:
Springer Nature Customer Service Center GmbH
Europaplatz 3, 69115 Heidelberg, Germany

Printed by Libri Plureos GmbH
in Hamburg, Germany